Henry Davis Hoskold

Engineer's Valuing Assistant

Being a Practical Treatise on the Valuation of Collieries and Other Mines...

Henry Davis Hoskold

Engineer's Valuing Assistant
Being a Practical Treatise on the Valuation of Collieries and Other Mines...

ISBN/EAN: 9783337069285

Printed in Europe, USA, Canada, Australia, Japan

Cover: Foto ©berggeist007 / pixelio.de

More available books at **www.hansebooks.com**

THE

ENGINEER'S VALUING ASSISTANT:

BEING

A PRACTICAL TREATISE

ON THE

VALUATION OF COLLIERIES AND OTHER MINES

INCLUDING

ROYALTIES, LEASEHOLDS AND FREEHOLDS, AND
ANNUITIES FROM OTHER SOURCES,

With Rules, Formulæ, and Examples.

ALSO

NEW SETS OF VALUATION TABLES

CALCULATED ON THE PRINCIPLE OF ALLOWING INTEREST TO THE PURCHASER
OF ANNUITIES AT ONE RATE, AND REDEEMING THE CAPITAL INVESTED
AT ANOTHER, AND PRACTICABLE RATE PER CENT.;

AND

TABLES OF VALUES

SHOWING THE DISCREPANCIES EXISTING IN THE ORDINARY TABLES OF PRESENT
VALUES, AND THE ERRORS CREATED BY THEIR USE;

SOURCES FOR THE REDEMPTION OF CAPITAL

AT DIFFERENT RATES PER CENT.;

REMARKS UPON HOME AND FOREIGN MINES

AS INVESTMENTS; ETC.

BY

H. D. HOSKOLD, F.R.G.S., F.G.S., M.Soc.A. & Inst.M.E., &c.

CIVIL AND MINING ENGINEER:

Late Mining Engineer to the Dean Forest Iron Co. for 16 years;
Author of 'A Practical Treatise on Mining, Land, and Railway Surveying and Engineering.'

LONDON:

LONGMANS, GREEN, AND CO.

1877.

THIS WORK

IS RESPECTFULLY DEDICATED

TO

J. PEASE, Esq., M.P.

HUTTON HALL, GISBOROUGH,

and

H. HUSSEY VIVIAN, Esq., M.P.

PARK WERN, SWANSEA,

BY THE AUTHOR.

PREFACE.

THE EFFORTS put forth in the literary world at the present time, and the facility of production and means of distributing the results of such labours, are truly astonishing, and without a parallel in past times, and for the multiplication of Books of all classes there seems to be a growing necessity, but although various subjects connected with Arts, Sciences and Manufactures have been largely treated upon, that of the Valuation of Mines has been entirely neglected.

This circumstance is the more surprising in as much as the subject is one of great national importance, affecting, as it does, —at least in some degree—the interest of all those who are connected with Mining and other branches of industry.

In these times, commercial prosperity in general seems to depend more or less upon its relations to honest and successful Mining adventure; and although Mining and other branches of trade have received and will still receive healthy stimulus, nevertheless there are periods of reaction, causing depression, the origin of which it is not always easy to trace and explain. However, in very many cases it may be referred to inflated notions of speculation, creating undue excitement, error in judgment, and an unwarrantable lavish expenditure of capital upon properties not capable of yielding profits compatible with the outlay.

Immense sums of money are frequently spent in the purchase of mineral properties, and it is a common occurrence that much more is paid in order to secure them than they are really worth, and cases coming within experience are by no means few where the estimated value has exceeded the true value by 40 or 50 per cent., due in many cases to the employment of an erroneous years' purchase. Table XII exhibits the *source* of such discrepancies.

True value, and the economic expenditure of money upon the purchase and development of any property, are therefore matters of such paramount importance, that it has been considered necessary for the general interests of the public to point out in this work, and illustrate by ample practical examples, how such discrepancies as those referred to have arisen, and the means necessary to be adopted in guarding against such an enormous waste of Capital.

Taking a rational view of the matter, it is desirable that any mode of deducing Values, having for its basis nothing better than an approximate rule, or a mere guess, should yield to more accurate treatment ; and as the interest at stake is so great, and almost universal, it seems to be most essential that the public should possess some standard work of reference—embodying information and easy rules of a reliable and practicable character, so that by mere consultation, the comparative merits and value of Mineral and other kinds of property that may come under consideration may be readily determined. Hitherto, however, no work of this description and specially devoted to this subject has appeared.

The present work is therefore an attempt to supply this need ; and it has been written chiefly with a view to facilitate such calculations as are required generally, and especially by those of the Profession on whom more particularly devolve the important and onerous duties connected with Valuation, to introduce a general system based upon equitable and scientific principles, and to assist in obtaining more uniformity and accuracy in general practice.

In past years when I was extensively engaged in valuing coal and other mines, the labour connected with the necessary and frequent calculations involving the use of rules derived from first principles became so tedious, that I determined once for all to prepare full and complete sets of Tables required, to be employed in Valuation as *labour savers*.

After much thought and labour this task has been completed, and the result supplied by the publication of this work, in which I have endeavoured throughout to render the treatment of the subject as simple and intelligible as its nature seemed to admit of, and, as I believe, free from all unnecessary mathematical surroundings.

Throughout the work will be found numerous examples of all the more important cases that can occur in practice, both in Simple and Compound Interest as applied to Valuation generally, including Annuities derived from Collieries or other Mines, Royalties, Leaseholds, Freeholds, and other sources.

These examples are derived from practice, and the utility and advantage of the Tables in expediting work are fully exemplified, and where, for the sake of illustration it has been found proper, or convenient, Logarithmic computations have been resorted to.

At the termination of the Third Part of the work a few pages have been specially devoted to Logarithmic calculations of a particular order, and the accuracy of the numbers selected from the tables has been rigidly tested, and in no instance has any error been discovered in them. In these calculations the great superiority of Mr. Gray's 12-figure Logarithms is made apparent ; and the readiness with which any Logarithm or Anti-logarithm can be found to 12 places of decimals is the great feature and recommendation of his Tables.

Plain rules and formulæ of a special character have been

author is of itself a sufficient guarantee of the accuracy of the principles involved in a work with which he may in any way be connected.

I have much pleasure in stating that I have received very great assistance from Mr. William Hewlett, M.E., one of my former Articled Pupils, and late Engineering Assistant. He *re-computed* and corrected nearly the whole of the Tables and Examples in this work, and for a period extending over a year and a half took a considerable interest in its progress.

I am also indebted to Mr. William A. Taylor, who has exhibited great kindness in assisting me, by reading, comparing and correcting the proofs—a labour of no small importance.

I now leave the work in the hands of an enlightened public, venturing to express a hope that it may prove as much a benefit to them as it has been a pleasure to me in writing it, and I take this opportunity of expressing my grateful acknowledgments for the liberal support and encouragement accorded to my former publications.

H. D. HOSKOLD.

LONDON: *May* 1877.

INTRODUCTORY NOTE

BY

PETER GRAY, F.R.A.S.

Honorary Member of the Institute of Actuaries, and Author of 'Tables and Formulæ for the Computation of Life Contingencies' &c.

INTRODUCTORY NOTE.

THE PRESENT WORK, on the subject of the Valuation of Mineral Property, contains matter of very great interest for both the Professional Valuer, and the Actuary:—for the former in its ample discussion of the principles which should guide him in the discharge of his duties; and for the latter in its treatment of the problems (of a somewhat unusual character) which arise in the practical application of those principles, as well as in the mass of original Tables it contains, specially adapted for the easy and exact solution of any case that may present itself. The Tables occupy no fewer than 225 pages, and of themselves form a standing monument to the perspicacity and industry of the author.

There is found to exist among professional valuators some diversity of opinion and practice in regard to certain points in the purely actuarial portion of their work; and upon these points I have been requested to give my opinion. I will do so as briefly and plainly as I can, supporting my views with the requisite amount of demonstration, occasionally diverging, perhaps, into cognate matters that may press themselves on the attention.

The course of proceeding in the Valuation of a Mine appears to be as follows:—The valuator, in the exercise of his professional skill and knowledge, names a sum and a term of years, the former to be considered as the annual income to be derived from the mine, and the latter as the number of years that this income is to last. It is further arranged between the parties, that the purchaser is to be allowed a specified rate of interest on his outlay, during the entire term. The required value is thus presented in the form of an annuity certain, the elements of which—the sum, the term, and the rate—are known; and there remains only the conversion of that value into a present sum.

One of the points on which I am requested to give my opinion is as to the correct method of valuing the annuity which forms the subject of the valuator's first determination.

Ordinarily the valuation of an annuity for a term of years, when the rate of interest to be allowed to the purchaser has been arranged, is a sufficiently simple matter. The well-known tables of Smart (reproduced by Jones in his *Treatise on Annuities*), and others, furnish, in the cases that usually arise, all the aid that can be required, even by the most inexperienced computer. But the cases with which we have here to do are somewhat complicated by the entrance of a consideration that does not present itself—in so pressing a way, at least—in general practice.

It cannot be doubted that the purchaser of an annuity for a term, on which he is to be allowed interest at a specified rate, ought, as regards this transaction, to be in the same position, pecuniarily, at the end of the term, as if he had lent his money during the term at the same rate. The lender receives his interest annually, and has the sum lent returned at the end of the term. But the purchaser of an annuity must recoup himself by investing the excess of his annuity over the annual interest on his outlay, at such a rate that at the end of the term his capital will be reproduced. The lowest rate at which this reproduction can be *assumed* by the vendor or purchaser to be effected, is the rate allowed in the purchase of the annuity, as will presently be shown. In the case of annuities purchased at current rates, but little inconvenience and loss will occur to the purchaser from this restriction as to the rate of re-investment, since *practicable* rates in respect of such will usually differ but little from the *stipulated* rates. In the cases with which we are here concerned, however, the state of matters is far otherwise. In the purchase of mining property the purchaser, for reasons with which we have nothing here to do—they are fully discussed in the following work—is usually, perhaps always, allowed a rate of interest on his outlay far exceeding that at which he can invest the surplus of his annuity, which is called with propriety the *Redemption Fund*; and hence, if the *ordinary tables* are used in the valuation of the annuity determined and assigned by the valuator, the result must be a loss to the purchaser, more or less heavy according to circumstances, since in them the difference between the two rates is ignored. In the present connexion, therefore, special methods must be employed.

I will show here, first, that to reproduce the capital at the end of the term, when the tabular value of the annuity is used, the redemption fund must be invested at the *stipulated* rate, that is, the rate allowed to the purchaser; and I will then show how, when the *practicable* rate is taken account of, the value of the annuity may be correctly determined.

Denote by a the annuity for n years, and by P_n the purchase money, which is to yield the purchaser r' per £ on his investment.

The tabular value of the annuity is, we know,

$$P_n = \frac{a(1 - v^n)}{r'}, \quad \text{where } v = \frac{1}{1 + r'} \; ; \; . \; . \; . \; . \; (A)$$

whence

$$a = \frac{P_n r'}{1 - v^n}.$$

Now, a year's interest on P_n, the purchase money, is $P_n r'$, and therefore, in accordance with what is above stated,

$$a - P_n r', \text{ or } \frac{P_n r'}{1 - v^n} - P_n r',$$

is the redemption fund; and it has to be shown that this, if invested as it accrues, at the rate r', will amount to P_n in n years.

$$\frac{P_n r'}{1 - v^n} - P_n r' = \frac{P_n r' - P_n r'(1 - v^n)}{1 - v^n} = \frac{P_n r' v^n}{1 - v^n}.$$

Multiplying numerator and denominator by $(1 + r')^n$, this expression becomes,

$$\frac{P_n r'}{(1 + r')^n - 1} \; ;$$

and this we know is the annuity which, at the rate r', will amount to P_n in n years. And it is thus shown that when the value of an annuity is determined by the *common tables* (for those tables consist of series of values of $\frac{1 - v^n}{r}$), it is necessary, in order that the capital shall be reproduced at the end of the term, that the redemption fund should be invested at the rate allowed to the purchaser.

a

I am now to show how, when the stipulated rate—that allowed to the purchaser—is r', and the practicable rate—that at which the redemption fund can be invested—is r, the correct value of the annuity may be determined.

Let, as before, a be the annuity for n years to be purchased, and P_n the purchase money.

The redemption fund is $a - P_n r'$; and if we denote by M_n the amount of an annuity of £1, for n years, at the rate r, (the *practicable* rate,) the amount of the redemption fund at the end of the term will be $(a - P_n r')M_n$. Hence, since this, by condition, is to equal the purchase money, we have the following equation :—

$$P_n = (a - P_n r')M_n ;$$

and from this we get,

$$P_n = \frac{aM_n}{1 + r'M_n} \quad \cdot \quad \cdot \quad \cdot \quad \cdot \quad (1)$$

This is the value required ; and it is in a form very convenient for calculation, either by logarithms or otherwise. The form, however, may be varied. Thus, dividing numerator and denominator by M_n, we have,

$$P_n = \frac{a}{\dfrac{1}{M_n} + r'} ; \quad \cdot \quad \cdot \quad \cdot \quad \cdot \quad (2)$$

and $\dfrac{1}{M_n}$ being the annuity which will amount to £1 in n years —in other words, the redemption fund necessary to produce £1 in that time—at the rate r, if for $\dfrac{1}{M_n}$ we write s_n, the expression assumes the more compact form,

$$P_n = \frac{a}{s_n + r'} ; \quad \cdot \quad \cdot \quad \cdot \quad \cdot \quad (3)$$

and this is the most convenient for use when, as in the present volume, we are furnished in Table V, with the values of s_n for all terms and rates that can present themselves in practice.

The form chiefly, for special reasons, used by Mr. ·Hoskold in the body of the work, is the basis of (3), by substituting in it for s_n its value, $\dfrac{r}{(1+r)^n - 1}$.

We thus have,

$$P_n = \frac{a}{\dfrac{r}{(1+r)^n - 1} + r'}, \ \text{or} \ \frac{a}{\dfrac{r}{R^n - 1} + r'}, \ \cdots \ (4)$$

writing R for $1 + r$.

I give now a numerical example, in further illustration of what precedes.

Let the annuity be £100 for 20 years, on which the purchaser is to be allowed 5 per cent., while the redemption fund can be invested only at 3 per cent. The present value—the purchase money—is required.

I shall solve this first by the formula (A), which ignores the difference between the stipulated and the practicable rates.

The formula is, for this case,

$$P_{20} = \frac{100(1 - v^{20})}{\cdot 05}.$$

This might be worked by Table IV, which gives the value of v^n for all required rates and terms. But it is easier to take at once the value of the annuity of £1 for 20 years, from Table XII, p. clxxvi. We thus have $P_{20} = £1246 \cdot 221$.

This value fulfils the condition of replacing the capital at the end of the term, *if* the redemption fund can be invested at 5 per cent.

Thus, a year's interest on the capital is $62 \cdot 311$, and hence the redemption fund is $100 - 62 \cdot 311 = 37 \cdot 689$. Now, the amount of £1 per annum in 20 years, at 5 per cent., being (Table III, p. xxxvi) $33 \cdot 0660$, that of $37 \cdot 689$ will be $33 \cdot 066 \times 37 \cdot 689 = £1246 \cdot 223$, establishing the theorem.

On the other hand, if the redemption fund can only be invested at 3 per cent., its amount at the end of the term will be no more than, (p. xxxiv,)

$$26 \cdot 8704 \times 37 \cdot 689 = £1012 \cdot 718,$$

showing a deficiency of £233·503.

I now give a correct solution by (3).

The formula is,

$$P_{20} = \frac{100}{s_{20} + \cdot 05}$$

a 2

$$s_{20} \text{ (p. liv)} \quad \cdot 03721571 \text{ at 3 per cent.}$$
$$\cdot 05 \qquad \cdot 05$$

$$
\begin{array}{ll}
\cdot 08721571 & \log \; \overline{2 \cdot 9405948} \\
& \text{colog } 1 \cdot 0594052 \\
100 & \log \; 2 \cdot \\
\hline
P_{20} \qquad 1146 \cdot 582 & \log \; 3 \cdot 0594052
\end{array}
$$

Hence, £1146·582 is the value sought, and it fulfils the pre-scribed condition as follows:—

A year's interest on P_{20}, at 5 per cent., is 57·329, and the redemption fund, therefore, is $100 - 57 \cdot 329 = 42 \cdot 671$. And $26 \cdot 8704 \times 42 \cdot 671 = £1146 \cdot 582$, as it ought to do.

It is needless to enter on an inquiry as to the comparative advantages of the expressions that have been given for the solution of the problem under consideration, for in truth almost every case under the problem that can present itself has been already solved, and the solution is recorded in the following work; so that it is very rarely indeed that there will be occasion to have recourse to any formula. Tables VI to IX, occupying pages lxv to cxi, give the years' purchase, that is, the value of P_n on the supposition that the annuity to be purchased is £1, for every *practical* combination of the stipulated and the practicable rates, with the element n, the duration of the annuity; so that to complete the valuation there only remains the multiplication of the proper tabular value by the annuity whose value is required. The process, in fact, is entirely assimilated to that requisite in the use of the *common tables*, with the important distinction in the results that, in Mr. Hoskold's tables, due account is taken of the disparity between the stipulated and the practicable rates, while in the common tables this disparity is altogether ignored.

Table XII is very instructive. It shows, for various com-binations of the *stipulated* and the *practicable* rates, the excess of value assigned by the *old* (the common) tables over the true value for every pound of annuity purchased. I leave this table to make its own impression.

I have now indicated with sufficient distinctness that the method of valuation which I have sought to illustrate, and which is that advocated and employed by Mr. Hoskold, is the correct one. But before leaving the subject I would call attention to a variety of the problem which presents itself to be dealt with when, as is sometimes the case, the annuity to be purchased is deferred; that is, which, while making the same number, n, does not commence its payments till after the lapse of, say, t years. The symbol for the value of the annuity, when subject to this condition, might with propriety be $P_{t|n}$.[1]

The value here, in accordance with a well-known principle, is,

$$P_{t|n} = v^t P_n, \text{ or } \frac{P_n}{(1+r')^t};$$

equivalent forms, since v^t and $(1+r')^t$ (in which r' is the stipulated rate), are reciprocal, each to the other; and hence, when the value of an immediate annuity is found (or known), that of the same annuity, when deferred, can be readily deduced.

A demonstration is given also, by Mr. Hoskold, on p. 34, founded on elementary principles. And I may remark that it is, as I believe, only now, that, for the first time, the value of the deferred annuity is correctly assigned.

Here, too, as in the case of the immediate annuity, the wants of the computer have been anticipated and supplied by Mr. Hoskold. Tables X and XI, occupying pages cxiii to clxxii, contain the values, for most practical rates, of a deferred annuity of £1 (the number of years' purchase), the periods of deferment ranging from 1 to 10 years. In consequence, the necessity for having recourse to a formula will very rarely occur, the value required in any particular instance being usually obtainable from the tables named by the merest inspection.

It is only such as have had some experience in the construction of tables who will be able to realise the great amount of labour involved in the formation of those that have been specially referred to—I mean Tables VI to XI; but it is very certain that everyone who may have occasion to use them for practical purposes will bear willing testimony to their great utility.

---

[1] Mr. Hoskold uses P_{t+n}. The form above suggested seems on the whole preferable: the suffix $t+n$, being the sum of the periods of deferment and duration together, is the entire term over which the transaction extends. [I will add that, having *instinctively* here written *deferment*, while Mr. Hoskold writes *deferrence*, I have been accustomed to do so on the authority of the late Prof. De Morgan (*Compan. to the Almk.*, 1840, p. 16). I find neither of the words in the dictionaries to which I have present access; probably therefore both may be equally legitimate—or illegitimate.]

On pages 30 to 32 Mr. Hoskold points out, demonstrates, and freely uses a relation that may be thus enunciated :—

The annuity for n years that £1 will buy, exceeds the annuity that will amount to £1 in n years, by r, the interest of £1 for a year.

This relation I find is not unknown to some actuaries; nevertheless, as it has not yet, so far as I know, found its way into the books, it may be worth while here to place it on record.

The proof is very simple. The annuity for n years that £1 will buy is,

$$\frac{r}{1-v^n},$$

which we may write thus,

$$\frac{r(1+r)^n}{(1+r)^n-1};$$

and the annuity that will amount to £1 in n years is

$$\frac{r}{(1+r)^n-1}.$$

Subtract now the second from the first, and we get

$$\frac{r(1+r)^n}{(1+r)^n-1}-\frac{r}{(1+r)^n-1}=\frac{r\{(1+r)^n-1\}}{(1+r)^n-1}=r;$$

and so the theorem is established.

The foregoing relation can be shewn to hold from other considerations than those adduced above. The formula (3), p. xiv, *ante*, when $a=1$, becomes

$$P_n=\frac{1}{s_n+r'},$$

and denotes the value of an annuity of £1 for n years, at the rate r', when the redemption fund is invested at the rate r; and if $r'=r$, the value indicated is that of the ordinary annuity. In this case then s_n+r will be the annuity for n years that £1 will buy, since this annuity and its present value are mutually reciprocal. And hence, since s_n is the annuity that in n years will amount to £1, we again see the relation to subsist.

I will just add by way of corollary, that, the value of the annuity which £1 will buy being of course £1, and that of the annuity which in n years will amount to £1 being v^n (since this is the value of £1 to be realised in n years), the difference of these values is $1 - v^n$. Now this must be the value of an annuity of r, (the quantity by which the annuities themselves differ,) for n years. And this is seen to be the case as follows :—

$$\frac{1 - v^n}{r} \times r = 1 - v^n.$$

There is another point in the valuation of Mining Property in regard to which diversity of opinion and practice exists among valuators; and on which also I have been requested to give my opinion. The point here referred to arises as follows :—

A mine is to be sold having a specified term to run. The valuator, in the exercise of his best judgment and technical skill, assigns the annuity on which the purchaser may probably reckon during the term of duration, with the rate of interest to be allowed him on the purchase-money. Here a new consideration sometimes—perhaps I should say frequently—arises. The sum named by the valuator as the probable annual return to the purchaser is that which he considers ought to be the return if the mine is fully developed. At the same time he may be of opinion that to bring the mine into this condition an expenditure of greater or less amount in the early years of the mine is necessary. In these circumstances he does not abate from his estimated annual return, but names a sum, as cost of development, to be expended by the purchaser in equal portions during the first few years of the mine, to bring the mine into the required condition : and which sum consequently, when valued subject to the conditions of its disbursement, will constitute a deduction to be made from the gross value of the mine, so as to determine the amount of the portion of this value payable to the vendor. And it is as to the manner in which this deduction is usually made that I am requested to give my opinion.

For illustration I quote a case given by Mr. Hoskold, p. 119.

The term of a colliery for the next 21 years is to be sold. It at present yields a net return of £8,000 per annum; and the valuator estimates that to maintain the return at this rate, during the term, it will be necessary for the purchaser to expend in works, &c., £12,000, in equal portions of £4,000 during each of the next 3 years. Also, the purchaser is to be allowed 20 per cent. per annum on his outlay, redeeming his capital at 3 per cent. Required the net amount now payable.

I will determine the required amount first in the customary way. .

By Table VII, p. xcii., the gross value of the mine is 4·25764225 × 8,000 =   £34.061·138

And the abatement is :—

Cost of works, &c. . . .	12,000	
Interest on the same, 3 years at 5 per cent.	1,800	13,800·000

Net amount now payable,   .   .   .   £20,261·138

Now this cannot be correct. The abatement here is the *amount* in 3 years, at 5 per cent. simple interest, of the disbursements *to be made* by the purchaser ; and it could be legitimate only if the entire £12,000 had been disbursed *three years ago*. The purchaser in fact *receives a bonus* for delaying payment of a portion of the purchase-money—a bonus, moreover, which *increases as the delay increases*.

Surely nothing further needs be said to establish the inadmissibility, in accordance with any rational or recognised principle, of the method just exemplified for determining the net amount now payable.

The following shews the manner in which I consider the required determination ought to be made :—

Gross value as before,   .   .   .   £34,061·138

Abatement for cost of works :—

Value of an annuity of £4,000 for 3 years, at 3 per cent. (Table XIII) 2·82861 × 4000 =   .   .   11,314·440

Net amount now payable,   .   .   £22,746·698   .

The annual payments to be made by the purchaser constitute an annuity, and there exists no conceivable reason why they should not be valued as such.

The only point in regard to which there may be thought to be room for question as to the validity of the method here employed, is the rate at which the annuity of £4,000 ought to be valued; and somewhat plausible reasons might be adduced for making the valuation at the rate of 20 per cent. I am quite satisfied, however, after full consideration, that any arguments in this sense that could be assigned are groundless. The purchaser is entitled to £20 per cent. on his outlay, which is the gross value of the mine. It is true that a portion only of this—in the present case the larger portion—goes at once into the pocket of the vendor; but the rest is expended in the amelioration of the property, whereby the purchaser is proportionally benefited.

I am pleased to find myself in regard to this second method of solution in entire accordance with Mr. Hoskold, who has largely attended to the subject; and whose remarks on pp. 120, 121, I commend to careful consideration.

P. GRAY.

LONDON:
June 11, 1877.

CONTENTS.

PART I.

REMARKS UPON THE VALUATION OF MINERAL PRO-PERTY AND MATTERS CONNECTED THEREWITH.

PART II.

CONSTRUCTION OF TABLES OF VALUES, WITH
RULES AND FORMULÆ.

Nature and value of annuities—Increase of principal; formula, rule and
example—Present value of a perpetuity of £1 payable once every
nth year, the first payment due n years hence; formula, rule and
example—Present value of a perpetuity of £1 deferred n years; formula, rule and example—Theory and mode of ascertaining the advantage of payment of annuity at varied intervals; values by yearly,
half-yearly, and quarterly payments—Amount of £1 per annum in n
years; formula, rule and example—Present value of £1 due n years
hence; formula, rule and example—The annuity which £1 will purchase, found by adding together the redemption fund and interest—
Redemption fund to reproduce £1 in n years; formula, rule and example—Present value of £1 per annum for n years, allowing one
rate of interest on purchase-money, and another rate of interest for
redemption; formulæ, rule and example—Recapitulation of formulæ
—Examples illustrative of, and demonstrating the accuracy of, the
formulæ—Deferred annuities, with two rates of interest; formulæ,
rule and example—Discussion of subject, with further examples;
formulæ and rules—Difference of opinion; analysis of ditto—Rule
established and confirmed by further practical examples—Redemption
of capital; examples—Conditions considered in forming the tables
of values—The tables calculated from first principles—Inaccurate
tables worse than useless—Tables great labour-savers, as values may
be found by mere inspection—The amounts proved by logarithmic
process—Tables have been frequently published, but have not been
calculated above 10 per cent.—Other writers, rules, formulæ, and
papers written on two rates of interest—Problem; solution of problem by other writers—*Mr. Peter Gray's* solution—Rates of interest
allowed on mining property 10 to 25 per cent.—Errors in the con-

---

PART III.

PRACTICAL EXAMPLES IN VALUING COLLIERIES, IRON AND OTHER MINES, ROYALTIES, LEASE-HOLDS, FREEHOLDS, LIFE INTERESTS, &c.

another rate, the duration and annuity being given—To find the redemption fund necessary to produce £1 within a given time; formula, rule and example—To find the present value of the unexpired term of the lease of a colliery subject to a lessor's royalty; the annuity, rate per cent. on purchase-money and for redemption, royalty, and value of the plant being given—To find the present value of a colliery with a certain annuity and duration, subject to a lessor's royalty to be paid quarterly; interest on purchase-money, royalty, and for redemption, being given—Difference in present value due to yearly and quarterly payments; rate of interest gained by quarterly payments—To find the present value of the lease of a colliery having a certain annuity and time to run, subject to a variation in the lessor's dues at fixed periods, and the expenditure of an additional sum of money to sustain the annuity during the period; interest on the purchase-money and for redemption being given, together with the estimated value of the plant—Detailed values—'Customary' interest—Valuation of the lease of a colliery under varying circumstances—Practical example; detailed values; redemption fund, &c.—Remarks on last preceding case—To find the present value of an unopened colliery with the overlying estate; time and cost of development, prospective annuity, rate per cent. upon purchase-money, for redemption, royalty, wayleave, &c., being given—Value under similar conditions as in preceding case, payments to be made half-yearly—Difference in value between yearly and half-yearly payments—Value under similar conditions, payments being made quarterly—Differences in value between yearly, half-yearly, and quarterly payments—To find the value of the lease of a mineral property with several seams of coal and iron ore; annuity, time of duration after opening the seams, interest on purchase-money and for redemption, wayleaves, royalties, &c., being given—Details of the whole of the valuations in preceding case—Reduction from gross to net values—Summary of preceding cases—Remarks on last preceding case—Lease of 21 years' duration—Comparison of values in preceding case, time of lease being shorter—Summary of values under shorter duration of lease—Deduction arrived at from comparison of values—To find the present value of the life interest of a person A, aged x, in the royalty of an iron mine; the annuity, interest on purchase-money and for redemption, being given; formulæ, rule and example—Correct method of valuing compared with 'customary' mode; detailed example,

---

PART IV.

TABLES.

ERRATA.

Page 12, line 16 from bottom, *reads*, 'il possible;' *should read*, 'if possible.' Page 18, line 6 from top, *reads*, 'Royal Forrester;' *should read*, 'Royal Forester.'

TABLE I.

					reads	*should read*	
¾ per cent.,	91	years,			1·9937856536		1·9737856536
3	,,	46	,,	,,	4·8950437169	,,	3·8950437169
13	,,	38	,,	,,	104·987432	,,	103·987432
15	,,	32	,,	,,	87·562068	,,	87·565068
15	,,	80	,,	,,	71750·979401	,,	71750·879401
17	,,	93	,,	,,	2194245·22623	,,	2194245·12623
22	,,	34	,,	,,	803·44413	,,	863·44413
25	,,	89	,,	,,	421687917·92926	,,	421687917·72926

TABLE III.

					reads	*should read*	
1¾ per cent.,	11	years,			12·0418439241		12·0148439241
3	,,	28	,,	,,	43·9309225246	,,	42·9309225246
5	,,	80	,,	,,	971·2288123372	,,	971·2288213372
5	,,	97	,,	,,	2251·9416156374	,,	2251·9146156374
8	,,	7	,,	,,	8·992803	,,	8·922803

Corrections to be made by the Pen

Page lxxii, for 63 years at 11 % reads 6·56986861 *should read* 8·56986861
„   lxxiii,   „   21   „   14 %   „   4.65651405   „   5·65651405
„   lxxxix, „   43   „   14 %   „   7·55203993   „   6·59203993
„   „   „   44   „   14 %   „   7·61245131   „   6·61245131
·,   „   „   45   „   14 %   „   7·63195168   „   6·63195168
„   „   „   46   „   14 %   „   7 65059272   „   6·65059272
„   „   „   47   „   14 %   „   7·66842240   „   6·66842240
„   xciii,   „   97   „   22 %   „   7·50839226   „   4·50830226
„   ccxix,   „   10¾d.   „   0.4470167   ·,   ·04470167

PART I.

PRELIMINARY REMARKS

UPON THE

VALUATION OF MINERAL PROPERTY

AND

OTHER IMPORTANT MATTERS

CONNECTED THEREWITH.

PART I.

VALUE has been defined as 'the quality in anything which fits it to be given and received in exchange;' the meaning of the term, however, has been much discussed and controverted. It frequently occurs in the writings of political economists, and has by them been employed in a modified sense, as '*value in use*' and '*value in exchange.*' Considered therefore relatively, the former may be defined as representing the *intrinsic*, and the latter as the *estimated* or market worth of an article.

The value of all exchangeable articles of utility must, however, be determined by the money worth set upon each commodity when brought into the market. The deduction then to be arrived at from the general order of things is that a *pound* sterling will command the purchase of a larger quantity of articles of commerce of one kind than it will of another, and the amount of each as compared with a pound sterling as a standard will also vary with the fluctuations of trade, which are dependent upon and regulated by the law of demand and supply. When, therefore, materials intended for commercial purposes are produced and brought into a proper marketable condition, they are said to possess a certain relative worth.

Value is as comprehensive as it is significant, constituting as it does a general standard of the comparative excellence or worth of all commodities necessary for the use, comfort, and maintenance of life.

Such articles as are of daily consumption cannot, however, be produced, distributed, and applied, without the expenditure of a considerable amount of labour, which in itself is the mainspring of creative wealth. The science of values, therefore, is one of immense public importance, and underlies political economy in all its branches.

I do not propose entering into this very important though

intricate subject further at present, except so far as relates to one of its leading branches, viz. Valuation of Mineral Properties.

Mining operations are now conducted over wide-spread fields, in which the most highly educated, the profoundest thinkers, as well as those in the possession of more moderate intellectual capacities, may enter and work, each in his own way and order contributing his mite of knowledge to the forwarding of the general interest, welfare, and intelligence of the age.

The great importance of legitimate home mining to the general support of the State, and the advancement of the interest and creation of wealth of the English nation, can hardly be over estimated; but that there are other than honourable speculations in mining matters, tending to produce opposite effects, is also true, and it is this latter class that should be avoided, discouraged, and exposed.

Millions of pounds sterling are annually sunk in worthless mining schemes, many of which never were or could be capable of yielding any good results, and some of them probably never had any real existence at all, except perhaps in men's imagination and on paper. It is therefore in this and many other ways of a similar nature that an honourable, generous, and unwary public are beguiled by the plausible representations of designing men, and led on to contribute year after year to the highly coloured but rotten plans and schemes for making money so frequently presented to their view, until a crisis is produced, resulting in the utter ruin of thousands.

In no branch of industry is there required so much real practical skill, sound judgment, indomitable perseverance, combined with undeviating integrity of purpose, and the knowledge of the application of scientific principles, as there is in Mining; indeed upon the possession of such qualities, exertion of abilities, and timely application of these elements to the various operations involved, the success of every mining enterprise mainly depends; but whether the general conduct of such undertakings is always entrusted to persons of this description, possessing abilities of the higher order, is quite another question, and the general experience of the mining world would probably go far to negative such an assumption.

Mining, when honestly conducted, is undoubtedly one of the most legitimate and lucrative undertakings that can possibly be entered into, as is evinced by the large fortunes realized by those who have entered into and carried on *bonâ fide* concerns for lengthened periods.

Of late years Mining has become a leading branch of industry, and upon the general success of such adventures depends very much that of nearly every other trade.

It is a subject of paramount importance, and has received and is still receiving great attention both from legislators and capitalists; but whether the Acts of Parliament recently provided, regulating the working of mines, will prove of great benefit, and justify public expectation, is a question time must solve.

The amount of capital invested in mines during the past three or four years was at least ten-fold, as compared to what was expended in any similar period during the past half century. During the former period the unprecedented stimulus given to trade at home, on the Continent, in America, and the Colonies, in articles of general commerce, has hurried on and produced the present reaction.

As time goes on, however, producing a certain firm balance in trade, and all illegitimate means of speculation have been checked, avoided by the public, and crushed when discovered, Mining will be placed upon a more solid foundation, and will receive still greater attention from men of wealth, and eventually become a strong arm of the nation.

The estimate of an exhaustion of minerals in our coal-fields will, I trust, fall very short of the truth; and as it is now known by recent discoveries that some of the English mineral fields extend much further than was formerly supposed, in time new boundaries will be assigned, giving to them a much larger area, and, of course, a larger quantity of minerals. I believe that other discoveries will also be made, upon further explorations being conducted, leading to results not now anticipated. Correct reasoning, therefore, should lead to the conviction that *Providence* has placed and hidden from immediate view immense wealth, at great depths in the earth's crust, which it is to be hoped will be brought forth as necessity requires until the end of time. Very little, if indeed anything definite, is known as to the condition and thickness of the deeper coal-formations below the *Permian* and the *Trias*; but however deep coal-seams may have to be encountered in the future, I am convinced they could not so exist without being of service to man, which implies the possibility of extraction; and that when their development is found to be necessary, practical science will be applied so as to overcome every difficulty which is now associated with great depths. We are well aware that the area of circles are to one another as the square of their diameters; it is therefore reasonable to infer that by increasing the size and number of down-cast shafts, also the in-take roads from them, and airways to similar up-cast shafts, in proportion to the depth, with the addition of more powerful ventilating machinery, a greater quantity of air may be conveyed, reducing the temperature to a fit condition for the support of life, and equal to that in mines of comparatively shallow depth, with ordinary sized shafts and air-ways. It therefore seems to me to simply resolve itself into a question of time and cost in the development of the deeper mines. With our present appliances for sinking shafts by machinery, and with the great improvements they will doubtless receive in the future, very great depths ought to be attained at a moderate outlay, and in a shorter period, as compared to the cost and time occupied by employing hand labour.

The great drawback at the present time to the immediate development of *virgin* properties, even of a good class, arises chiefly from the fact that the minerals contained in them exist at a considerable depth, and from opinions expressed by persons who really have no experience or valuable knowledge with regard thereto, but who nevertheless take occasion to insinuate their advice with the view to damage the interests of some and serve those of others. It is, no doubt, an undeniable fact that capitalists are frequently misled as to the nature and value of mineral properties.

Numerous mines of all kinds are constantly introduced to the notice of capitalists, and each offer to sell is generally accompanied with some statement of the merits of the property, or a general report compiled by some local engineer. The documents so produced are sometimes from men of great ability, experience, and integrity, but in many other cases they are concocted for a purpose, which in a general way may be seen by men of judgment upon the face of them.

Unless, therefore, the writer of any such report is a man of some *status*, it is customary for an intending purchaser to instruct some other engineer to proceed to the *locus in quo*, examine into all matters connected with such property and report thereon, the value and reliability of such report depending entirely upon the ability and truthful purpose of the engineer so *engaged* and *trusted*. Capitalists should therefore use due caution in employing persons upon whose very *word* so *much* depends.

I consider that gentlemen embarking their capital in extensive mining undertakings should well understand the position they are about to occupy, whether voluntarily entered into, or by the introduction of others, and that persons entrusted with their absolute confidence should make a sacrifice of their own interests, if needs be, rather than deviate one iota from a strictly honourable course.

The best advice should be given, whether each individual proprietor is a business or financial man, or otherwise; but that parties are constantly let into worthless concerns is too obvious a fact to admit of doubt or denial.

It appears to me there is great need in London and other large cities for the establishment of firms of mining engineers, comprising persons of practical ability from each mining district, who would conduct their business in an honourable way, and upon terms that would receive the countenance and co-operation of bankers, brokers, financiers, and other gentlemen seeking *bonâ fide* properties as investments.

All the mines of importance for sale in each district would consequently be known, and in a short time would flow into the offices of such establishments. Such a business, conducted upon high class principles, must undoubtedly find support from the general public, and would exert a healthy influence, and produce results of incalculable benefit to the legitimate mining community and public at large.

All advice upon mining business comes within the province of the Mining Engineer, who should not only be a scientific man, but a financier as well. It is, however, true that the latter acquisition is not necessary to the actual conduct of the mere operations of a mine, except so far as relates to its cost, but it is an additional knowledge very necessary to be possessed in order to be able to make any mine pay a dividend upon the capital invested in it.

Mining operations conducted on virgin tracts, in comparatively old, well-known, and proved fields, are not so much questions of speculation as of sufficient capital to develope the minerals contained therein, as the number, thickness, depth, quality, and condition of the several veins or seams of coal contained in such fields are generally known. But when it is intended to open up a colliery or other mine in a field wholly unexplored, the case is different, and frequently becomes very perplexing, and creates considerable toil and anxiety in arriving at a just and reliable conclusion with regard thereto.

To the experienced mining engineer, who is, or at least should be, a geologist, there are always certain distinctive features in the strata protruding to the surface of the earth in many places, in every district, or laid bare by quarries and cuttings, pointing out the probable existence, and also the kind of minerals likely to be discovered; even in an entirely new field the indications are sufficient generally to lead to the exact site of the outcroppings of the minerals, unless covered by a newer formation, as is the case of various seams of coal in the Somersetshire, Leicestershire, some portions of Staffordshire, and other coal-fields. This was notably the case at the Sandwell Park Colliery in Staffordshire, a spot where no coal was believed to exist by the generality, being outside the known field, but a recent sinking there has however turned out an immense success, through the knowledge, energy, and persevering skill brought to bear upon it by Mr. Henry Johnson. To him therefore alone is due the credit for the discovery of the extension of this field.

In cases where the outcroppings of mineral veins or coal-measures are overlaid by strata of a newer formation, explorations become more difficult and expensive, and can only be carried on successfully by resorting to boring operations, which are in doubtful cases preferable to sinking a shaft; but should it be decided to sink a trial shaft in the first instance, and a considerable sum is to be so expended, dependent upon the recommendation and knowledge of any individual, it becomes a question for mature deliberation for all parties concerned, before entering upon it.

When the minerals are known to exist, and it is only a question of development, the annual income likely to be derived, and consequently the value of the mine, will be *affected* more or less by the mode of operating, and by the amount of skill brought to bear upon it, for if the plan proposed to be adopted has been well thought out and laid down, failure in any one important point in the execution, resulting from defective knowledge of the district in general—leading to a large expenditure upon inadequate machinery—or other mischance, the value may be considerably diminished, or indeed entirely lost.

I am aware of a similar case where a party came into a certain mining district with which I am perfectly familiar, but with which they had no acquaintance whatever, and after purchasing a small colliery property, they erected plant and commenced sinking two shafts, and carried them down I believe about 150 yards, without making any provision for a pumping engine or pumps, and although they were repeatedly informed by persons conversant with the district that a considerable though not excessive quantity of water existed, nevertheless they entirely disregarded such advice, and upon continuing their shafts down to the watery strata the pits were soon flooded. The outcrop of the coal-seams and general measures were elevated considerably above the top of the pits. They afterwards erected a pumping engine, but the machinery was cumbersome and inadequate, and to the present day the water remains, although it is believed that over £100,000 has been wasted. Now here is a most glaring case, although perhaps not an isolated one, for the site selected was to the dip of the field of coal, and it might have been anticipated that all the water in the rise area would flow downwards. The outcroppings of this field were well defined and thoroughly known; several land works also existed, surrounding the site of the new winning, but no attempt whatever was made to ascertain the quantity of water pumped from each mine to the rise, although it was susceptible of computation.

Putting a *problematical* case, and assuming that a valuation had been made of this particular property—which was small in area—and that it was £60,000, which without actually going into figures I believe to be too much for it, by spending £100,000 in vain attempts, would leave a loss of the entire sum,

minus the present value of the plant and machinery; but supposing that all the coal had been extracted, and that the profit
per ton upon which the valuation was made had been realized,
there would be a clear loss of £40,000, and the only set-off
against this would be, as before stated, the present value of the
machinery.

It is very probable that many such cases as this have
occurred, but it is high time that the expenditure of large sums
of money should be entrusted to men of better judgment.

Before commencing operations for the development of any
mine, it is very necessary to examine all the valley outcroppings,
or low levels, if such exist, natural or artificial outfalls or free
drainage, existing old adits and pits from which water is or
may have been pumped, surrounding the entire area to be
developed, also the probable effect produced from the average
rainfall due to the district, the quantity of water likely to be
delivered from the rise area may then be closely computed, and
the size of the pumps necessary to raise a similar quantity of
water from a given depth be determined. Of course in such a
case ample allowance should be made in the size to provide
against sudden inflow of water through porous strata, occasioned
by excessive wet seasons, and other contingencies. The allowance to be made must depend upon the requirements of the
case, and the judgment and capabilities of the engineer in charge
of the execution of the works; but it is not unfrequently the
case that the hands of a good man are completely tied by the
control exercised by a board of directors, who perhaps for the
first time may have engaged in mining. Such interference is
most absurd, and occasionally proves very ruinous to the shareholders, because a really good and efficient man could not work
under such restrictions.

I believe it will always be found, as a rule, that to err on
the side of *excess* of size for machinery and pumps is far better
than defect.

It is clearly the work of the Mining Engineeer in charge of
getting up a Report, Estimate, or Valuation, to ascertain every
fact, and to bring out every point bearing upon any property
under consideration, whether in favour of or against it; and it
is only by such proceedings that a satisfactory conclusion can
be arrived at, but it is very important that all facts should be

ascertained by personal attention, not taking for granted or using the information supplied by others, unless it were to agree with independent investigation.

The characteristics of any adjoining property will generally form a good guide as to the condition of the minerals in the estate in question; but as there are faults and denuded parts existing in every coal-field, it would be very difficult to discover if any such existed, and how far the seams of coal might be affected thereby, if the property or estate is situate at a long distance from any well-known underground workings.

The strike of any general disturbances may, however, be determined pretty clearly if they have been found to exist in any neighbouring colliery.

It is not possible, when property is so circumstanced, to determine with absolute accuracy, the exact quantity of minerals contained within any given *area not explored*, and when faults and other disturbances are suspected, but not defined, it frequently becomes a very complicated question, and then an approximately correct estimate only can be expected.

A great many points involved and relating to each distinct property, will present themselves to the engineer for consideration and analysis, and there will always be found some special and distinctive features and circumstances connected with each property which will tend to affect the value, which can only be determined by the persevering skill and judgment of the engineer.

When a colliery or other mine is opened up and partially exhausted, the mode of procedure is very different and more direct, as all the seams are laid open, and everything in connection with them can be satisfactorily determined, and it only becomes a question of accuracy in surveying the underground workings of the colliery and those adjoining it, and the surface boundaries, in order to determine the reserve area, and consequently the quantity of coal, presuming, however, such area is free from faults, and the seams of coal of uniform thickness.

In a general way, the quantity of coal per statute acre may be accurately determined by taking the average specific gravity of several samples from different parts of a seam, and then deducting a certain proportion for waste. The quantity to be allowed will vary with different seams, and under different circumstances in each district, and sometimes in different collieries in the same district.

The proportion of large and small, and the marketable quantity of coal to be obtained from any seam, will depend upon the uniform thickness and condition throughout, the system of extraction employed, and the general management.

My practice has been to allow $\frac{1}{5}$th upon the quantity as determined by the specific gravity, when the seams are found, or at least believed to exist, in a healthy condition, leaving about 1,200 tons per statute acre of one foot thick, to be realised by extraction. I have, however, known the yield to differ from this, both in excess and defect.

My experience of Hematite Iron Mines, taken throughout a district, is that they are capable of yielding about from 5,000 tons to 10,000 tons per acre, and in a few instances as much as 20,000 tons per acre. This refers to general deposits, existing in the Carboniferous Limestone, and not to surface or mere accidental and isolated patches.

In making an estimate of the quantity of minerals to be expected from any unopened mine, such a quantity should be assigned per acre as would be justified by the experience of the general yield of a whole district.

Of course barren, unhealthy, and denuded portions exist in most stratified mines in every locality, and these should be discovered, il possible, and due allowance made for them in the final result.

Great attention should also be given to the nature of the strata to be passed through in sinking, the cost of labour and materials, and, in fact, to everything connected with a general estimate of the cost of winning, including plant and machinery of every description necessary to produce certain results. These are points demanding very full investigation, involving considerable experience and judgment in the execution of works of a similar nature to those under consideration.

The cost of establishing an extensive winning, including the conduct of all the present and future operations, affects the value of a property to a present purchaser very considerably, although it does not alter the original condition of the property. It is therefore of the first importance to ascertain the outlay likely to cover the cost of the whole of the development, not forgetting to make ample allowance for contingencies, or any unforeseen difficulties which may be encountered.

The position of the property in relation to railways, markets, and to surrounding collieries, competition in trade, demand for the produce, cost of labour and production, and the net profit per ton, are among some of the principal points which are of very great importance to be determined.

The cost of production will be very much affected by, and depend upon, the state of the labour market, the nature and inclination of the measures immediately over and underlying the coal—involving a small, moderate, or large quantity of timber—quantity of noxious gases to be encountered, uniform thickness and quality of the seams, existence of faults, or denudations, and whether any of the coal has to be left in order to support any part of the estate or royalty, upon which any portion of a town or other buildings may have been erected, as also the amount of capital, if any—in the case of a going concern—required to extend any present or future operations which may be necessary to support a given yield of coal, or other minerals, per annum, during the remainder of its duration.

The accessibility of any other seams of coal or minerals in the royalty from existing winnings, and if the property is extensive, and the lease of short duration, the probability as to its renewal, amount of dead rent and royalty, and power to assign the lease at any time, with its responsibilities, are all points which must be duly weighed by the valuer on behalf of the vendor, and the incoming tenant. There are also other points too important to be lost sight of, viz. facilities for the extension of surface arrangements, such as new erections, railway branches, areas in reserve for tipping colliery refuse; also any trespass committed upon the royalty at remote points by the workings of adjacent collieries, all tending to produce an effect upon the cost per ton of the minerals raised, and all of which must enter into the calculation, in order to arrive at a just and reliable valuation.

Every purchaser of mining property, should have ample allowance made upon his purchase, but the amount of such an allowance, as a *percentage,* must depend upon a point difficult to calculate, viz. the attendant risk to be incurred in mining matters, in making a certain annuity or annual income during

the existence of the mine, and to be placed in a position to recoup the capital invested within the period of duration or time
of the lease. /

All things being considered upon a fair basis, and assuming
the property to be a good one, no one would be in a better position
or qualified to ascertain the attendant risk, than an experienced
mining engineer, but whether from caprice, fear, or doubt as to
certain results, the question is too frequently left to the decision
of an intending purchaser; it would therefore appear to resolve itself into a question of agreement between vendor and
purchaser, and no doubt it is a safe plan of throwing all
responsibility upon the shoulders of the purchaser, and would
save the reputation of the engineer, assuming the property in
question did not eventually come up to the expectations which
sanguine persons may have entertained with regard to it.
Hitherto, therefore, in very many cases, valuation has been considered more as arbitrary means, dependent upon mere opinion,
than that of a system based upon correct and scientific principles.

The income derived from the working of mines may be
ascertained from the general accounts, if they have been strictly
and truthfully kept, and the value deduced therefrom, coupled
with all existing circumstances connected with the mine, but it
would be necessary to employ in the process the average annuity
that may have resulted over a series of years in the past.

With regard to the amount of percentage to be allowed, or
years' purchase a mineral property is worth to a present purchaser, much difference of opinion has existed and still exists,
as will appear from the following quotations:—

In 1829 a Committee of the House of Lords examined Mr.
Buddle, a mining engineer in the county of Durham, upon the
valuation of mines, and he stated that, 'Having considered
what the risk is worth according to the nature of the colliery,
the profit is estimated as an annuity, and that annuity is purchased at so much per cent., varying according to the probable
risk from 8 to 20. In some instances it would be a safer
purchase at 10 per cent., and redeem the capital, than in other
cases of great risk it would be at 20 per cent.; but then, in
these valuations, if it is for a purchaser, I generally submit a
scale of purchase at such a rate as would be worth so much, and
at such a rate so much. You may take my advice as to the

risk, but you must decide for yourself. One man may be satisfied with 10 per cent., while another less adventurous might
expect fully 15; therefore it altogether depends upon the
opinion of the person purchasing at what rate per cent. he would
purchase.'

Mr. Dunn states on page 82 of his work on the Coal Trade,
that Mr. Buddle in his evidence asserted that ' 5 per cent. was
the average profit of collieries, after returning the capital. The
highest rate of profit he knew of was 14 per cent., including
redemption of capital, viz., 5 per cent. profit, and 9 *per cent.
redemption.*' Some error must exist in this statement, and it
is most probable that 9 per cent. profit and 5 per cent. redemption was meant, but it is difficult to see how a mistake could
have been made, inasmuch as Mr. Buddle says 5 per cent. profit
in two cases.

The report of the Select Committee of the House of Commons, published in 1857, on the Rating of Mines, presents the
opinions of several engineers and others called to give evidence
upon the mode of valuing mines. J. Pease, of Darlington, said
' he would calculate his purchase on about 10 years, as applied
to coal-mines.' Mr. S. Dobson, Glamorganshire, said that he
' thought coal-mines should be valued at an average of 8 to 10
years' purchase. Land is worth about 30 years' purchase, dwel-
ling-houses from 20 to 25 years', manufactories perhaps 15
years', and in respect to coal-mines, you may take the average
at from 6 to 8 years' purchase, and you may take ironstone
mines at much the same. He thought there was no more difficulty in fixing the annual value of a colliery than there is in
fixing the annual value of a factory. You must always take
into account the quantity of minerals raised, because the annual
value of the property in all cases (or nearly so) depends on the
quantity raised.' J. H. Coterell, surveyor, Bath, ' had settled
the value of mines in arbitration, and fixed them at from 6 to
8 years' purchase; they were very short of railway accommoda-
tion to their collieries; when they got that, he thought the
mines in his district would be worth a little more.' T. J.
Taylor, of Earsdon, Northumberland, upon being asked ' How
many years' value do you calculate you ought to give if you
were going to open a mine?' said, ' There are two distinct
circumstances which arise for consideration in answer to this

question. The first is that where the freehold of a mine is purchased it is usual to allow 8 per cent. upon the perpetuity—that would be $12\frac{1}{2}$ years' purchase. The duration of a mine is less than a perpetuity—say 10 or 11 years' purchase; the allowance for that depends entirely upon the length of time the mine has to last. The other case is the case of the purchase of a lessee's interest in a mine; the purchase of the interest of the occupier of the mine in distinction from that of the lessor. Then an annuity has to be purchased, subject not only to the mining risk but also to the occasional risk; it is calculated as an annuity for the term of the lease. It varies from 12 to 18 per cent., and gives from 8 to 5 years' purchase.' H. W. Schneider, M.P., said: 'From my knowledge of the subject of the value of mines, taking iron-mines and coal-mines of every description—taking the whole of England from one end to the other; I would not give 10 years' purchase for all the mines in the country—including royalties. If anyone would give me 10 years' purchase for my own best mine (Park Iron Mine, in Dalton, Lancashire), I would very gladly take it; indeed, in stating in general terms 10 years' purchase, I have gone beyond the mark. Public opinion, which is the best criterion in such cases, shows that from 7 to 8 years is about the average with respect to mining property generally throughout the country. If you look to the value of any great mine, and see the dividend it is paying, and multiply that dividend by 7, you will find that that is very nearly the value of that mine. As regards the royalty—supposing the amount of the royalty is £10,000 a-year, the question would be, what would that £10,000 produce in land? Taking it at 3 per cent., it would give you £300 a-year, and that would be somewhat equal to 10 years' purchase for the royalty.'

In Dunn's work on the Coal Trade, at pages 208–9, he says: 'If [the mine is] unopened or unproved, its value must be necessarily dubious, especially if the prospective period of its being brought into productiveness be uncertain. These various data, therefore, must be calculated, and suitable allowance made for time and uncertain value in the winding up of the moneyed consideration. The rental, then, being once assumed, the value will be the present worth of an annuity during the expected term of its duration, minus the number of years' duration which

it is expected to lie dormant; the rate of return being varied
by the valuator according to certain or uncertain data, and the
probable profit to be realised under all the circumstances of the
case. The customary course of valuing the lessor's interest in
mining property in Scotland has been 10 years' purchase upon
the ordinary rental, unless some prospective increase of value
present itself; but in the North of England it is constructed
after a more detailed principle. First, then, the prospective
annual value must be assumed, as also its duration, and if it
amount to a perpetuity it will be valued as a freehold; but as
this description of property is liable to uncertain or suspended
return, a percentage of 8 to 10 per cent. rebate is taken to be
equitable. For instance, a landlord's interest in a coal property,
say £500 per annum for 30 years as a perpetuity, is worth, at
8 per cent. rebate, = 11·25 years' purchase, or £562,500. The
lessee's interest is treated in a similar manner, but is subject to
still greater uncertainty, inasmuch as it involves consideration
of stock and other expenditure, and even the duration of the
lease itself, which might be given up or brought to a termination by policy or by some catastrophe. The first and main
consideration is the probable profit to be derived amongst all
the varying circumstances of the cost of working, the amount
of selling price, the probable yearly quantity to be produced,
and probable expenditure necessary from time to time to keep
up the said contemplated quantity.

'These, therefore, are data which must à priori be assumed,
after which the valuation resolves itself into the following principle :—Assuming the annual profit during the lease to be
£1,000, and the unexpired term to be 15 years, then it is an
annuity, the purchase value of which, under all the uncertainties
of the case, ought to repay a purchaser 14 per cent. per annum,
with a return of capital = 6·14 years' purchase, or £6,140;
then, taking the colliery stock as valued, in a working state, at
£6,000. But, to be sold off by auction at the end of the term,
including expenses for £2,500. The value of the said £2,500
to be received by the purchaser, at the end of the term of 15
years, is worth, in ready money, at 5 per cent. discount, ·48
purchase, or £1,200. Leaving net value £7,340. The rate of
purchase value differs from 12 to 18 per cent., according to the

degree of risk and uncertainty of the profits, whether from mine accidents or the fluctuations of trade.'

We shall test Mr. Dunn's statements in another portion of the work.

In a published Report, made in 1872, on the Cannop Bridge Level, parts of Royal Forrester, Speculation, and Rose-in-Hand Colliery Gales, in the Forest of Dean, Mr. Marcus Scott states that an annuity of '£3,000 for 28 years, at 20 per cent. per annum, is worth £14,909 in present money;' or in other words, that the present value of £1 per annum at 20 per cent. for that period is 4·96967 years' purchase. Also that the 'published (and unpublished) tables by which the calculations are facilitated are compiled on the assumption that a purchaser can re-invest annually at compound interest (and at the same rate of interest) the surplus money above the rate of interest he calculates on making on his purchase money. As, for instance, suppose we take, under Table 3 (of his Report), that a purchaser is going to pay down £62,893 (for an *immediate* annuity of £7,000 per annum for 24 years), on which he calculates he will realise 10 per cent. (on his purchase money), and at the end of 24 years he will have redeemed or recouped the sum of £62,893. Now, 10 per cent. on that amount is £6,289·3; the difference between which and the annuity or profit of £7,000 is £710 7s., which amount, invested annually at 10 per cent. compound interest, will, at the end of 24 years, amount to within a fraction of the purchase money, £62,893; but, suppose a purchaser can only invest for the purpose of redemption, at the rate of 3 per cent. compound interest, then, instead of realising 10 per cent. on his purchase money, he would only realise £8·179, or £8 3s. 6d.'

This is certainly a most unintelligible passage, and a mistake as well, for, if the 10 per cent. is allowed for, or added to, the redemption fund at 3 per cent., and then unity divided by it, we get the present value. Thus, if a purchase were made, allowing interest on the purchase at 10 per cent., and redeeming the capital at 3 per cent. at the expiration of 24 years, we have 7·74909 years' purchase, and the present value = £54,243.

Then, 10 per cent. upon this . . = £5424·363.
And the redemption fund, at 3 per cent. = £1575·637.
_____
Annuity . . = £7000·000.

Mr. Scott calculated the whole of his values upon the principle—if it may be so called—of redeeming capital at the same rate per cent. as that allowed upon the purchase money; but he does not say anywhere in his Report where money could be placed in order to accumulate at 10 per cent. interest, but he does value up to 20 per cent. upon the same assumption.

He, however, refers to the redemption of capital at 3 per cent., and proceeds to remark: ' To calculate the whole of the Tables I have given you' (*i.e.* in his Report) 'on the latter mode of present value and investment, would entail an enormous amount of figures, as there are no published Tables that I am aware of which give the *years' purchase* on investment for redemption at 3 per cent., except Willich's, which only go as high as for the purchaser to realise 5, 6, and 7 per cent. with investments at 3, 3½, 4, and 5 per cent.'

He also states that ' the calculations of annuities for any number of years deferred, or any number of years with redemption at 3 per cent., are very intricate.'

The discrepancies that have arisen in valuations made by the inaccurate mode practised by Mr. Scott and many others will be fully treated of hereafter.

In the case of unopened mines it has been my practice, in deducing the present value deferred, to allow 20 per cent. to a present purchaser, and redeem the capital at 3 per cent. per annum; which I consider in a general way is a safe mode of dealing with any mine with average prospects; although, in special cases, where mines had a more certain character, I have allowed a percentage as low as 14, and in some of less certainty as high as 25.

A rule cannot be laid down expressing the attendant risk of mining adventure, as nearly all mines exist under circumstances differing widely from each other. It is a matter of experiment: each mine must, therefore, stand upon its own merits, and the amount of percentage to be allowed must also be varied according to the circumstances of each particular case.

In working up a valuation, after the number of tons are ascertained in the given area, a reasonable and practicable output per annum must be assumed—such as would be justified by the probable state of the market, continuance of supply from the surrounding collieries, and other circumstances—which, multiplied into the profit per ton, will give the yearly income or annuity expected.

The annuity so determined has to be purchased upon an agreed or allowed percentage, and resolves itself into a question of compound interest, or the present value of £1 per annum at a certain rate on the purchase, and to redeem the capital—not in an *imaginary way*—but at another practicable rate per cent., and during a defined period, multiplied into the annuity expected per annum for the present value.

If the mine is not opened, the annuity must be considered as deferred during the period the mine is unproductive; thus, if the time necessary to win a mine.is 3 years, and its duration afterwards 50 years, allowing 20 per cent. to a present pur-. chaser, and redeeming the capital at 3 per cent. per annum, the present value deferred would be 2·77070179 years' purchase, which would accumulate during the time occupied in winning the mine to 4·78777025, which, in point of fact, is the present value of £1 per annum or years' purchase immediate, at the rates of interest stated.

PART II.

ONSTRUCTION AND USE OF VALUATION TABLES,

RULES AND FORMULÆ.

PART II.

EVERY beneficial interest or sum of money accruing, or to accrue, and to be paid at the end of a year, or portion of a year, may be considered as an annuity, and may be either terminable with the life of an individual or perpetual. Any sum of money left unpaid for a certain number of years is called an annuity in arrear, and when not payable until after a fixed number of years it is said to be a reversionary or deferred annuity.

In either case the annuity is transferable, and may be purchased on certain agreed terms; each class of annuities must, however, receive a particular mode of treatment, adapted to, and peculiar to the nature of the circumstances connected with each particular case.

If money could not be employed, and a marketable rate of interest obtained for its use, the value of any sum of money or annuity would be equal to that to be paid at the end of one year, multiplied by the whole period or number of years the annuity has to run; but as compound interest is involved in all these cases, it is clear that if A desires to sell an annuity to B, and which has to last a certain number of years, a certain agreed interest or discount must be allowed to B upon the whole sum to be purchased, and received by him for the fixed period.

The Increase of the Principal at compound interest may be illustrated by the following mode of expression :—

Putting r = interest on £1 for one year or other integral period,

   ,, R = amount of £1 with one year's interest,

   ,, n = any integral number of years,

   Then

(1)   .   .   . $R^n = (1 + r)^n$.

Supposing the rate of interest to be 3 per cent., then $r =$ ·03, and $R = 1 + ·03 = 1·03 =$ the principal of £1, and simple interest on it at the above rate for one year. If improved in a similar manner during the second year, it would amount to $(1 + ·03)^2$ or $= 1·0609$, and so on until $(1 + ·03)^{100}$ would amount to 19·2186319809.

In words, the rule may be thus expressed, *Add to unity or 1 the interest due upon it at the end of the first year; involve the sum, to the power whose index is the number representing each successive year, in the given period.*

It is manifest that the present value of £1, at 3 per cent., must be such a sum less than £1 as would, if improved by a year's interest, amount to it. Thus the principal of £1, and interest, ·03, thereon $=$ £1·03, the amount; and $\dfrac{1}{1·03} =$ ·9708738, the present value of £1. For, ·9708738 × 1·03 $=$ £1.

Similarly, the present value of £1, due 6 years hence, at 3 per cent., would be $= \dfrac{1}{1·194052} =$ ·837484. It therefore follows that if £1 is raised to any amount resulting as shown from its improvement at compound interest, at a certain rate per cent., during any number of years, and unity or 1 is divided by it, the resulting number or quotient will represent the present value of £1, due at the end of the same periods the amounts were raised for. The value or years' purchase of perpetuities may be found by dividing the annuity by the rate of interest on £1 for one year. Thus $\frac{1}{3} = 33·3333$, $\frac{1}{4} = 25$, $\frac{1}{5} = 20$, $\frac{1}{7} = 14·2857$, and $\frac{1}{10} = 10$ years' purchase respectively.

The Present Value of a Perpetuity of £1 payable once in every *n*th year, the first payment due *n* years hence, will be denoted by V_n; thus we have,

$$(2) \qquad \cdot \qquad \cdot \qquad V_n = \frac{1}{R^n - 1}.$$

And, for the value of such a perpetuity payable every 10 years, at 4 per cent. we have,

$$V_{10} = \frac{1}{R^{10} - 1} = \frac{1}{1·48024 - 1} = 2·0823.$$

The present value of £1 to be paid annually in perpetuity, at 4 per cent. is, as stated above, $= 25$ years' purchase; but if, instead of being annual, the payments are only made at intervals of say 2, 3, or 4 years, or other periods, by taking the amount of R^n from the Tables for the variable periods, the formula $V_n = \dfrac{1}{R^n - 1}$ will of course continually represent the present value of the perpetuity.

If the perpetuity is *deferred* for say 5 years, so that the first payment is to be made 15 years hence, the value found as above must be multiplied by v^5; and if the perpetuity is *anticipated* 5 years, the value found must be multiplied by R^5.

In the former case v^5 being ·82192711, we have

$$2\text{·}0823 \times \text{·}82192711 = 1\text{·}71149882;$$

and in the latter case R^5 being 1·21665290, we also have $2\text{·}0823 \times 1\text{·}21665290 = 2\text{·}53344909$, the value of the perpetuity.

Again, putting s_n = redemption fund, we also have

$$(3) \qquad \qquad V_n = \frac{s_n}{r}.$$

Thus $\dfrac{\text{·}083292}{\text{·}04} = 2\text{·}0823$, as before.

Also, the Present Value of a Perpetuity of £1, deferred n years, may be deduced as follows:—

Putting D = deferred value,

,, v^n = value of £1 due n years hence,

we have

$$(4) \qquad \qquad D = \frac{v^n}{r}.$$

Thus $\dfrac{\text{·}6756}{\text{·}04} = 16\text{·}8891$, value at 4 per cent. for 10 years.

$$\text{Also } \frac{1\text{·}00}{\text{·}04} = 25,$$

and $\text{·}6756 \times 25 = 16\text{·}8891$, the value deferred as before.

When large sums are invested at compound interest, a certain advantage would accrue to an investor if interest on capital were to be paid at several equal intervals during the year, instead of one single payment at the end of the year.

It does not come within the scope of this work to enter into a theoretical investigation of the subject, but the practical mode generally adopted in solving problems of this nature may be exemplified as follows :—

If interest were to be realised m times in a year, at the rate $\dfrac{r}{m}$, the expression becomes

$$(5) \qquad . \qquad . \qquad . \qquad \left(1+\frac{r}{m}\right)^{mn}.$$

Assuming the principal to be £1, and $r = \cdot06$ per £1 for one year, for half-yearly payments we have

$$\left(1+\frac{\cdot06}{2}\right)^{2} = 1\cdot0609.$$

Payments being made quarterly, we also have

$$\left(1+\frac{\cdot06}{4}\right)^{4} = 1\cdot0613635505.$$

By the same rule, for monthly payments the amount would be $1\cdot06167781$, and for weekly payments it would be $1\cdot06179981$.

On the same principle, if it were possible for payments to be made momentarily, the amount of all the increments would depend upon, and be expressed by, the well-known principle of the 'Binomial Theorem,' and if the series are continued to a sufficient extent, would $= 2\cdot718281828459$, which is the base of the *Napierian* logarithms. The log of this number is $0\cdot434294481903$, and $\cdot06 \times 0\cdot434294481903 = \cdot02605766891418$, the natural number of which is $1\cdot061836546557$, or the amount.

Thus, it is evident from the nature of the above formula, that if payments were made on the assumption that a year could be divided into an indefinite number of periods, the resulting amount of all the increments, at the end of the year, would, according to this hypothesis, be in excess of that derived from the employment of periods of time having greater duration, as a day, week, or month, &c. &c.

The Amount of £1 per Annum, if invested and improved at compound interest, in n years, may be determined by the following expression :—

Let r = interest on £1 per annum.
„ M_n = amount of £1 per annum for n years.
„ R = $(1+r)$, *as in last case.*

Then

$$(6) \quad . \quad . \quad . \quad M_n = \frac{R^n - 1}{r}.$$

Assuming the rate of interest to be 3 per cent.,

Then

$$M_1 = \frac{R - 1}{\cdot 03} = \pounds 1$$

for the first year's amount, and if improved for the second year,

$$\frac{R - 1}{\cdot 03} = \frac{1 \cdot 0609 - 1}{\cdot 03} = 2 \cdot 03 \, ;$$

and so on, until

$$\frac{19 \cdot 2186319809 - 1}{\cdot 03} = 607 \cdot 28773269,$$

the amount at the end of 100 years.

In words the rule is thus expressed :—*Deduct unity or 1 from the amount of £1 in n years, and divide the remainder by the rate per £1.*

The amounts may also be found thus :—Multiply the first year's amount, *i.e.* $1 \times 1 \cdot 03 + 1 = 2 \cdot 03$, the second year's amount ; then $2 \cdot 03 \times 1 \cdot 03 + 1 = 3 \cdot 0909$, the third year's amount. The same results will be obtained by adding the amount of £1 in n years, *i.e.* R^n, to the amount of £1 per annum in n years, or M_n; thus $1 + 1 \cdot 03 = 2 \cdot 03$, then $2 \cdot 03 + 1 \cdot 0609 = 3 \cdot 0909$, the third year's amount.

When interest can be realised m times in a year, the expression becomes

$$(6a) \quad . \quad . \quad M_n = \frac{\left(1 + \frac{r}{m}\right)^{mn} - 1}{\frac{r}{m}}.$$

Therefore, for half-yearly payments, the interest being at the rate of 4 per cent. per annum, for 10 years, we have

$$M_{10} = \frac{\left(1 + \frac{\cdot 04}{2}\right)^{20} - 1}{\frac{\cdot 04}{2}} = 12 \cdot 1486848994.$$

And for quarterly payments we also have

$$M_{10} = \frac{\left(1 + \frac{\cdot04}{4}\right)^{40} - 1}{\frac{\cdot04}{4}} = 12\cdot221593339.$$

The Present Value of £1, due n years hence, may be determined from the following data :—

Putting v^n = present value of £1 due n years hence,

    ,, R^n = $(1+r)^n$ as before,

(7) .   .   .   then $v^n = \dfrac{1}{R^n}$.

Supposing 5 per cent. to be the rate of interest, we have

$$v = \frac{1}{R} = \frac{1}{1\cdot05} = \cdot9523809523,$$

the first year's present value, and the 6th year's present value would be equal to

$$\frac{1}{1\cdot340095641} = \cdot7462153964.$$

In words the rule may be thus expressed :—

Divide unity or 1 *by the amount of £1 in* n *years; the quotient will then represent the present value of £1 due at the end of* n *years.*

The same results may also be obtained by first dividing unity or 1 by the amount of £1 in one year, and afterwards to constantly divide the successive quotients by the same amount.

$$\text{Thus } \frac{1}{1\cdot05} = \cdot9523809523\,;$$

$$\text{then } \frac{\cdot9523809523}{1\cdot05} = \cdot9070294784,$$

the second year's present value.

If it were possible to realise interest m times in a year, the expression becomes

(7a) .   .   . $v^n = \dfrac{1}{\left(1 + \dfrac{r}{m}\right)^{mn}}.$

Therefore, for half-yearly payments, at 4 per cent., and for 5 years, we have

$$v^5 = \cfrac{1}{\left(1 + \cfrac{\cdot04}{2}\right)^{10}} = \cdot8203483,$$

And for quarterly payments we also have

$$v^5 = \cfrac{1}{\left(1 + \cfrac{\cdot04}{4}\right)^{20}} = \cdot8195444.$$

The Redemption Fund that will amount to £1 in n years may be computed from the following expression :—

Putting s_n = redemption fund,

,, R^n and r = the elements as previously assigned.

Then

$$(8) \quad . \quad . \quad . \quad . \quad s_n = \frac{r}{R^n - 1}.$$

Allowing the rate of interest to be 3 per cent, and to redeem £1 at the end of 3 years, we have

$$s_3 = \frac{\cdot03}{R^3 - 1} = \frac{\cdot03}{1\cdot092727 - 1} = \cdot3235303633,$$

or the redemption fund ; and for redemption in 20 years, we also have

$$s_{20} = \frac{\cdot03}{1\cdot8061112347 - 1} = \cdot0372157076.$$

The rule for finding the redemption fund may be written in words thus :—

Divide the rate of interest per £1 by 1 less than the amount of £1 in the time.

Assuming interest to be convertible m times in a year, the expression becomes

$$(8a) \quad . \quad . \quad . \quad s_n = \frac{r}{\left(1 + \cfrac{r}{m}\right)^{mn} - 1}.$$

Therefore, for half-yearly payments, at 4 per cent., and for 10 years, we have

$$s_{10} = \cfrac{\cdot04}{\left(1 + \cfrac{\cdot04}{2}\right)^{20} - 1} = \frac{\cdot04}{1\cdot485947396 - 1} = \cdot08231343;$$

And for quarterly payments we also have

$$s_{10} = \frac{\cdot 04}{\left(1 + \frac{\cdot 04}{4}\right)^{40} - 1} = \frac{\cdot 04}{1 \cdot 4888637336 - 1} = \cdot 08182239,$$

the redemption fund.

We may also deduce similar results from

$$(9) \quad . \quad . \quad . \quad . \quad s_n = \frac{1}{M_n}.$$

Thus

$$s_n = \frac{1}{12 \cdot 221593} = \cdot 08182239, \text{ the redemption fund, as before.}$$

It may be remarked here, that for very nice work, *i.e.* to make the simple interest at a certain rate per cent. on the deduced value and redemption fund balance the annuity exactly, it is necessary to employ a table computed to eight or ten places of decimals.

Putting A = the Immediate Annuity which £1 will purchase, we have

$$(9a) \quad . \quad . \quad . \quad A = s_n + r'.$$

The annuity is therefore readily obtained by adding to the redemption fund necessary to produce £1 at the end of any given period, the interest allowed upon the investment.

Thus, the redemption fund necessary to produce £1 at the end of 3 years, at 3 per cent. = ·3235303633; then, the interest allowed on the investment being 3 per cent. we have,

$$\cdot 3235303633 + \cdot 03 = \cdot 3535303633,$$

or the annuity. This rule applies to all percentages; for, assuming the interest to be allowed on the outlay to be ·20 per £ instead of ·03 per £, for a period of 40 years' duration, we have the redemption fund necessary to produce £1 at the end of the assigned period = ·01326238; then

$$\cdot 01326238 + \cdot 20 = \cdot 21326238,$$

or the annuity which £1 will purchase.

Table V. is therefore well adapted for determining the annuity, without having a special Table for that purpose.

The Present Value of £1 per Annum, deduced by the old rule for n years, may be computed as follows:—

$$(10) \quad . \quad . \quad \text{Present value } p_n = \frac{R^n - 1}{R^n r} \text{ or } \frac{1 - v^n}{r}.$$

Assuming the rate of interest to be 3 per cent. per annum for 5 years, we have

$$\frac{1 - \cdot 8626087846}{\cdot 03} = 4 \cdot 57970719, \text{ the present value.}$$

The value deduced by either of the last preceding rules is erroneous, when it is necessary to employ rates of interest above those which can be realised in the money market for the redemption of capital. See Table XII. for discrepancies in the old table of present values.

The Present Value of £1 per Annum for n years, allowing to a purchaser of annuities one rate of interest on his purchase money, and to redeem his capital at the expiration of the time by annually investing the overplus at another practicable rate, may be deduced as follows:—

Putting P_n = present value,

  ,, R, s_n, and r = the elements as previously assigned,

           and r' = the interest allowed on capital.

We have

$$(11) \quad . \quad . \quad . \quad P_n = \frac{1}{\dfrac{r}{R^n - 1} + r'} \quad \text{or} \quad \frac{1}{r' + s_n}.$$

Assuming the rate of interest on capital to be 5 per cent., and to redeem it at 3 per cent., at the expiration of 3 years, we have

$$\frac{1}{\cdot 3235303633 + \cdot 05} = \frac{1}{\cdot 3735303633} = 2 \cdot 677158534.$$

Assuming interest to be convertible m times in a year, the expression becomes

$$(11a) \quad . \quad . \quad P_n = \frac{1}{\dfrac{r}{\left(1 + \dfrac{r}{m}\right)^{mn} - 1} + r'}.$$

Therefore, for half-yearly payments, interest on capital being 10 per cent., and redemption 4 per cent., and for 10 years, we have

$$P_{10}= \cfrac{1}{\cfrac{\cdot04}{\left(1+\cfrac{\cdot04}{2}\right)^{20}-1}+\cdot10} = \cfrac{1}{\cfrac{\cdot04}{1\cdot485947396-1}+\cdot10} = 5\cdot484859137,$$

And for quarterly payments, we also have

$$P_{10}= \cfrac{1}{\cfrac{\cdot04}{\left(1+\cfrac{\cdot04}{4}\right)^{40}-1}+\cdot10} = \cfrac{1}{\cfrac{\cdot04}{1\cdot4888637336-1}+\cdot10} = 5\cdot499842456,$$

the present value.

It will be observed that the purchase money being P_n, it is evident from $\dfrac{1}{r'+s_n} = P_n$, that the interest r' allowed or expected to be realised for investing a sum P_n, would be equal to $P_n r'$, and s_n, invested at another rate per cent., r, which being accumulated at compound interest, will reproduce the original capital P_n at the expiration of a certain defined period.

The annuity being unity or £1, is consequently made up of two distinct parts, that is, r' per cent., a years' interest on P_n, and s_n, which being invested at another rate of interest per cent., r per annum, will produce P_n.

The annuity of £1 is therefore equal to $r'+s_n$, which may probably be more clearly seen by the following mode of working:—

Putting the period of duration, n, of the annuity　　．　= 55 years

　　,,　　the rate of interest, r, to reproduce P_n within the period of 55 years ．　．　．　．$\Big\}$ = ·03

　　,,　　the rate of interest r' to be realised on the purchase money P_n　　．　．　．　．$\Big\}$ = ·20

Then from (11) we have

$$P_n = \cfrac{1}{\cfrac{r}{R^n-1}+r'}.$$

Also $R^n = R^{55} = 5\cdot082148592$,

$$\text{and } P_{55} = \cfrac{1}{\cfrac{\cdot03}{R^{55}-1} + \cdot20} = \cfrac{1}{\cfrac{\cdot03}{5\cdot082148592-1} + \cdot20}$$

$$= \cfrac{1}{\cdot00734907104 + \cdot20} = \cfrac{1}{\cdot2073497104} = 4\cdot82278505, \text{ or}$$

present value.

To insure, therefore, the purchase of an immediate annuity of £1 under these conditions, the purchaser must pay down a sum of £4·82278505 = P_{55}, the present value, or years' purchase.

Again, s_{55} at r per cent. = s_{55} at ·03 = ·00734907104, which is the redemption fund necessary to reproduce £1 in the given time

Then $P_{55}\,r' = 4\cdot82278505 \times \cdot20$ . . = $\cdot9645570100$

And $P_{55}\,s_{55} = 4\cdot82278505 \times \cdot00734907104 = \cdot0354429899$

Also $r' + s_{55} =$ the annuity receivable, or £1·000000000

If, therefore, £·0354429899 is annually invested at the rate of ·03 per cent. compound interest, it will reproduce P_{55}, the original purchase money, or capital, at the expiration of the term of 55 years.

Thus the amount of £1 per annum for 55 years = 136·0716197, which, multiplied by £·0354429899 = £4·82278505, the original capital, or P_{55}.

What has been hitherto advanced relates more particularly to formulæ, and rules, employed in the construction of the Tables necessary for determining the present value of immediate and deferred annuities, realised under certain conditions; but when annuities are deferred, and the present value required to be tabulated, special treatment must be adopted; and the construction of Tables of this nature becomes very tedious.

In calculating the Tables in this work of the present value of £1 per annum for n years after t years, allowing a purchaser interest on his purchase money at a certain agreed rate per cent., also such a surplus as, invested at another practicable rate per cent., would reinstate the capital at the end of the term, the following conditions were necessary to be considered.

If instead of an annuity of p pounds being entered upon immediately, it can only be realised at the end of the t-th year, and to continue n years thereafter, the purchaser will expect to realise r' per cent. on his outlay P_{t+n}, during the whole term of $t+n$ years; and here, as was shown in the last preceding case, he can invest the surplus annuity only at the rate of r per cent.

It is necessary, therefore, to determine the relation existing between P_{t+n} and p, and, as it may be seen that no annuity can be paid during the deferred term of t years, P_{t+n} would accumulate or amount at the end of the t-th year to $P_{t+n}(1+r')^t$. When, however, the annuity is entered upon, which as a matter of course it would be at the end of the t-th year, it is, as previously shown, separable into two parts, that is to say, first, a year's interest on the amount which the purchase money P_{t+n} has now attained, namely, $P_{t+n}r'(1+r')^t$; and, secondly $p-P_{t+n}r'(1+r')^t$, the sum which must be invested at the rate of r per cent., and which will reproduce $P_{t+n}(1+r')^t$ at the end of $t+n$ years. Then, by condition, we have,

$$\{p-P_{t+n}r'(1+r')^t\}\,M_n = P_{t+n}(1+r')^t.$$

Solution of this equation gives

$$(12)\quad . \quad P_{t+n} = \frac{p\,M_n}{(1+r')^t(1+r'M_n)},\text{ and}$$

$$(13)\quad . \quad p = \frac{P_{t+n}(1+r')^t(1+r'\,M_n)}{M_n};$$

in both of which M_n denotes the amount of an annuity of £1 in n years.

If in (12) p be put $=1$, we have for the value (the years' purchase,) when the annuity is £1,

$$(14)\quad . \quad P_{t+n} = \frac{M_n}{(1+r')^t(1+r'M_n)};$$

and if the value, i.e., the sum invested, be £1, we have from (13) by making $P_{t+n}=1$, the annuity which £1 will purchase, viz.:—

$$(15)\quad . \quad p = \frac{(1+r')^t(1+r'\,M_n)}{M_n}.$$

The value of an annuity to continue 55 years after 3 years deferrence, r' being $= \cdot 20$, and $r = \cdot 03$, may be deduced from (14).

Thus,

$$P_{3+55} = \frac{M_{55}}{(1 + \cdot 20)^3(1 + \cdot 20\,M_{55})}.$$

Table (III.) gives $M_{55} = 136 \cdot 07161972$, at 3 per cent. Therefore,

$$P_{3+55} = \frac{136 \cdot 07161972}{1 \cdot 728(1 + \cdot 20 \times 136 \cdot 07161972)} = \frac{136 \cdot 07161972}{48 \cdot 75435056} =$$

$2 \cdot 7909633636$, value of deferment required. Again, if the purchase money P_{t+n} is made $= \pounds 1$ for the same continuance and period of deferment, and at the same rates, the annuity $\pounds 1$ will purchase may be deduced from (13):—

Thus,

$$p = \frac{(1 + \cdot 20)^3(1 + \cdot 20\,M_{55})}{M_{55}};$$

and by substituting the numerical quantities we have

$$p = \frac{1 \cdot 728(1 + \cdot 20 \times 136 \cdot 07161972)}{136 \cdot 07161972} = \frac{48 \cdot 75435056}{136 \cdot 07161972} =$$

$\cdot 3582991858$, or the annuity deferred which $\pounds 1$ will purchase; and it is, as it ought to be, the reciprocal of the value, when the annuity is $\pounds 1$.

$$\text{For, } p = \frac{1}{P_{t+n}},$$

thus:—

$$\frac{1}{2 \cdot 790963636} = \cdot 3582991858,$$

the deferred annuity which $\pounds 1$ will purchase, as before.

The value of the annuity when deferred, may be readily derived from the value when immediate, by virtue of the following relation,

$$P_{t+n} = P_n v^t,$$

where n is the term of continuance, and t the term of deferment. Applying this to the last example, we have,

$$P_{3+55} = P_{55} v^3.$$

P_{55} is $= £4\cdot82278505$, and v^3 (at $\cdot20$ per £1) is $= £\cdot57870370$ (See Table IV.) Hence,

$P_{3+55} = 4\cdot82278505 \times \cdot57870370 = 2\cdot790963636$, the same as before.

In order to illustrate the power of the Tables, and to give an additional method of obtaining the deferred value, we have,

$\checkmark$ (16) .   . $P_{t+n} = \dfrac{1}{\dfrac{(1+r')^t - 1}{P_n} + r' + s_n}$.

Then, by substitution, we also have,

$$P_{t+n} = \dfrac{1}{\dfrac{(1+\cdot20)^3 - 1}{4\cdot82278505} + \cdot20 + \cdot00734907} = \dfrac{1}{\cdot3582991937} =$$

$2\cdot790963578$, or value of deferrence, practically the same as above.

There is nothing in the amount of work involved in this method to frighten a student—on the contrary, I consider it simpler than when employing M_n. But for practical purposes, and in order to get over a larger amount of calculation in a given time, no doubt $P_n v^t$ should be employed, which is the simplest possible form the formula can be made to assume. Tables X. and XI. were computed by this rule.

Again, for obtaining the deferred annuity which £1 will purchase, we have the following expression :—

$\checkmark$ (17) .   . $p = \dfrac{(1+r')^t - 1}{P_n} + r' + s_n$. Or,

$$p = \dfrac{(1+\cdot20)^3 - 1}{4\cdot822785051} + \cdot20 + \cdot00734907 = \cdot3582991937.$$

It may be here remarked that it is not necessary in practice to work up any of the elements involved in the solution of these problems, as they are tabulated in this work, and may be immediately obtained by reference.

It is to be observed that when working the numerical quantities represented by the formulæ, (14), (15), the operation should be taken from right to left, thus :—

$$M_n \times r' + 1 \times (1+r')^t,$$

i.e., $136\cdot07161972 \times \cdot20 + 1 \times 1\cdot728.$

If t is made equal to o, that is to say, if the annuity can be made available on present entry, then $(1+r')^t = 1$, and the formula deduced, becomes for this case

$$(18) \quad . \quad . \quad P_n = \frac{pM_n}{1+r'M_n};$$

$$(19) \quad . \quad . \text{ also } p = P_n \frac{(1+r'M_n)}{M_n}.$$

Putting p therefore $= £1$, we have from (18)

$$P_n = \frac{M_n}{1+r'M_n};$$

and by substituting the numerical quantities we also have

$$P_{55} = \frac{136\cdot0716197}{1+(\cdot20 \times 136\cdot0716197)} = \frac{136\cdot0716197}{28\cdot21432394} = 4\cdot822785051,$$

which is the present value, or years' purchase immediate.

Again, putting $P_n = £1$, we also have from (19),

$$p = \frac{1+r'M_n}{M_n};$$

and by substitution we also have

$$p = \frac{1+(\cdot20 \times 136\cdot0716197)}{136\cdot0716197} = \frac{28\cdot21432394}{136\cdot0716197} = \cdot20734907104,$$

which is equal to the redemption fund necessary to produce £1 in the given time, *plus* the interest allowed to a present purchaser. See (9a.), page 30.

The results deduced from the last two preceding formulæ for immediate annuities, prove the accuracy of the plan upon which the Tables of this class have been computed for this work.

The subject of *Deferred Annuities* has been considered by some to be very complicated, and by many avoided altogether—when two rates of interest are involved—as something unapproachable. The great difference of opinion that exists in relation to the proper mode of treating the question, as applied to Mines, has led me to investigate it thoroughly, and I believe the conclusions arrived at are such as are not to be controverted.

The resulting number deduced from (14) and (16), that is to say 2·790963578, is the sum necessary to be paid down by a present purchaser, in order to secure an annuity of £1 for 55 years (which is not to commence, however, until the expiration of 3 years), which would yield him 20 per cent. during the entire period of 58 years, and redeem the purchase money, that is to say £2·790963578, and its amount during the 3 years of deferment, together equal to £4·822785051, by investing the surplus annuity at 3 per cent. compound interest.

Again, under similar conditions, if, instead of £2·790963578, *one pound* only had been invested, then an annuity of £·3582991858 would have been secured by the purchaser. Generally, therefore, in cases of *deferred* annuities of this kind—that is, when two rates of interest are involved—a certain sum, P_{t+n}, has to be paid down immediately ; but as no annuity is or can be payable under the circumstances during the deferred period, the purchase money, P_{t+n}, accumulates at the rate allowed to the purchaser on his capital, or r' per £, to a certain sum $= P_{t+n} (1 + r')^t = P_n$; but, at the expiration of t years, the deferred period closes, and the annuity commences or is then entered upon, and its payments have to yield interest at the rate agreed upon between the parties to the business, or r' per £ on the accumulated purchase money $P_n = P_{t+n} (1 + r')^t$, and also a sum sufficient to reinstate the sum P_n, to which the purchase money has accumulated at the end of the assigned term of $t + n$ years, at another rate per £, or r. In the present case the deferred period t is equal to 3 years, and the term n to run afterwards is equal to 55 years.

Then,

$$P_{t+n} (1 + r')^t = P_{t+n} (1 + ·20)^3 = P_{t+n} \times 1·728 = £2·790963578$$
$$\times 1·728 = £4·822785051 = P_n,$$

the amount to which the purchase money has accumulated at 20 per cent. at the end of the deferred period.

The interest on $P_n = P_n r' = 4·822715051 \times ·20 = £·964557010$, or that part of the annuity due to the agreed percentage. P_{t+n} being the present gross value to be paid down $= £2·790963578$, and the redemption fund required to produce £1 at the expiration of 55 years at 3 per cent. is equal to £·007349071.

Then,

$P_n s_n$ at r per cent. $= P_n \times {\cdot}007349071 = £4{\cdot}822785051$
$\qquad \times {\cdot}007349071 = £{\cdot}03544299$,

the amount necessary to be annually set aside and to accumulate at 3 per cent. for the assigned term of n years.

Also the interest on P_n, or $P_n r'$, *plus* that
on $P_n s_n$ at r per cent., is equal to
$P_n \times {\cdot}20 =$ to that portion of the annuity enjoyed for present use . . $= \quad {\cdot}964557010$
And $P_n s_n$ at r per cent. $=$ the other part
set aside for redemption within the period $= \quad {\cdot}035442990$

And $P_n r' + P_n s_n$ at r per £ . . . $= £1{\cdot}000000000$
or the annuity to be received by the purchaser under the proposed conditions.

If further proof of the accuracy of the foregoing mode of working were required, it is only necessary to multiply the amount of an annuity of £1 in 55 years at 3 per cent. by the surplus annuity set aside to reproduce the capital at the expiration of the given time.

Thus, the amount of £1 per annum for n or 55 years

$\qquad = 136{\cdot}0716197 \times £{\cdot}03544299 = £4{\cdot}822785051$,

the original capital invested, with accumulated interest.

When the sum invested is £1, the annuity purchased, as previously shown, is equal to £$\cdot$3582991858, and if treated as above, $P_{t+n} (1 + r')^t = £1 (1 + {\cdot}20)^3 = 1{\cdot}728 = P_n$, the accumulated amount during the deferred period of 3 years;

Then $1{\cdot}728 \times {\cdot}20$ . . . . $= \quad {\cdot}3456000$
And $£P_n = £1{\cdot}728 \times {\cdot}0073490$. $= \quad {\cdot}0126991$

$\qquad\qquad\qquad\qquad\qquad\qquad £{\cdot}3582991$

the annuity as previously determined.

Then, if we multiply the amount of an annuity of £1 as before, we have $136{\cdot}07161970 \times {\cdot}00734907104 = £1$, the original capital, or purchase money paid down.

If further proof of the principle involved in the return of the capital were required, we may select an example embracing a short duration, and proceed in detail as follows:

The present value of £1 per annum, allowing 20 per cent. and to reproduce it at 3 per cent. within a period of 5 years after 3 years = £1·490142634, which accumulates to £2·574966472 in 3 years.

The redemption fund to produce this

sum is . . . . . = ·485006705689
And £2·574966472 × ·20 . . = ·514993294311
—————————
The annuity . . £1·000000000000

And in detail thus :—

·485006705689 = 1st year's redemption fund.
30·1 inverted.
—————————
·485006705689
14550201170
—————————
·499556906859 = amount at end of 1st year.
·485006705689 = 2nd year's redemption fund.
—————————
·984563612548
30·1 inverted.
—————————
·984563612548
·29536908376
—————————
1·014100520924 = amount at end of 2nd year.
·485006705689 = 3rd year's redemption fund.
—————————
1·499107226613
30·1 inverted.
—————————
1·499107226613
44973216798
—————————
1·544080443411 = amount at end of 3rd year.
·485006705698 = 4th year's redemption fund.
—————————
2·029087149100
30·1 inverted.
—————————
2·029087149100
60872614473
—————————
2·089959763573 = amount at end of 4th year.
·485006705689 = 5th and last year's redemp. fund.
—————————
2·574966469262 = the accumulated present value.

The first year's redemption fund to
be invested   .   .   .   = £·485006706689,

And at 3 per cent., at the end of
the year becomes   .   .   = £·499546906859.

The second instalment of the redemption fund.   .   .   = £·485006705689,

Is again invested, and at the end of
the second year the fund is   = £1·014100520924,

To which, at the end of the third year, £·485006705689 is
again added, and so on to the end of the fifth year, when the
original purchase money, £1·490142634, and its accumulation
during the deferred period, by multiplying it by

$$(1 + r')^5 = 1·728 = £2·574966470.$$

Care must, however, be taken that no delay is occasioned in
investing the annual instalment at the proper time, otherwise
a discrepancy will exist in the account at the end of the period.

The Tables introduced into this work have been carefully
calculated from data deduced from first principles, and involved
in the doctrine of interest and annuities. The formulæ and
rules which were employed in their construction, are laid down
in the most simple form, so as to be readily understood, and
applied by those who may not have either time or inclination to
investigate, and employ rules containing algebraical combinations of a more complicated nature. I have strenuously endeavoured to divest the subject of all intricate formulæ and elaborate
mathematical reasoning, that would, in my opinion, tend in
any way to confuse it. I trust, therefore, that this has been
effected so far as it was considered to be convenient and beneficial, and consistent with the nature of the enquiry. And it is
presumed that any person having occasion for calculations of
this nature may, by merely consulting the Tables, obtain at
sight any years' purchase for a given time and rate of interest,
and consequently arrive at a reliable conclusion as to the value
of any annuity in a much more satisfactory manner, in less
time, and with greater ease than could be expected to result
from a tedious process of direct calculation. The same remark
applies to all the other Tables.

Those who are sufficiently expert, and object to the use of
tables as labour savers, will find that the rules laid down are suffi-

cient for the calculation of values in a specific and direct manner, or for the production of tables similar to those I have referred to.

Inaccurate tables are worse than useless, and without employing some special means for the correction of error, it certainly could not be expected that tables involving so many figures and direct computations could be entirely free. Considering this, therefore, and being aware from long experience of the trouble and difficulties that are created by the employment of incorrect tables of various kinds, I was led to adopt means to the end in view. I have, therefore, every reason to believe that the result is, Tables free from error, and which may absolutely be relied upon.

With regard to the Tables of Amounts, an additional test as to the accuracy was applied to the final number in the column of each rate per cent. The mode of calculating an extreme number by a logarithmic process in a series having no ratio, will be fully illustrated in another portion of the work.

Tables of the value of leases and annuities have frequently been published: that of Mr. Ward was written as far back as 1710; but Mr. Smart's celebrated five Tables of Compound Interest, which appeared in 1726, far excelled all that had been done previously to that time: indeed, his tables have been incorporated more or less into the works of many writers to the present time.

The tables specially referred to are—

1. The amount of £1 in any number of years.
2. The present value of £1 due at the end of any number of years.
3. The amount of £1 per annum for any number of years.
4. The present value of £1 per annum for any number of years.
5. The annuity which £1 will purchase for any number of years.

None of the tables of this class that I have seen (and I have examined a large number of works upon the subject), are computed to rates of interest higher than 10 per cent., and many of them extend only to 5 per cent.

The *fourth* and *fifth* tables, previously described, must

necessarily be inaccurate for rates of interest higher than from $4\frac{1}{2}$ to 5 per cent. This will be fully demonstrated further on.

Tables I., II., and III. of the Amounts in this work were originally calculated to 15 decimal places, with a view to print them to 10 places; but on account of the great expense of publishing, I determined to reduce all the other Tables to their present condition. The ordinary table of the present value of £1 per annum is the same in the works of all writers upon annuities; and, as the basis upon which it has been computed is in error, it follows that the annuity which £1 will purchase is also in error, because the latter is dependent for its formation upon the former. That is, p_n being the present value, and A = the annuity £1 will purchase, we have $A = \dfrac{1}{p_n}$. Thus for 60 years at 10 per cent., in the old table, $p_n = 9{\cdot}967157$; and $\dfrac{1}{9{\cdot}967157} = {\cdot}1003295122$, or the annuity.

For the same period of time, and rate per cent., but redeeming capital at the rate of 3 per cent., $P_n = 9{\cdot}42214381$. Also, we have $A = \dfrac{1}{P_n} = \dfrac{1}{9{\cdot}42214381} = {\cdot}1061329587$, or the annuity. Thus it is evident that the years' purchase upon the old basis, is in excess of the truth, whilst the annuity which £1 will purchase, derived from it, is in defect.

The reverse is the case in the Tables calculated for this work. For $9{\cdot}967156 - 9{\cdot}42214381 = {\cdot}54504319$, the difference in excess of a years' purchase; and ${\cdot}1061329587 - {\cdot}1003295122 = {\cdot}0057034465$, the difference in defect.

Mr. Peter Hardy states it as his belief that Mr. Griffith Davies was the first to compute a table showing 'the value of an annuity on a single life, which was to pay the purchaser 5, 6, or 7 per cent. on his outlay, and to replace the original capital at 3 per cent., according to the Northampton rates.' This table seems to have been published in 1825. Mr. Benwell also appears to have written a small pamphlet on the subject, containing a table of limited extent (similar to the one appended to Mr. Hardy's paper), and published it in 1831. Between the years 1837 and 1850, a rather cumbrous rule was introduced into the Appendix of Inwood's Tables, relating to two rates of interest. I give it here from my copy of that work published in 1850:—

'Let a = amount of clear improved rent.

$b = \begin{cases} \text{amount of } \pounds 1 \text{ per annum at } n \text{ per cent. com-} \\ \text{pound interest for } r \text{ years.} \end{cases}$

$c = \begin{cases} \text{rate of interest per cent. required on purchase} \\ \text{money.} \end{cases}$

x = amount of purchase money.

$y = \begin{cases} \text{sum to be annually laid by at } n \text{ per cent. com-} \\ \text{pound interest, to replace capital at expiration} \\ \text{of lease.} \end{cases}$

Then from this statement we shall have

$by = x$ $\begin{cases} \text{for the amount of } \pounds 1 \text{ per annum for the num-} \\ \text{ber of years of the lease at the given rate of} \\ \text{compound interest, multiplied by the number} \\ \text{of pounds annually laid by, must equal the} \\ \text{purchase money.} \end{cases}$

$\dfrac{cx}{100} + y = a$ $\begin{cases} \text{for } \dfrac{cx}{100} = \text{the annual interest on purchase} \\ \text{money, since the annual interest on any sum} \\ = \text{that sum multipled by the rate of interest,} \\ \text{and divided by 100, and the annual interest} \\ \text{on the purchase money added to the sum an-} \\ \text{nually laid by to replace capital, must} = \text{clear} \\ \text{improved rent.} \end{cases}$

From the first of these equations, $y = \dfrac{x}{b}$, which, substituted in the second, gives

$$\frac{cx}{100} + \frac{x}{b} = a,$$

$$\therefore \quad x = \frac{100ab}{100 + bc},$$

$$\text{and } y = \frac{x}{b} = \frac{100a}{100 + bc}.$$

This rule supposes that an annuity may be purchased, securing interest on the purchase money at one rate, and reinstating it at another rate, per cent.

Mr. William Morgan gave a solution to this problem at page 321 in the Appendix to his work on Annuities and Assurances, published in 1821, and I give it here verbatim :—

'Problem IV.—To determine the sum which should be paid for any given annuity for n years, so as to secure to the purchaser the return of his capital at the expiration of the term, supposing him to have the means of reproducing that capital at ρ per pound, and that the value of the annuity is computed at r per pound.

'Solution.—Let a be the given annuity, and x the capital to be reproduced at the end of n years, or the sum which should be paid for the annuity on the above conditions.

Since $\dfrac{(1+\rho)^n-1}{\rho}$ is the amount of £1 per annum at ρ interest in n years,

$$(a-rx) \times \frac{(1+\rho)^n-1}{\rho} \text{ will be equal to } x ;$$

from which equation x is easily found

$$= \frac{a \times \overline{(1+\rho)^n-1}}{\rho + \overline{(1+\rho)^n-1} \times r}.$$

'Example.—A purchases an annuity of £65 for 10 years, and is to be allowed £9 per cent. in the purchase, but being unable to improve the difference between £65 and the interest at £9 per cent. on the capital at a higher rate than £3 per cent.; it is proposed to make him such allowance in the purchase money as shall enable him to replace his capital at the end of the term by improving it at this reduced interest. In this case ρ is ·03, $r =$ ·09, $n =$ 10, and $a =$ 65; and x will therefore be

$$= \frac{65 \times \overline{(1·03)^{10}-1}}{·03 + \overline{(1·03)^{10}-1} \times ·09} = \frac{65 \times ·344}{·03 + ·344 \times ·09} = 366·710.$$

'In other words (and this is meant for a proof); 366·710 × ·09 = 33·004, and 65 − 33·004 (= 31·096) multiplied into 11·464, the amount of £1 per annum in 10 years, gives 366·710.'

Taking the annuity at £1 instead of £65, the result comes

out equal to the years' purchase. It is thus deduced from Mr. Morgan's figures;

$$\frac{366\cdot710}{65} = 5\cdot641692307,$$

but by taking the amount for 3 per cent to 8 places of decimals, and working his problem, we get a little difference in the result. Thus:

$$\frac{65 \times (1\cdot03)^{10} - 1}{\cdot03 \times (1\cdot03)^{10} - 1 \times \cdot09} = \frac{65 \times \cdot34391638}{\cdot03 + \cdot34391638} = \frac{22\cdot3545647}{\cdot0609524742}$$

$$= 366\cdot7540160, \text{ and } \frac{366\cdot7540160}{65} = 5\cdot642369476,$$

which is the present value of £1 per annum under the conditions.

In the year 1850, Mr. Peter Hardy, a well-known writer on annuities, wrote and introduced a paper on this subject to the Institute of Actuaries, which created considerable interest at that time. His mode of treating the question is here given :—

'Problem.—To determine the present value of an annuity certain of £1 for n years, which is to pay during its continuance a given rate of interest on the original purchase money, and to replace that purchase money at the expiration of the term at a different rate of interest.

'Solution.—The payments of the annuity being each = £1, let i' = the rate of interest which the purchaser intends to make on each £1 of his investment, or, as it may be termed, *the remunerative rate.*

'Let $(r - 1)$ = the rate of interest at which the purchaser expects to accumulate the surplus annuity, in order to replace the original capital, or, as it may be termed, the *accumulative rate.* Let $\frac{r^n - 1}{r - 1}$ = the amount of an annuity of £1 for n years forborne and accumulated at $(r - 1)$ rate of interest, and put V = the required value.

'Now it is obvious that

Vi' = the purchaser's annual interest,

$1 - Vi'$ = the surplus annuity to be accumulated,

so that in n years it may reproduce V.

If £1 per annum in n years will accumulate into

$$\frac{r^n - 1}{r - 1};$$

then

$$V = (1 - Vi')\frac{r^n - 1}{r - 1};$$

and if, for the sake of simplicity, we represent

$$\frac{r^n - 1}{r - 1},$$

by a single symbol, say A, we shall have

$$V = (1 - Vi')A,$$
$$V = A - Vi'A,$$
$$V + Vi'A = A,$$
$$V(1 + i'A) = A,$$
$$V = \frac{A}{1 + i'A}.$$

'Example.—Required the present value of an annuity of £1 per annum for 20 years, the purchaser to make 5 per cent. per annum interest of his outlay, and to replace his capital by the investment of his surplus annuity at 3 per cent.

'Here the annuity = 1, i' = ·05, $(r - 1)$ = ·03 and A = 26·8703 at 3 per cent., and

$$\log 26·8703 = 1·4292677$$
$$\underline{·05}$$
$$1·343515$$
$$\underline{1}$$
$$\log 2·343515 = \frac{0·3698650}{1·0594027} = \log 11·466 = \text{value.'}$$

The rule proposed by Mr. Hardy, was intended to effect the same purpose as that of Mr. Morgan, but at the time of its introduction it was considered by some to be very diffuse, and that the subject admitted of simpler and more lucid treatment.

Peter Gray, Esq., F.R.A.S. and F.R.M.S., an eminent mathematician, and author of several valuable works, took the

matter up in the *Assurance Magazine*, and, signing himself
'A Subscriber,' published a mode of solving the problem in a
much more simple, lucid, and satisfactory manner than that
adopted by his predecessors. In my opinion it is most elegantly
constructed, and admirably adapted to the purpose. I give it
here according to the author's own version, and for the original
letter see pages 101 and 102 of the *Assurance Magazine,*
vol. i. 1851 :—

'Call the sum advanced, the present value, m

Annuity   .   .   . p

The realised rate   .   . r per £.

Investing rate   .   . r'

And we have to find the relation between m and p when r and
r' are given.

'The annual interest on the sum advanced is mr, whence
the sum to be annually invested is $p-mr$; and if A denotes
the amount of an annuity of £1 at the investing rate during
the term, we have by condition

$$(p-mr)A = m.$$

'From this we get

$$m = \frac{pA}{1+rA}, \text{or} \frac{p}{\frac{1}{A}+r} \quad . \quad . \quad . \quad . \quad . \quad (1)$$

and $$p = \frac{m(1+rA)}{A}, \text{ or } m\left(\frac{1}{A}+r\right) \quad . \quad . \quad . \quad (2)$$

'If the annuity is £1 we have from (1)

$$m = \frac{A}{1+rA}, \text{or} \frac{1}{\frac{1}{A}+r}.$$

'Again, if the sum advanced or present value is £1, we get,
from (2),

$$p = \frac{1+rA}{A}, \text{or} \frac{1}{A}+r.$$

'Once more, if the arrangement as to the annuity and
the sum to be advanced is made between the parties without

reference to rates, we find for the realised rate from the first expression

$$(p - mr)A = m,$$

$$r = \frac{p - \dfrac{m}{A}}{m}, \text{ or } \frac{p}{m} - \frac{1}{A}.$$

The rule laid down in (1) is the same as that which I have employed, wherein $\frac{1}{A}$ is what I have termed the redemption fund, and called by French writers the amortizing annuity.

Attached to Mr. Gray's letter, is a very important note by the Editor of the *Assurance Magazine*, in which he states that his correspondent ' did not seem to have anticipated that Mr. Hardy's paper would appear in that magazine,' and that the subject had been previously investigated by Mr. Morgan, and further that it had not ' occurred to any of these writers that by far the most simple way to treat the question would be to construct tables showing the annual payments required at practicable rates to produce £1 at the end of n years. Calling these results r', we shall have the relation between p and m (to use our correspondent's notation) by inspection, and

$$m = \frac{p}{r + r'} \quad p = m\,(r + r'), \text{ and } r = \frac{p}{m} - r'.\text{'}$$

Although rules have thus been supplied by a few mathematicians, nevertheless there is but little to be found upon the subject in books, neither can I discover any table computed to any extent by their means. Mr. Hardy calculated and attached a small table to his paper previously referred to, extending to one 8vo. page only, for rates of interest of 5, 6, and 7 per cent., and to redeem capital at rates of interest of 3, 3½, 4, and 5 per cent. carried to 3 decimal places.

This table was introduced into Inwood's Tables, by the permission of Mr. Hardy, in 1853, but it is cut down to 2 places of decimals, and extends to two 12mo. pages only. This original table by Mr. Hardy, was also cut down to two places of decimals, and incorporated into Mr. C. M. Willich's book, as may be seen in the edition published in 1871. It is less perfect than the original, as the years' purchase is only given for every 10th year after 50 years.

Mr. Downing Biden also published in 1864, two 8vo. pages of tables of this kind for rates of interest from $3\frac{1}{2}$ to 8 per cent. carried to 3 places of decimals.

There is also a table of the same class, in Hurst's 'Architectural Surveyors' Hand Book,' published in 1866, computed from $3\frac{1}{2}$ to 10 per cent., and to 4 places of decimals, and is the best that has come under my notice. All these tables, however,— especially the last-named—are very useful *within* the limits assigned; but I am not acquainted with any tables of this class that are sufficiently extensive to be of any real practical use to persons engaged in valuing Mineral properties, where high rates of interest are expected to be realised on the purchase money, or capital invested.

The rate of interest allowed to a purchaser of mineral property, such as Collieries, Iron Mines, and others, frequently ranges between 10 and 25 per cent., but more generally between 14 and 20 per cent., depending of course upon the character of the property. It is evident, therefore, that tables calculated for rates of interest no higher than 8 or 10 per cent., and to 2 or 3 places of decimals, could not be employed for ascertaining the true value of annuities derived, or to be derived, from high rates.

It is stated on page 2, in all the editions of 'Inwood's Tables of Annuities' that I have seen—that is to say, those published from 1837 to 1866—that 'A lease or annuity for 14 years, to make 3 per cent. and get back the principal, is worth 11·296 years' purchase of the clear annual rent,' and this rule is repeated as a foot-note as far as page 9, as being true for all the rates of interest up to 10 per cent. The table goes no higher than 10 per cent., but it is identical with Mr. Smart's table—and that of all subsequent writers—of the present value of £1 per annum, for any number of years. This table, and others of its kind, to be found in most works on Annuities, is constructed correctly according to the mode laid down; but as that mode is based on incorrect principles, its application to the valuation of annuities, where interest is allowed at higher rates per cent. than can possibly be found for reproducing capital, is entirely fallacious, for the principle upon which it is based assumes that we can reproduce capital which may have been invested, at the same rate of interest as that allowed and expected to be realised on the purchase money invested.

Tables of this class are, therefore, limited in their use to cases where the rate of interest on the capital invested, is the same as that which may be practically obtained in the funds for redeeming the capital. I have, however, good reason for concluding that many well-known Engineers, and others, still employ tables of this kind in valuations connected with mineral properties, even when the rate of interest ranges between 10 and 20 per cent. A clear proof of this assertion may undoubtedly be found on examining the evidence published as having been given before a Committee of the House of Commons, on the Rating of Mines in 1857, and fully quoted in the foregoing pages of this work.

I desire it to be understood, however, that I am antagonistic to none, but feel great respect for those gentlemen who gave the evidence referred to, as I conceive they believed they were right. I certainly cannot, however, consent to pass over a matter so vastly important, and which is, in my opinion, at variance with reasoning based on correct principles.

Referring therefore again to Mr. Taylor's statement, 'that a perpetuity at 8 per cent. is worth $12\frac{1}{2}$ years' purchase,' that is to say, the present value of £1 per annum according to the old Tables is 12·4943 years' purchase, and it must be evident that it is this class of table Mr. Taylor employs in arriving at the value as stated.

The value by Mr. Taylor's statement = 12·4943
„   by correct tables   .   . = 12·24789
Difference   .   .   =   ·24641

or 4s. 11d. too much, equal to 24·641 per cent. lost on every £1 annuity purchased according to his rule. Mr. Dunn also states (see his work on the Coal Trade) 'that 30 years' duration at 8 per cent. is 11·25 years' purchase' that is, the present value of £1 per annum at 8 per cent. for the period stated is 11·2578 years' purchase.

The value by Mr. Dunn's statement = 11·2578 years' purchase.
„   by correct tables   .   . =   9·8991   „   „
Difference   .   =   1·3587

or £1 7s. 2d. too much, equal to 135·868 per cent. lost on every
£1 annuity purchased. But to be clear that no mistake has
occurred in Mr. Dunn's statement, he further adds, 'that for a
duration of 15 years at 14 per cent.,

$$
\begin{aligned}
\text{The years' purchase is} \quad &= 6\cdot14 \\
\text{The value by correct tables} &= 5\cdot16084867 \\[-2pt]
\hline
\text{Difference} \quad\quad\quad\cdot &= \cdot97915133
\end{aligned}
$$

or 19s. 6¼d. too much, equal to 97·915 per cent. lost on every
£1 annuity purchased by the use of the old tables. (See Table
XII.)

Now, as these gentlemen must have believed their method
of deducing the value to be true, it would be very much out of
place to pass any severe stricture on them; but as correct rules
were in existence, by which the accurate value in years' purchase
at a given rate could have been ascertained, before they gave
their evidence, it is, to say the least, a great pity they did not
avail themselves of them.

Furthermore, I apprehend that such evidence was obtained for
the purpose of recommending the passing of some enactment as
to the Rating of Mines; it was, therefore, placing the Committee
in a wrong position; as all the evidence collected and published
by them tends to show that coal and other mines were of greater
value than could, in strictness and on just and equitable principles, be assigned them.

The question is, as I have before said, of vital importance,
and may, I think, be so far demonstrated as to put it beyond a
doubt or mere matter of opinion; and by employing the proper
rule previously referred to and laid down in this work, the truth
may possibly appear more clear and convincing by deductions
arrived at from numerical examples.

Taking 3 per cent. as interest to be realized on capital
invested, and redeeming that capital at the same rate within
14 years, we have, the redemption fund to reproduce the
capital at 3 per cent. within 14 years = ·05852634.

Then ·05852634 + ·03 = ·08852634 and $\dfrac{1}{\cdot08852634}$

= 11·29607314 years' purchase, practically the same as in

Inwood, and in all other writers on Annuities, but correct to more places of decimals.

Now, assuming a purchase was effected by allowing 20 per cent. instead of 3 per cent., and by the same rule also to recoup at 20 per cent., we have, the redemption fund to recoup at 20 per cent. within 14 years = ·0168930552.

$$\text{Then } \cdot 0168930552 + \cdot 20 = \cdot 2168930552 \text{ and } \frac{1}{\cdot 2168930552}$$

= 4·610567171 years' purchase.

Suppose the annuity to be purchased equals £20,000, its present value, immediate, would be £4·610567171 × 20,000 = £92211·34342.

Then 20 per cent. upon this sum =   18442·268684
And £92211·34342 × ·0168930552 =   1557·731316

The annuity   .   . £20,000·000000

It is, therefore, evident that if a purchaser were to invest £92211·34342 in purchasing an annuity of £20,000, derived from some mineral property, he could not invest annually in any funds so small a sum as £1557·731315 for 14 years, that would yield him 20 per cent. It would, therefore, be impossible to realise or reproduce the original capital invested, within the time, under the circumstances.

Now, if we apply the proper rule, which is founded upon the principle that an investor realises a certain rate on his capital, and can reproduce that capital at another, but lower, and more practicable rate, we shall find that a serious discrepancy exists in the last preceding mode of ascertaining the value.

Taking 20 per cent. as interest to be realised on capital, and to redeem that capital within 14 years at 3 per cent. compound interest, we have the redemption fund necessary to replace £1 within the time = ·058526339.

$$\text{Then } \cdot 058526339 + \cdot 20 = \cdot 258526339,$$

$$\text{and } \frac{1}{\cdot 258526339} = 3 \cdot 868077828 \text{ years' purchase,}$$

which is the true sum that must be given, in order to secure £1 annuity for 14 years, allowing 20 per cent. upon it, and to replace it at 3 per cent. within the time.

To secure an annuity, therefore, of £20,000, there must be invested a sum equal to £77361·55656; and to get it back in 14 years at 3 per cent. an annual redemption fund of £4527·6887446 would be required to be set aside to accumulate at compound interest.

$$\text{For 20 per cent. on } £77361\cdot55656 = 15472\cdot31131$$
$$\text{And } £77361\cdot55656 \times \cdot058526339 = 4527\cdot68869$$

$$\text{The annuity as before} = £20000\cdot00000$$

The present value obtained by Inwood's rule, and endorsed by many others (see pages 2 to 9 of his book) .   .   .   . $= 92211\cdot34342$
The present value found by correct method, viz. to realise at one rate per cent., and to redeem at another rate, say 3 per cent. . $= 77361\cdot55626$

$$\text{Difference}   .   .   . = £14849\cdot78716$$

It is conclusive, therefore, that a present purchaser would be paying too much by £·742489358, in order to secure £1 annuity, or a total of £14849·78716.

For the difference between the incorrect and the true years' purchase

$$" = £4\cdot610567171 - 3\cdot868077813 = £\cdot742489358,$$
$$\text{and } £\cdot742489358 \times 20000 = £14849\cdot78716,$$

being the difference in error as before, or a loss of 74·2489 per cent. on the annuity purchased. (See Table XII.)

The practice, therefore, of valuing upon tables constructed on the assumption of reproducing capital at the same high rate of interest, as that which may be realised on it, is opposed to the truth, and calculated to mislead and injure a purchaser to a very large extent.

The subject of *Deferred* Annuities, embracing *two* different rates of interest per cent., has not, in my opinion, hitherto received so much attention as some other of its branches, although deferred annuities involving *one* rate of interest have been frequently and ably discussed.   Nevertheless, after diligent search and enquiries that I have instituted, I cannot discover anything which appears to me to bear directly upon the case—when *two* rates of interest are considered—in any of the published works devoted specially to Annuities.

The rules already in use may give approximations, but I have deemed the question of sufficient importance to call for further investigation; and with a view to establish more uniformity in practice, appropriate formulæ, and practical rules, have been devised, peculiarly adapted to the construction of tables of this nature.   These rules have previously been laid down in a former portion of this work, and a proof of their accuracy, demonstrated.   They are, I believe, important ones, and the Tables calculated by their aid are now introduced for the first time.

When one rate of interest is considered, the general principle applied to deferred annuities is, that when the value at a specific rate of a benefit with reference to a specified epoch has once been determined, the value at the same rate with reference to any other epoch is assigned by multiplying that first determined, by the power of v whose index is the number of years in the period of deferment, or the interval through which the value is in a sense transferred.

Thoman's definition is '*that the present value of a deferred annuity is equal to the difference between two immediate annuities of the same yearly income, one for the whole term, the other to continue until the time of entering on the deferred annuity.*

This rule, however, embraces but one rate of interest in the present value of £1 per annum, but it has, I believe, been followed by all writers on Annuities, and by many valuers since Thoman's time.   It is thus illustrated :—

Assuming the annuity to be deferred t years, and to continue n years afterwards, that is, say 55 years after 3 years, we have

The present value of $3 + 55 = 58$ years
at 3 per cent. . . . $= 27\cdot33100549$
The present value of $58 - 55 = 3$ years
at 3 per cent. . . . $= 2\cdot82861138$

$$\text{Present value deferred 3 years} = \pounds24\cdot50239411$$

Now if we suppose the interest allowed on an investment is 20 per cent., and also to reproduce the capital at the same rate, employing the above rule, we have for a duration of 14 years after 3 years:

The present value of $3 + 14 = 17$ years at
20 per cent. . . . . . $= 4\cdot7746338$
The present value of $17 - 14 = 3$ years at
20 per cent. . . . . . $= 2\cdot1064815$

$$\text{Present value deferred 3 years} . = \pounds2\cdot6681523$$

But allowing a purchaser 20 per cent. upon his investment, and to reproduce the capital at 3 per cent. for a similar period, that is to say 14 years after 3 years, we have from (14),

$$P_{3+14} = \frac{M_{14}}{(1 + r')^3 (1 + r' M_{14})}.$$

Or,

$$\frac{17\cdot08632416}{1\cdot728 \times (1 + \cdot20 \times 17\cdot08632416)} = \frac{17\cdot08632416}{7\cdot63303363} = 2\cdot238470965$$

years' purchase, or value deferred three years.

The deferred value by Thoman's or Inwood's rule $= 2\cdot668152300$
The deferred value by correct method . $= 2\cdot238470965$

$$\text{Difference} = \pounds\cdot429681335$$

A purchaser would, therefore, be paying too much for each £1 annuity, by £·429681335, or 8s. 7d.; and if an annuity of £20,000 were purchased, the gross overpaid sum would amount to £8593·6267, or £8593 12s. 6½d., or a total loss of 42·968 per cent. upon the annuity.

For 2·668152300 × 20000 =   53363·0460
And 2·238470965 × 20000 =   44769·4193

Difference as before = £ 8593·6267

Again, taking another case under Thoman's and Inwood's rule, as generally adopted, and allowing 20 per cent. to a present purchaser, and to reproduce the capital at the same rate, we have for a duration of 55 years after 3 years :

The present value of £1 per annum for
  58 years at 20 per cent. .   .   . = 4·999872221
The present value of £1 per annum for
  3 years at 20 per cent.   .   . = 2·106481481

Present value deferred 3 years = £2·893390740

But, by allowing to the said purchaser 20 per cent. upon his investment, and to reproduce the capital at 3 per cent, the period of time being as in the last preceding case, or 55 years after 3 years, and adopting the correct rule for such a case, we have

$$\frac{136\cdot07161972}{1\cdot728 \times (1 + \cdot20 \times 136\cdot07161972)} = \frac{136\cdot07161972}{48\cdot75435056} = 2\cdot790963639,$$

the correct value deferred.

The deferred value by Thoman's or Inwood's
  rule .   .   .   . = 2·893390740
The deferred value by correct method .   = 2·790963639

Difference = £·102427101

The difference, therefore, is equivalent to 2s. 0½d. per £1, and, if as before, an annuity of £20,000 were purchased, the overpaid value or total loss would = £·102427101 × 20000 = £2048·54202; that is to say, every £1 annuity purchased under such conditions would cost too much by £·102427101, or 10·243 per cent.

When *one* rate of interest only is considered in the purchase of an annuity, and redemption of the capital invested is made at the same rate per cent, the rule as employed by Inwood, which may be found from pages 2 to 9 of his Tables, and which has been the subject of investigation in this work, would undoubtedly be correct, and the value derived, *tolerably reliable for percentages up to 4 or 4½*; but it has been shown that when *two* different rates of interest enter into the question, its application to the valuation of mineral or any other property would produce erroneous results.

In the purchase of mineral properties, the rate of interest allowed on the investment—that is to say, purchase money or capital—should not be fixed at the same rate as that proposed, or which may be found practicable, for the redemption of the capital, otherwise a difficulty would be created in obtaining a return of the large sums annually laid out in gigantic mining concerns; and this applies to Collieries as well as other mines.

It has been previously stated that in the purchase of Mines, the rate of interest ranges from 14 to 20 per cent.; but it is evident that such rates could not be realised for the purpose of redeeming capital invested in a mine by available means, such, for instance, as in land, houses, Consols, or other similar well-known securities. What I mean by the redemption of capital, is being put in a position to find the original sum expended, in safe keeping, at the time the mine and annuity ceases.

From a consideration of these circumstances, it is difficult to conceive how those persons accustomed to make valuations—dependent upon tables which assume that the profit rate is identical with the reproductive rate of interest, no matter how high the former may be—are justified in adopting and continuing such a system. It has, however, been strongly urged to me, as a sort of defence of the system, that a proprietor having an open colliery or other mine, equal in duration to a perpetuity, or say 100 years, may profitably reinvest, so to speak, his surplus annuity in the extension of his mining operations, and thereby produce an annual increase in minerals, and consequently in annuity, and so gain upon it the same rate of interest as that realised on the original capital already brought into productive action. The surplus annuity, which should have been invested in some good security at either 2½ or 3 per cent., thus becomes charged as original capital, simply written off the ledger accounts, or paid to shareholders as dividends year by year, until the mine is unproductive ; but, at the same time, there has been no special means adopted on this hypothesis in providing for the redemption of the accumulated capital.

Only upon this suggested, but really impracticable mode, can an attempt ever be made of reinvesting the surplus annuity at the same rate as that realised on the capital, unless, indeed, a purchaser is willing to accept 3, 4, or 5 per cent. on his investment.

But to realise the idea fully, the mine operated upon in such a way as that suggested, must be made to yield constant results, and the state of the market must also be such, as to produce uniform profits upon the minerals annually produced, in the case of a going concern, but in the case of a deferred annuity, expected to arise from an undeveloped property, the rate per cent. upon which the mine was purchased and expected to be obtained during its continuance, must not only be *guaranteed* by an engineer's report, but, to be actually realised, the general state of trade must not fluctuate so as to depreciate the value, as previously determined, when the mine is brought into productive condition. Again, at the expiration of 20 years from the present time, the result to be obtained from any mine cannot be absolutely guaranteed; and, although the value may have been arrived at with very great care and judgment, nevertheless unforeseen contingencies, arising from some particular and unavoidable circumstance connected with a mine, coupled with a downward tendency of trade, may depreciate the value of that particular mine.

Then, if it be granted that from such causes the annuity derived from any mine is not a constant amount, but that it may suffer from the fluctuations of trade, it is also granted, or at least it would follow, that as the annuity is a variable quantity, although it had formerly been fixed, or guaranteed, by allowing a certain rate per cent., as *profit* on the purchase, a valuation made upon the assumption that the capital could be redeemed at a uniform rate, as high as that fixed for profit, would be altogether unreliable.

I have gone into this matter rather fully, because I am aware that the kind of tables I have mentioned as being erroneous in principle, are much employed in Valuation, and that if a rule becomes established as applicable to one mine, it would be equally competent to apply it to all other mines. I should, however, discard the principle involved, and reject any valuation made upon it.

To such considerations as those enumerated, earnest attention must be given, and in order that large capital sums should not be lost, it is undoubtedly the safest, as well as the wisest plan to anticipate, as far as possible, every contingency, and out of the general annuity derived from a certain realised rate per cent. on the capital invested in any mine, set aside such a sum as may be determined by calculation, and which, if invested in Consols, or some other fund equally secure, will, at the normal rate of 3 per cent. interest, reproduce the original capital at the time when the annuity ceases.

I intended to have concluded this subject here, but having communicated my views to a friend, he forwarded to me a pamphlet, entitled 'An Investigation of the Errors of all Writers on Annuities,' by William Rouse, published by Lackington, Allen, & Co., 1816. It extends to 40 small 8vo. pages, and, as I conclude it is scarce, I take the liberty of making a quotation from it. He says, commencing at page 36, 'As to the tables published at rates of interest *above* 5 per cent. per annum, when the principle on which they are formed is considered, they will be found both impracticable and illusive to purchasers.

'The principle on which all the tables hitherto published, for the valuation of terminable incomes, whether for years or lives, have been formed, is that the yearly income will not only be equal to the interest per cent. named in the tables, but as *much more* as will replace (at the end of the term of life) the capital employed. For instance, if a person pay £802 for an income or £100 per year for 17 years, he employs his money, or capital, at 10 per cent. interest (according to the tables); that is, he is supposed to receive £10 for every £100 advanced, and as much more as will replace the £802 at the end of the 17 years. 10 per cent. on the capital employed is £80 4s. 0d., but as he will receive £100 each year, the difference, or £19 16s. 0d. per year, is the sum to replace the £802 at the end of the term; this it will certainly do *if* a man can make 10 per cent.

interest on £19 16s. 0d. every time he receives it; but at 5 per cent. the same sum will only amount to £511 in 17 years. Now I appeal to the common sense of every man, if it be practicable to improve small sums of money at a greater interest than 5 per cent.?—indeed, beyond this rate it is illegal to lend money, and no leases, annuities, or government securities, can be purchased with small sums of a few pounds each, which in general form the excess of interest to replace the capital with when the income ceases.

'Such being the principle on which all the tables of compound interest for the valuation of leases and annuities for *years* or lives are formed, they must be *practically* wrong where they exceed the rate of 5 per cent., which being the legal interest of the country, all calculations to replace the capital at the end of the term ought to be made in this rate; and as these tables *form the basis of all the calculations for annuities on lives*, they must follow the same fate; for the present value of an annuity *certain*, for any number of years, is the several *present* values of the several sums to be received at the end of the first, second, third, &c., years to the end of the term, *added together*; and the present value of a life annuity is nothing more than the amount of the said several values, *each diminished* in proportion as the respective probabilities of the person being alive at the end of the several years to receive them, are below certainty, and continued to the most probable extent of life. Now, as the values of annuities for lives depend on the combinations of these two sets of tables, if one requires new modelling, and the other is practically wrong, all the results at rates exceeding 5 per cent. must be doubly incorrect.'

Biden also states that 'the ordinary tables (present value of £1 per annum) at high rates are erroneously applied in valuing leases, &c., because they assume the possibility of making annually an investment of surplus at those high rates, which is impracticable.'

The question of the redemption of capital is of as much importance to the landlord or lessor, as it is to the lessee. For, in the case of collieries and other mines, the lessee removes annually so many acres of the minerals contained in the estate. Unless, therefore, the lessor invests annually a certain sum derived from the royalty dues at the termination of the lease, or exhaustion of the minerals in the royalty area, the 'Fee Simple' in the mineral estate would be entirely lost.

On the contrary, if provision had been made for redemption at the end of the term of the lease, or when the mineral estate is exhausted, the annual investment would accumulate to the original value of the royalty, and the landlord, or lessor, would be in possession of a sum which could be invested in land or other property of a permanent character.   Thus, the original value of the mineral estate would be continued in another form.

All terminable annuities, no matter from what source derived, should be purchased upon a principle, which would allow a portion of such benefit to be annually invested, and capable of yielding back the original capital at the termination of the income.

PART III.

PRACTICAL EXAMPLES

IN THE

VALUING OF COLLIERIES,

IRON AND OTHER MINES,

ROYALTIES, LEASEHOLDS, FREEHOLDS, LIFE INTERESTS, &c.

ALSO

ANNUITIES DERIVED FROM ANY SOURCE, EITHER
IMMEDIATE OR DEFERRED.

3. **In What Time** will £650 amount to £858 at 4 per cent. simple interest?

Here £650 × ·04 = 26·00, and 858 − 650 = 208.

Then $\dfrac{208}{26\cdot00}$ = 8 years.

4. **What is the Rate** per cent., simple interest, when £650 amounts to £858 in 8 years?

Here 650 × 8 = 5200, and 858 − 650 = 208.

Then $\dfrac{208\cdot00}{5200}$ = ·04.   And ·04 × 100 = 4 per cent.

5. **What Discount** should be allowed for the present payment of a bill of £920 due at the end of three months, interest being at the rate of 5 per cent.?

Here 3 months = $\dfrac{3}{12}$ ths of a year = ·25.

And ·25 × ·05 = ·0125.   Then ·0125 + 1 = 1·0125.

And $\dfrac{920}{1\cdot0125}$ = £908·64197.

Then 920 − 908·64197 = £11·35803, the discount required.

6. **What Will an Annuity** of £650 amount to in 12 years at 5 per cent. simple interest?

Putting a = annuity, or £650,
r = interest, or 5 per cent.,
t = years, or 12,
M = amount.

Then $M = \left(\dfrac{t\,(t-1)\,r}{2}+t\right) a,$

And $M = \left(\dfrac{12\,(12-1)\,\cdot05}{2}+12\right) 650$ = £9945.

7. **What Annuity** will amount to £9945 in 12 years at 5 per cent. simple interest?

Here $a = \dfrac{2M}{t\,(t-1\;r)+2t},$

Or $a = \dfrac{9945 \times 2}{12\,(12-1)\cdot05+24}$ = £650, the annuity.

8. In What Time will an Annuity of £650 amount to £9945 at 5 per cent. simple interest?

$$\text{Here } t = \frac{\sqrt{8r\dfrac{M}{a} + (2-r)^2} - (2-r)}{2r},$$

$$\text{And } t = \frac{\sqrt{8 \times \cdot 05 \times \dfrac{9945}{650} + (2-\cdot 05)^2} - (2-\cdot 05)}{2 \times \cdot 05}$$

$$= 12 \text{ years, the time required.}$$

9. At What Rate per cent. simple interest, will an annuity of £650 amount to £9945 in 12 years?

$$\text{Here } r = 2\frac{\left(\dfrac{M}{a}-t\right)}{t\,(t-1)},$$

$$\text{Or } r = 2\frac{\left(\dfrac{9945}{650}-12\right)}{12 \times (12-1)} = \cdot 05, \text{ and } \cdot 05 \times 100 = 5 \text{ per cent.,}$$

the rate required.

COMPOUND INTEREST.

10. **What will £6500 Amount** to in 40 years at 5 per cent. per annum compound interest?

$$\text{Here } (1\cdot05)^{40} = 7\cdot039988712. \text{ (See Table I.)}$$

Then,

$7\cdot039988712 \times 6500 = £45759\cdot926628$, the amount required.

By logarithms:

log 1·05 . . . . . . = 0·021189299070

40

0·847571962800

„ 6500 . . . . . = 3·812913356643

„ 45759·9266290 . . . = 4·660485319443

11. **What Principal will Amount** to £45759·926629 in 40 years, at 5 per cent. per annum, compound interest?

Here $\dfrac{45759·926629}{7·039988712169}$ = £6500, the principal required.

By logarithms:

log 1·05 × 40 .	.	.	.	= 0·847571962800
„ 45759·926629	.	.		= 4·660485319443
„ 6500 .	.	.	.	= 3·812913356643

the principal, as previously determined.

It is evident that by employing Mr. Gray's Logarithmic Tables to 12 places of decimals, we obtain better results than could be supplied from the common 7-figure table.

12. **What is the Rate per cent.** when £6500 amounts to £45759·926628 in 40 years?

Here we have

$$\frac{45759·926628}{6500} = 7·0399887121,$$

the amount of £1 in 40 years, and this number will be found in the column of the amounts under 5 per cent., which is the rate required.

Or thus:

$$\sqrt[40]{7·0399887121} = 1·05,\text{ and } 1·05 - 1 = ·05.$$

Then ·05 × 100 = 5 per cent.

By logarithms:

log 45759·926628	.	.	.	= 4·660485319443
„ 6500	.	.	.	= 3·812913356643
			40)	0·847571962800
„ 1·05	.	.	.	. = 0·021189299070

And 1·05 − 1 = ·05, then ·05 × 100 = 5 per cent., as above.

13. **In what Time will £6500 amount** to £45759·926628 at 5 per cent. per annum compound interest?

Here $\dfrac{45759·926628}{6500}$ = 7·039988712,

the amount of £1 at 5 per cent. per annum. Then, by inspecting the Tables under 5 per cent., the number 7·039988712 will be found opposite 40 years, which is the time required.

By logarithms:

log 45759·926628   .   .   . = 4·660485319443
„   6500   .   .   .   . = 3·812913356643
$$\overline{0·847571962800}$$

and log 1·05 = ·02118929907.

Then $\dfrac{·847571962800}{·02118929907}$ = 40 years, as above.

AMOUNT OF ANNUITIES AT COMPOUND INTEREST.

14. **What will an Annuity** of £6500 amount to in 40 years, at 5 per cent. interest being payable annually?

The amount of £1 per annum in 40
years, at 5 per cent., see Table (III). = 120·7997742425
$$\underline{\hspace{3cm}£6500}$$
£785198·5325762500

By logarithms:

log 1·05 = ·02118929907
$$\underline{\hspace{3cm}40}$$ Natural Number.
0·84757196280 = 7·039988712169.
$$- 1·$$
$$\overline{6·039988712169}$$

Then,

log 6·039988712169 = 0·781036126991
„   6500   .   . = 3·812913356643
$$\overline{4·593949483634}$$
„   ·05   .   . = $\overline{2·698970004336}$
5·894979479298 = £785198·5325821,

the amount as above.

Also $\dfrac{6·039988712169 \times 6500}{·05}$ = £785198·53258197.·

15. **What Annuity will Amount** to £785198·53258197 in 40 years, at 5 per cent. per annum?

Here £785198·53258197 × ·05 = 39259·9266290985.

The amount of £1 in 40 years, Table (I) = 7·039987712169,

And 7·039988712169 − 1 = 6·039988712169,

Then $\dfrac{39259\cdot9266290985}{6\cdot039988712169}$ = £6500, or annuity.

By logarithms:

log 1·05 = 0·02118929907

$$\dfrac{\qquad\qquad 40}{0\cdot84757196280}$$ = 7·039988712169

− 1

$$\overline{6\cdot039988712169}$$

log (785198·53258197 × ·05) = 4·593949483634

  ,,   6·039988712169   = 0·781036126991

                         3·812913356643

                        = £6500 as above.

By the Tables—

The amount of £1 per annum in 40 years at 5 per cent., see Table (III) = 120·7997742425,

Then $\dfrac{785198\cdot5325821}{120\cdot7997742425}$ = £6500, the annuity.

16. **What will an Annuity** of £500 amount to in 10 years, at 4 per cent. compound interest, the annuity and interest being payable half-yearly?

Here, as the time is 10 years, and the rate of interest 4 per cent. for half-yearly payments, it becomes 10 × 2 = 20 half years, and $\dfrac{4}{2}$ = 2 per cent.

Then (1·02)²⁰ = 1·4859473960,

And 1·4859473960 − 1 = ·4859473960, see Table (I) for 2 *per cent.* at 20 years.

Also $\dfrac{\cdot4859473960 \times 500}{\cdot04}$ = £6074·34245.

Again, taking the amount of £1 per annum for 20 years, at 2 per cent., see Table (III) = 24·2973697989.

$$\text{And } \frac{500}{2} = £250;$$

Then 24·2973697989 × 250 = £6074·34245.

If, however, payments were made quarterly, then we should have 10 × 4 = 40 quarter years, and $\frac{4}{4}$ = 1 per cent.

Then $(1·01)^{40}$ = 1·4888637336, the amount, see Table (I);

And 1·4888637336 − 1 = ·4888637336;

$$\text{Then } \frac{·4888637336 \times 500}{·04} = £6110·79667.$$

By employing the amount of £1 per annum for 40 years, at 1 per cent., see Table (III) = 48·8863733588.

$$\text{And } \frac{500}{4} = £125.$$

Then 48·8863733588 × 125 = £6110·7966985, practically the same as above.

17. **Required the Time** in which an annuity of £6500 will amount to £785198·53257625, interest being at 5 per cent.

Here 785198·53258197 × ·05 = 39259·9266290985,

$$\text{And } \frac{39259·9266290985}{6500} = 6·039988712169,$$

And 6·039988712169 + 1 = 7·039988712169

Then log 7·039988712169 = 0·847571962800

And „ 1·05 = 0·021189299070

$$\text{Whence } \frac{0·847571962800}{0·021189299070} = 40 \text{ years, the time required.}$$

By the Tables—

$$\frac{785198·53258197}{6500} = 120·7997742425,$$

the amount of £1 per annum, at 5 per cent.; and in Table (III) under 5 per cent., this number corresponds to 40 years, the time required.

18. **An Annuity** of £6500 amounts to £785198·53258197 in 40 years; required the rate per cent.

Here $\dfrac{785198\cdot53258197}{6500} = 120\cdot7997742425$,

or the amount of £1 per annum in 40 years, at the rate required ; and upon finding this number in the Tables opposite 40 years, the rate per cent. will be found at the head of the column in which the number $120\cdot7997742425$ is found. In this case the rate is 5 per cent.

Much has been written upon the theory of this problem, but, after all, such investigations have only led to approximate results by a no very direct method, which is neither convenient or facile. Dual Arithmetic, however, according to Mr. Byrne, is said to furnish direct means for its solution.

PRESENT VALUE OF SUMS AT COMPOUND INTEREST.

19. **What is the Present Value** of £500 due at the end of 30 years, allowing interest at the rate of 5 per cent.?

By Table (I), $(1\cdot05)^{30} = 4\cdot3219423752$, the amount of £1 in 30 years, at 5 per cent.

Then $\dfrac{500}{4\cdot3219423752} = £115\cdot6887243266$,

the value required.

Again $\dfrac{1}{4\cdot3219423752} = \cdot2313774486$,

and $\cdot2313774486 \times 500 = £115\cdot6887243$, the present value as before. The number $\cdot231373486$ may be obtained direct from Table (IV) opposite 30 years and under 5 per cent, to 8 places of decimals.

By logarithms :

log 500	.	.	.	.	$= 2\cdot698970004336$
„ 1·05 = ·02118929907 × 30					$= 0\cdot635678972100$
„ 115·6887243181	.	.	.		$= 2\cdot063291032236$,

practically the present value, as before.

20. **What Sum may be Secured** at the end of 30 years by a present payment of £115·68872433, interest to be allowed at the rate of 5 per cent.?

By Table (I), $(1·05)^{30}$ = 4·3219423752, and 4·3219423752 × 115·68872433 = £500, the sum required.

Also $\dfrac{115·6887243266}{·2313774486}$ = £500, as before.

By logarithms :

$$\log 1·05 \quad . \quad . \quad . \quad . \quad . \quad = 0·021189299070$$

$$30$$

$$\overline{0·635678972100}$$

$$,, \quad 115·6887243181 \quad . \quad . \quad = 2·063291032236$$

$$,, \quad 500 \quad . \quad . \quad . \quad . \quad . \quad = 2·698970004336$$

21. **At the End of a Certain Term** the sum of £500 has to be paid in discharge of a debt, but allowing 5 per cent. discount from the sum then due, a settlement may be effected by a present payment of £115·6887243181 ; what was the number of years at the expiration of which the £500 should have been paid?

$$\log 500 \quad . \quad . \quad . \quad . \quad . \quad = 2·698970004336$$

$$,, \quad 115·6887243181 \quad . \quad . \quad = 2·063291032236$$

$$\overline{0·635678972100}$$

$$\text{Log } 1·05 = 0·021189299,$$

And $\dfrac{0·6356789721}{0·02118929907}$ = 30 years, the time required.

22. **If £500 is due at the end** of 30 years, and may be discharged by a present payment of £115·6887243181, what rate of interest per cent. was allowed ?

$$\log \quad 500 \quad . \quad . \quad . \quad . \quad = 2·698970004336$$

$$,, \quad 115·6887243181 \quad . \quad . \quad = 2·063291032236$$

$$30 \,)\, \overline{0·635678972100}$$

$$,, \quad 1·05 \quad . \quad . \quad . \quad = 0·03118939907$$

Then $1·05 - 1 = ·05$, and $·05 × 100 = 5$ per cent., the rate required.

PRESENT VALUE OF ANNUITIES AT COMPOUND INTEREST.

23. **The Lease of an Estate** has 30 years to run, the annual value of which is £805, but is held subject to the payment of £270 per annum ; what is the present value of the title, allowing interest at the rate of 5 per cent. ?

By Table (IV) the present value of £1 due 30 years hence, at 5 per cent. = ·23137745, and 1 − ·23137745 = ·76862255.

$$\text{Then } £805 - £270 = £535,$$

$$\text{And } \frac{·76862255 \times 535}{·05} = £8224·261285,$$

the present value of the title.

Again, a similar result may also be obtained by employing the present value of £1 per annum for 30 years, at 5 per cent. Thus, the present value of £1 per annum in 30 years, assuming redemption of capital can be effected at the same rate, = £15·372451.

$$\text{And } £15·372451 \times 535 = £8224·261285,$$

the present value, as before. Table (XII) gives values of this class to 5 places of decimals.

24. **What Annuity** to be continued 30 years, may be purchased for £8224·261285, allowing interest to the purchaser at 5 per cent. per annum ?

The present value of £1 per annum, at 5 per cent., redeeming capital at the same rate, for 30 years = 15·372451.

$$\text{And } \frac{8224·261285}{15·372451} = £535, \text{ the annuity required.}$$

We may also determine the annuity by employing the present value of £1 due 30 years hence. Thus by Table (IV), the present value of £1 due 30 years hence = ·23137745, and the arithmetical complement of this quantity = 1 − ·23137745 = ·76862255.

$$\text{And } \frac{8224·261285 \times ·05}{·76862255} = £535, \text{ as above.}$$

25. **An Annuity** of £535 was purchased for £8224·261285, interest being allowed at the rate of 5 per cent.; required the duration of the annuity.

$$\text{Here } 8224 \cdot 261285 \times \cdot 05 = 411 \cdot 21306425,$$

$$\text{And } 535 - 411 \cdot 21306425 = 123 \cdot 78693575.$$

$$\text{Then } \frac{\log 535 - \log 123 \cdot 78693575}{\log 1 \cdot 05}$$

$$= \frac{2 \cdot 728353782021 - 2 \cdot 092674812446}{\cdot 02118929907} = \frac{\cdot 635678969575}{\cdot 02118929907}$$

$$= 30 \text{ years, the time required.}$$

PERPETUITIES.

Perpetual Annuities are those which are to continue for ever, and are consequently treated in a different manner from annuities which are to continue for determined periods.

26. **What is the Present Value** of an estate in fee simple of £1200 per annum, interest of money being at the rate of 5 per cent.?

$$\text{Here } \frac{100}{5} = 20, \text{ and } 1200 \times 20 = £24000$$

the present value required, or

$$\frac{1200}{\cdot 05} = £24000, \text{ the value as before.}$$

27. **What Perpetuity** will £24000 purchase, interest of money being at the rate of 5 per cent.?

$$\text{Here } £24000 \times \cdot 05 = £1200,$$

the annuity required; or

$$\frac{100}{5} = 20 \text{ and } \frac{24000}{20} = £1200, \text{ as above.}$$

28. **What Rate of Interest** is realized when £24000 will purchase an annuity in perpetuity of £1200?

$$\text{Here } 1200 \times 100 = 120000,$$

$$\text{And } \frac{120000}{24000} = 5 \text{ per cent., the rate of interest required.}$$

REVERSIONS.

Reversionary or Deferred Annuities, are those which are not to be entered upon until after the expiration of a certain defined period. This subject has been fully entered into in the foregoing part of this work; it is therefore unnecessary to enlarge upon it here, except by example.

29. **What is the Present Value** of a deferred annuity of £650, to continue 20 years, but not to be entered upon until after the expiration of 6 years, allowing interest at the rate of 4 per cent.?

The present value of £1 per annum for 26 years, at 4 per cent., assuming capital can be redeemed at the same rate per cent., $= 15\cdot982769$, and for 6 years $= 5\cdot242137$.

$$\text{Then } 15\cdot982769 - 5\cdot242137 = 10\cdot740632,$$

$$\text{And } 10\cdot740632 \times 650 = £6981\cdot4108,$$

the present value deferred. This example is given to show the mode generally adopted in solving the problem.

Again, the present value of £1 per annum for 20 years, at 4 per cent., $= 13\cdot590326$, and the present value of £1 due in 6 years, at 4 per cent. $= \cdot79031453$.

$$\text{Then } 13\cdot590326 \times \cdot79031453 \times 650 = £6981\cdot4108,$$

present value as before.

Also, the present value of £1 per annum in 6 years, at 4 per cent. $= \cdot79031453,$

$$\text{And for 26 years } = \cdot36068923,$$

$$\text{And } \cdot79031453 - \cdot36068923 = \cdot4296253.$$

$$\text{Then } \frac{\cdot4296253 \times 650}{\cdot04} = £6981\cdot4108.$$

There is, therefore, an agreement in the present value deferred, deduced by three independent processes; but this could not have occurred if two different rates of interest had been involved in the years' purchase.

30. **What Annuity,** to continue 20 years after the expiration of the next 6 years, may be purchased for £6981·4108, interest being allowed at the rate of 4 per cent.?

Here $(1\cdot04)^{-6}$ . . . . . $= \cdot7903145257$

And $(1\cdot04)^{-26}$ . . . . $= \cdot3606892329$

$$\cdot4296252928$$

Then £6981$\cdot$4108 × $\cdot$04 $= 279\cdot256432$,

And $\dfrac{279\cdot256432}{\cdot4296252928} = £650$, the annuity.

31. **The Sum** of £6981$\cdot$4108 is expended in the purchase of an annuity of £650, commencing after the expiration of 6 years. What length of time will the annuity continue when the rate of interest is 4 per cent. ?

Here the amount of £1 in 6 years $= (1\cdot04)^6 = 1\cdot2653190185$,

And $1 - \dfrac{£6981\cdot4108 \times \cdot04 \times 1\cdot2653190185}{£650}$

$= 1 - \dfrac{£353\cdot34847486693664}{£650}$

$= 1 - \cdot5436130378 = \cdot4563869622.$

Then $\dfrac{-\log \cdot45638696}{\log 1\cdot04} = \dfrac{\cdot340666770752}{\cdot017033339299} = 20$ years,

the time required.

It may be here remarked that the logarithmic tables generally employed in these cases are those before referred to, by means of which the natural number can be easily found to 12 places of decimals.

The French mathematician, Callet, has a limited but very valuable Table of Logarithms in his work, to 20 decimal places, but the natural number can only be obtained to a few figures from it.

32. **What is the Present Value** of the reversion of a perpetuity of £650 per annum, after 6 years' deferment, interest allowed being at the rate of 5 per cent ?

By Table (IV), the present value of £1 due in 6 years

$$= \dfrac{\cdot74621540 \times 650}{\cdot05} = £9700\cdot8002.$$

Or thus:

$$\dfrac{100}{5} = 20, \text{ and } 20 \times £650 \times \cdot74621540 = £9700\cdot8002,$$

as before.

By logarithms:

log 650 . . . . . = 2·812913356643

„ ·05 . . . . . = $\overline{2}$·698970004336

„ £13000 . . . . = 4·113943352307

value of perpetuity.

Then,

log of perpetuity . . . = 4·113943352307

„ 1·05 × 6 . . . . = 0·127135794420

„ £9700·8002 . . . . = 3·986807557887

value as before.

33. **The Reversion** of an estate in fee simple, after 6 years' deferrence, is sold for £9700·8002 ; what annuity should it produce, so as to allow the purchaser 5 per cent. upon his purchase money ?

By Table (IV), the present value of £1 due in 6 years = ·7462154.

Then $\dfrac{£9700·8002 × ·05}{·74621540} = \dfrac{485·04001}{·74621540} = £650$, the annuity.

34. **If a Perpetual Annuity** of £650 is purchased for £9700·8002, allowing interest at the rate of 5 per cent., what period of time must the annuity be deferred before being entered upon ?

Here

$$\frac{£9700·8002 × ·05}{£650} = ·74621540.$$

Then

$$\frac{-\log ·74621540}{\log 1·05} = \frac{·127135792462}{·02118929907} = 6 \text{ years,}$$

the deferred period.

35. **Thirty years** having expired of a lease having 40 years' duration, what sum should be paid for renewing such lease for the lapsed period, supposing the estate to produce a clear rental of £200 per annum, interest being allowed at the rate of 5 per cent. per annum ?

Here the case is that of a deferred annuity, commencing 10 years hence, and to continue 30 years afterwards. If it were possible to redeem capital at the rate of 5 per cent., the following is the usual mode of treating the question :—

The old present value of £1 per annum for 40 years is 17·159086, and that for 10 years is 7·721735 ;

Then 17·159086 — 7·721735 = 9·437251, or years' purchase.

Also 9·437251 × £200 = £1887·450200,

the present value or sum to be paid down.

The old present value of £1 per annum for 30 years, at 5 per cent. = 15·372451 ; and this deferred 10 years = 9·43730.

Then 9·43730 × £200 = £1887·4600, the present value.

If, however, the capital can only be redeemed at 3 per cent. per annum, the present value of £1 per annum, allowing 5 per cent. upon the capital for a duration of 30 years = £14·080688, but deferred 10 years = £8·644417. (See Table X.)

Then 8·644317 × £200 = £1728·8634,

the present value or sum to be paid down.

It will be observed, that in the examples given in this section, no rate of interest has been employed higher than 5 per cent., and then only upon the assumption that any capital sum expended may be redeemed at the same rate per cent. as that allowed on the purchase.

It is a question of the value of money at the time of purchase, or the highest possible rate of interest available to a purchaser for the redemption of his capital for the future period, taking into consideration, however, the extra attendant risk always incurred, when any sum invested is believed to be capable of being redeemed at high rates of interest.

Most monetary transactions should undoubtedly be governed by the average rate of interest realised from a fluctuating market, over a series of years ; at least this would be the wisest course. But if higher rates are required, and accepted upon any transaction, the probability of an eventual realisation is much further removed. Generally, therefore, the higher we ascend the scale from the normal rate of interest, or 3 per cent., so is the risk of the redemption of capital increased proportionally.

VALUATION OF MINES.

36. **What is the Present Value** of a Colliery extending over 1200 acres, and yielding 160,000 tons of coal per annum, to continue 60 years? The average annuity derived from the Colliery during the last 10 years has amounted to £16,520, and that arising from the surface, let as a farm, is £2400. The interest allowed on the purchase of the Colliery to be at the rate of 14 per cent. per annum, and to redeem the capital at the rate of 3 per cent. per annum. Working plant to be included in the purchase. Interest allowed on the purchase of the rough farm land to be at the rate of 4 per cent.

Here, as previously laid down, we have

P_n = present value, or purchase money.
r' = rate of interest allowed on ditto.
r = rate of interest allowed for redemption.
M_n = the amount of an annuity of £1 at r per cent.
p = the annuity.

Then, from (19), we have

$$P_n = \frac{M_n}{1 + r' M_n}.$$

Substituting the numerical values for these symbols, we also have,

$$r' = 14 \text{ per cent.}$$
$$M_n = 163\cdot05343680$$
$$p = £16520.$$

Then,

$$\frac{163\cdot05343680 \times 16520}{1 + (\cdot14 \times 163\cdot05343680)} = \frac{2693642\cdot776}{23\cdot82748115} = £113047\cdot73505193,$$

or £113047 14s. 8d.,

the present value.

The present value of an annuity of £1, at 14 per cent. for 60 years, and to redeem the capital at the rate of 3 per cent. is 6·84308324, or, in other words, the years' purchase. (See Table VII, correct to 8 places of decimals.)

Then $6·84308324 \times £16520 = £113047·73512480$,
or $£113047\ 14s.\ 8d.$,

practically the same as before.

To find the Redemption Fund, we also have

M_n = the amount of an annuity of £1 at r per cent. for n years.

p = the annuity.

r' = the interest allowed on purchase money.

s_n = redemption fund.

Then

$$s_n = \frac{100\,p}{100 + (r'M_n)},$$

and by substitution we also have

$M_n = 163·05343680$,
$p = £16520$,
$r' = 14$ per cent.

Then

$$\frac{100 \times £16520}{100 + (14 \times 163·05343680)} = \frac{1652000}{2382·748115} = £693·31709449,$$
or $£693\ 6s.\ 4d.$, or s_n.

Again, from (8) we have

$$s_{60} = \frac{r}{R^{60} - 1} = \frac{·03}{5·891603 1041 - 1} = ·0061329587,$$

which is the same as found in Table (V), and which will produce £1 in 60 years at 3 per cent.

Then $£113047·73512480 \times ·0061329587 = £693·31709066$,
or $£693\ 6s.\ 4d.$,

practically the same as before.

Then, for the disposal of the annuity, we have

The yearly interest on $£113047·7351248$, at 14 per cent. per annum . . . = 15826 13 $7\frac{3}{4}$

And the annual redemption fund to replace the purchase money within the 60 years would be . . . . . = 693 6 $4\frac{1}{4}$

Together equal to annuity as above . . $£16520$ 0 0

Then, if we multiply the amount of £1 per annum for 60 years by the annual redemption fund, the original capital will be reproduced.

Thus 163·05343680 × 693·31709066 = £113047 14s. 8d., the purchase money or capital invested.

The land being in perpetuity, and 4 per cent. being allowed to a purchaser, it is worth 25 years' purchase.

Then £2400 × 25 = £60000, the present value.

The present value of the Colliery	. =	113047	14	8
The present value of the estate .	. =	60000	0	0
Total present value	. = £	173047	14	8

37. **What is the Present Value** of the unexpired term of a lease of a Colliery of 40 years, subject to a royalty to the lessor of 6d. per ton upon all coal raised? The present output is 90,000 tons per annum, and the average gross annuity derived, £10125. Interest on the purchase money to be allowed at the rate of 16 per cent., and to redeem the capital at the rate of 3 per cent. The rate of interest upon the royalty to be allowed at 8 per cent. The estimated worth of permanent plant and stock is £45,000, to be sold at the end of the term for say £12,000, upon which a discount of 5 per cent. is to be allowed.

Here by Table (VII), the present value of an annuity of £1 for 40 years, so as to allow a purchaser 16 per cent. upon his purchase money, and to redeem the same within the time at 3 per cent. compound interest . =		5·77159342
Annuity . . . . . . = £		10125
Total present gross value . . . =	£58437·38337750	
	or £58437 7s. 8d.	

For proof we have the interest on £58437·383375, at 16 per cent. per annum . . =	9349	19	7½
And the annual redemption fund to replace the gross value within the 40 years would be . . . . =	775	0	4½
The gross annuity as above . . . = £10125	0	0	

The annuity resulting from the royalty on 90,000 tons per annum, at 6*d.* per ton = £2250.

And by Table (VII), the present value of an annuity of £1 for 40 years, at 8 per cent., and to redeem the capital at 3 per cent. =	10·72243731
Annuity . . . . . . =	£2250

Present value . . . . . = £24125·48394750
or £24125 9*s.* 8*d.*

Also, for the value of the plant, we have

The present value of £1 due at the end of 40 years, at 5 per cent., by Table (IV) =	·14204568
	£12000

Present value . . . . . = £1704·54816000
or £1704 10*s.* 11½*d.*

From the present gross value of the Colliery lease . . . . . . =	58437	7	8
Must be deducted the present value of the royalty . . . . . . =	24125	9	8
	£34311	18	0
To which must be added the present value of the plant . . . . =	1704	10	11½
Total present net value of the Colliery lease =	£36016	8	11½

For proof of the value of the royalty we have,

The interest on £24125·4839475 at 8 per cent. per annum . . . . =	1930·038715
And the annual redemption fund to replace the value at the end of 40 years = £24125·4839475 × ·0132623779 . =	319·961285
Lessor's gross annuity. . . =	£2250·000000

For proof of the lessee's value we also have,

The interest on £34311·89943 at 16 per cent. per annum . . . . =	5489·9039088
And the annual redemption fund to replace the lessee's value at the expiration of 40 years = £34311·89943 × ·0132623779 =	455·0573767
Lessee's gross annuity . . . =	£5944·9612855

Then for proof of lessor and lessee's gross annuity we have,

£1930·038715 × 2 . . . . =	3860·077430
Lessor's redemption fund . . =	319·961285
Lessee's gross annuity . . . =	5944·961285
The gross annuity as deduced on page 80 =	£10125·000000

Again we also have,

£319·961285 + £445·0573767 . . =	£775·018662

Or £775 0s. 4½d. as before.

38. **What is the Present Value** of a Colliery yielding 60,000 tons of coal per annum, subject to a royalty to the lessor of 8d. per ton upon all coals raised? The estimated duration is 25 years, and the annuity accruing £6000. Interest on the purchase money to be at the rate of 16 per cent. per annum, and to redeem the capital at 3 per cent. per annum, plant and stock included. The annuity arising to the lessor and lessee, however, to be paid quarterly.

Here the rate of interest to redeem being 3 per cent., we have $\dfrac{·03}{4}$ = ·0075 for the quarterly ratio.

Then $\dfrac{·0075}{(1·0075)^{100} - 1} = \dfrac{·075}{1·11108384} = ·00675065676$,

the quarterly redemption fund (amount taken from Table II).

And $\cdot006750165676 \times 4 = \cdot027000662704$,

the annual redemption fund, with increase due to the quarterly increments.

Also $\cdot027000662704 + \cdot16 = \cdot187000662704$,

Therefore $\dfrac{1}{\cdot187000662704} = 5\cdot347574631$,

the years' purchase, or present value of an annuity of £1, for 25 years, when paid quarterly.

Then we have . . . . . 5·347574631
Annuity . . . . . . = £6000

Present gross value of the Colliery . = £32085·447786000
or £32085 8s. 11¼d.

The annuity accruing to the lessor = 60,000 tons at 8d. = £2,000, and the years' purchase being as above,

We have . . . . . . 5·347574631
Annuity . . . . . = £2000

Present value of the royalty . . = £10695·149262000
or £10695 2s. 11¾d.

The present gross value of the Colliery . = 32085 8 11¼
The present gross value of the royalty . = 10695 2 11¾

Present net value of Colliery. . . = £21390 5 11¼

The present value of an annuity of £1, at 16 per cent. per annum, and to redeem the capital at 3 per cent. per annum, for 25 years, when the annuity accrues annually . . . . = 5·335385792
Ditto ditto, when the annuity is paid quarterly = 5·347574631

Difference . . . . . . = £·012188839

or nearly 3d. in every £1 annuity purchased, an excess in value due to quarterly payments.

39. **What is the Present Value** of a Colliery Lease having 44 years to run, and producing 200,000 tons per annum? But

in order to continue this yield during the whole term, it will be necessary to expend £40,000 in additional works, extending over a period of 3 years. The average annuity derived from the Colliery during a series of years in the past has been, and still is £20,000, and the lease is held subject to a royalty to the lessor of 6d. per ton during the ensuing 21 years, and 9d. per ton for the remainder of the term, or 23 years. Interest on the purchase to be allowed at the rate of 12 per cent. per annum, and to redeem the capital at 3 per cent. per annum. The interest allowed to a present purchaser of the royalty to be at the rate of 7 per cent. per annum, and to redeem the capital at 3 per cent. The plant is estimated to have cost £100,000 when the Colliery was opened, but to be sold at the end of the term for say £16,000, and upon this sum a discount is allowed at the rate of 5 per cent per annum.

The present value of an annuity of £1 per
annum for 44 years, at 12 per cent., and
to redeem the capital at 3 per cent. per
annum, is by Table (VII)   .   .   =   7·62021768
Annuity   .   .   .   .   .   =   £20000

Total present gross value .   .   .   =   £152404·3536000
or £152404 7s. 0¾d.

Then it is customary for a valuer to say :

'From this gross value of £152404   7   0¾
  Must be deducted the cost of ad-
      ditional works   .   .   .   40000
  And interest thereon at 5 per
      cent. for 3 years'   .   .   6000   46000   0   0

Present gross value of Colliery, after deducting outlay, and interest as determined
by the customary mode   .   .   .   = £106404 7s. 0¾d.

Further on, special reference will be made to the customary mode of allowing interest at the rate of 5 per cent. upon any sum of money set apart, or estimated by a valuer for maintaining a certain yield from mines for a definite period. An independent mode of solution will also be introduced.

For the royalty we have 6*d*. per ton for the first period of 21 years, and 9*d*. per ton for 23 years afterwards; the correct value may therefore be more readily determined by *assuming* that the royalty is fixed at 9*d*. per ton for the whole period of 44 years, and deducting therefrom the present value due to the excess of royalty, or 3*d*. per ton for 21 years. This is evident, as an average of the two royalties could not give the correct value, neither could it be obtained accurately by two separate valuations, that is, *first* upon that due to the annuity arising from 6*d*. per ton for 21 years, and *secondly* to that due to the annuity arising from 9*d*. per ton for 23 years afterwards.

The reason for this is obvious.

The years' purchase due to 21 years at 7 per cent. = 9·53545399; and if we treat that number as the basis of a distinct valuation, and then proceed to value the second period of 23 years similarly, we should find the years' purchase = 9·919266819, which is only removed in point of time 2 years from the former period, *i.e.* 21 years, whereas the termination of the second period of 23 years is removed 44 years from the commencement of the first period, and the years' purchase for the period of 44 years = 12·31074584.

Therefore, by Table (VII), the present value of an annuity of £1 per annum for 44 years, allowing interest at the rate of 7 per cent. per annum, and to redeem the capital at the rate of 3 per cent. per annum . . . . . = 12·31074584

£7500

Present value if the royalty were at 9*d*. per ton for the whole period of 44 years = £92330·59380000

And by Table (VII), the present value of an annuity of £1 for 21 years, allowing interest at 7 per cent., and to redeem the capital at 3 per cent. . . = 9·535453993

£2500

Present value of excess of royalty for 21 years at 3*d*. per ton . . . = £23838·634982500

From the present value of the royalty at 9*d.*
  per ton for 44 years . . . = 92330·59380000
Must be deducted the present value of the
  royalty at 3*d.* per ton for 21 years . = 23838·63498250
  _____
Present net value of royalty . . . = £68491·95881750
  or £68491 19*s.* 2*d.*
Then for the machinery, we have the present
  value of £1 due at the end of 44 years at
  5 per cent. . . . . . = ·11686133
  £16000
  _____
Present value of machinery . . . = £1869·78128000
  or £1869 15*s.* 7½*d.*

REDUCTION FROM GROSS TO NET VALUES.

To the present gross value of the Colliery,
  after deducting outlay and interest . = 106404 7 0¾
Must be added the present value of thé
  machinery . . . . . = 1869 15 7½ .
  _____
  £108274 2 8¼

From which must be deducted the present
  value of the royalties . . . 68491 19 2
Present net value of Colliery Lease . = £39782 3*s.* 6¼*d.*

40. **What is the Present Value** of a Colliery, the lease of
which has 21 years to run, subject to a royalty to the lessor of
4*d.* per ton on all coals raised, but which royalty is now worth
8*d.* per ton during the whole term ? The output from the Colliery is 170,000 tons per annum, and the annuity derived is
£17,000, and that due from the royalty (which is to be deducted
from the lessee's gross annuity) is £2833 6*s.* 8*d.* Interest allowed
on the purchase of the Colliery at the rate of 14 per cent. per
annum, and to redeem the capital at the rate of 3 per cent. per
annum. The interest allowed on the royalty to be at the rate of
8 per cent. per annum, and to redeem at 3 per cent. The excess

of royalty to be at the same rate. Plant and stock included in the sale of the Colliery.

The present value of an annuity of £1 for
21 years, so as to allow a purchaser 14 per
cent. upon his purchase money, and to
redeem the capital at 3 per cent. per
annum . . . . . . = 5·718475674
Annuity . . . . . . = £17000

Present gross value of Colliery . . = £97214·086458000
or £97214 1s. 8½d.

And for the lessor's royalty, we have,

The present value of £1 per annum at 8 per
cent., and to redeem at 3 per cent. for 21
years . . . . . . = 8·705358535
Annuity due to lessor . . . = £2833·333

Present value of lessor's royalty . . = £24665·1796140472
or £24665 3s. 7d.

Also, for the excess of the value of the royalty, at 4d. per ton, we have

The years' purchase as above . . = 8·705358535
Annuity due to excess of royalty . = £2833·333

Present value of excess of royalty . = £24665·1796140472
or £24665 3s. 7d.

From the present gross value of the Colliery
lease . . . . . . = 97214 1 8½
Must be deducted the present value of the
lessor's royalty . . . . = 24665 3 7

£72548 18 1½
To which must be added the present value
of the excess of royalty . . . = 24665 3 7

Present net value of Colliery lease . = £97214 1s. 8½d.

Proof of the accuracy of the valuation of the Colliery may be thus obtained :—

The yearly interest at 14 per cent. upon
£97214·086548 . . . . =   13609 19 5¼
And the annual redemption fund that will produce £1 in 21 years = ·0348717765;
then 97214·086548 × ·0348717765 . =   3390 0 6¾

Together equal to annuity . . . = £17000 0 0

And the amount of £1 per annum for 21 years, at 3 per cent., by Table (III) = 28·67648572, and if this number is multiplied into the annual redemption fund, the original purchase money would be reproduced. Thus, 28·67648572 × £3390·027899 = £97214 1s. 8½d., the original capital invested.

For proof of the valuation of the royalty, we also have,

The yearly interest at 8 per cent. upon
£24665·1796140472 . . . =   1973 4 3½
And the annual redemption fund that will produce £1 in 21 years is, by Table (V),
= ·0348717765; then 24665·1796140472
× ·0348717765 . . . . =   860 2 4½

Annuity derived from royalty . . = £2833 6s. 8d.

Also, £860·11886 × 28·67648 = £24665 3s. 7d., the original present value of the royalty, as previously deduced. That is to say, if £860 2s. 4½d. were laid by annually, at 3 per cent. compound interest, the original sum paid for excess of royalty would be reproduced.

The last preceding case assumes an incoming tenant, who, upon purchasing the lease of the Colliery and everything therewith connected, subjects himself to all the conditions entered into by the lessee. At the onset, therefore, he is entitled to have a deduction made from the present value of the Colliery lease. In this case it is taken at 4d. per ton, and the resulting annuity upon the output is treated in the usual way; the question being, what is its present value, presuming it were about to be sold? This must be taken as a *minus* quantity, inasmuch as the purchaser of the Colliery, or representative of the lessee, subjects himself to the payment of the royalty to the lessor. On the other hand, the lessee has possessed himself of a valuable lease, the royalty of which, as fixed for the ensuing 21 years, is under its real value ; that is, it is considerably less than that charged upon the surrounding collieries. The lessee is, therefore, entitled to sell his Colliery lease at an enhanced value, equivalent to what is due to the difference existing in royalty between his and the surrounding collieries. Certain questions, however, would arise, such as whether the Colliery would be exhausted in 21 years ?—and if not, what would be the probable amount of royalty for the next term of extension of lease ? This should be provided for as far as possible in the lease ; but, if left an open question, then the incoming tenant may fairly raise objections, and seek to effect a compromise, which probably would result in diminishing the value of the excess of royalty. Of course all such questions involve a consideration of the basis upon which royalties are determined, which may always be open to dispute and reference; and here experience and judgment would weigh materially in settling the matter. Then, again, as to the determination of the amount of royalty of any particular Colliery, from that of the surrounding ones, the question as to whether such collieries are working under similar conditions must undoubtedly be taken into consideration.

41. **What is the Present Value** of an undeveloped freehold Colliery extending over an area of 1000 acres, containing several workable seams of excellent coal, capable of yielding 420,000 tons per annum for a period of 80 years, and producing by estimation an annuity of £42,000 ? The time occupied in developing the Colliery is estimated at 4 years, at the expiration of which time it is expected the above yield will commence. The Colliery is obtained under favourable circumstances, there being very little water, and good rock roofs exist over the different seams of coal. The interest allowed is 18 per cent. per annum, and to redeem the capital at the rate of 3 per cent. per annum, and the estimated cost of developing, with plant, is £80,000, with the customary rate of 5 per cent. interest thereon for 4 years, or during the time of development. The overlying estate, held in fee simple, is also to be sold with the minerals, the rate of

interest allowed being 5 per cent., and is let out as farms, producing an annuity of £3000.

Here the present value of an annuity of £1 for 80 years,
allowing a purchaser 18 per cent. upon his purchase money, and
to redeem the same at the rate of 3 per cent. per annum
= 5·46114612. But as this annuity is deferred 4 years, from
(14) page 34, or (16) page 36, we have

The deferred value = 2·816798415.

The deferred value may be more easily obtained from :—

$$P_{t+n} = P_n v^t.$$

Thus $P_{80} v^4 = 5\cdot46114612 \times \cdot5157888751 = 2\cdot816798415$, as
before.

Again, by Table (X), the years' purchase = 2·816798, true
to 6 places.

Also	2·816798415
Annuity . . . . . . =	£42000
Present gross value of Colliery . .	= £118305·53343000

Then a valuer would say :

'From which must be deducted the estimated cost of development	80000	
And the customary interest thereon at 5 per cent. for 4 years' .	16000	96000·00000000

Present net value of Colliery, after deducting outlay, and customary interest thereon	= £22305·53343000
	or £22305 10s. 8d.

The interest allowed on the purchase of the estate being
5 per cent., it is worth 20 years' purchase. We therefore have

£3000 × 20	=	60000 0 0
Present net value of Colliery . . .	=	22305 10 8

Total present net value of Colliery and Estate = £82305 10s. 8d.

The accuracy of the calculations referring to the last preceding case may be further corroborated thus:—

By Table (I), the amount of £1 in 4 years at 18 per cent. = 1·93877776.

Then 1·93877776 × £118305·5334 = £229368·13704,

or that sum to which the deferred years' purchase will amount during the deferred period.

Then the annual interest at 18 per cent. on
£229368·13704 . . . . = 41286 5 3½
And the annual redemption fund to replace
the same within the time at 3 per cent. compound interest . . . . = 713 14 8½

Together equal to the annuity . . . £42000 0 0

If, however, the payments of the annuity in the last preceding case were half-yearly, or quarterly, in order to ascertain the present value deferred 4 years, we must proceed as follows:

The interest allowed to redeem the capital being at the rate of 3 per cent per annum, the *pro ratâ* half-yearly and quarterly rates would be represented by

$$\frac{·03}{2} = ·015 \text{ half-yearly,}$$

and $\frac{·03}{4} = ·0075$ quarterly.

Then $\dfrac{·015}{(1·015)^{160} - 1} = \dfrac{·015}{10·828461 - 1} = \dfrac{·015}{9·828461}$

= ·00152626992, the half-yearly redemption fund.

Then ·00152626992 × 2 = ·00305253984, the yearly redemption fund, and ·00305253984 + ·18 = ·18305253984,

and $\dfrac{1}{·1830525398} = 5·462912457,$

the years' purchase immediate. The redemption fund for half-yearly and quarterly payments may be obtained direct from Table (V).

As the annuity is deferred 4 years, and payments are made half-yearly, the problem must be subjected to the principle involved in (5) and (7a.), pages 25 and 28.

The rate of interest being 18 per cent. per annum, the half-yearly and quarterly ratios

$$= \frac{\cdot 18}{2}, \text{ and } \frac{\cdot 18}{4} = \cdot 09 \text{ and } \cdot 045.$$

Then,

$$(1 + \cdot 09)^8 = 1 \cdot 99256264;$$
$$\text{and } (1 + \cdot 045)^{16} = 2 \cdot 02237015, \text{ or the}$$

amounts due to half-yearly and quarterly payments.

These numbers are readily obtained from Table (I), under 9 and $4\frac{1}{2}$ per cent. for 8 and 16 years.

Then for the present value of £1 due 4 years hence, at 18 per cent., for half-yearly payments we have,

$$v^4 = \frac{1}{1 \cdot 99256264} = \cdot 50186628;$$

which is, as it should be, less than the value found in Table (IV) for the same rate per cent. and period of deferment. The present value deferred, is now readily deduced from

$$P_{t+n} = P_n v^t;$$

the relation of which has been fully explained on page 35. Thus,

$$P_{80} = 5 \cdot 462912457; \text{ and } v^4 = \cdot 50186128, \text{ and}$$

$$P_{80+4} = P_{80}v^4 = 5 \cdot 462912457 \times \cdot 50186628 = 2 \cdot 741651553,$$

or the years' purchase deferred.

For proof, we have,

$$1 \cdot 99256264 \times 2 \cdot 741651553 = 5 \cdot 462912457,$$

or the immediate value or sum to which £2·741651553 would

have accumulated at ·18 per cent during the 4 years of deferment.

Then . . . . 2·741651553
Annuity . . . . . = £42000

Present gross value of Colliery . . = £115149·365226000

giving, for the gross value, a smaller sum by £3156·168204, when the payments are made half-yearly.

Again, for the quarterly payments, we also have

$$\frac{·0075}{(1·0075)^{320}-1} = \frac{·0075}{10·924902-1} = \frac{·0075}{9·924902} = ·00075567493.$$

Then ·00075567493 × 4 = ·0030226998,

the annual redemption fund due to quarterly payments,

And ·0030226998 + ·18 = ·1830226998,

Therefore $\dfrac{1}{·1830226998} = 5·463803129,$

the years' purchase immediate, due to quarterly payments.

But, being deferred 4 years, we have the amount of £1 in that period at ·18 per cent. due to quarterly payments = 2·02237015.

Then $v^4 = \dfrac{1}{2·02237015} = ·4944693235$, or the

present value of £1 due 4 years hence, accruing from quarterly payments.

Here we have $P_{80} = 5·463803129$, and $v^4 = ·4944693235$; Then 5·463803129 × ·4944693235 = 2·701683037, the present value deferred 4 years, and due to quarterly payments.

The proof is 2·701683037 × 2·02237015 = 5·463803129, or the immediate value or sum to which £2·701683037 would have accumulated at ·18 per cent. during the deferred period of 4 years.

Then . . . . 2·701683037
Annuity . . . . . = £42000

Present gross value . . .. . = £113470·687554000

Here the difference between the values, when the payments are made yearly and half-yearly, = £3156·168204; that between the yearly and quarterly payments = £4834·845876; and that between the half-yearly and quarterly payments, = £1678·677672.

The case, thus treated, is of considerable importance as applied to the Valuation of Mines, being greatly in favour of a purchaser. The principle upon which it is based has been formerly illustrated, and should always be applied when payments are made half-yearly and quarterly.

42. **What is the Present Value** of a mineral property, upon which two shafts have been sunk within a short distance of the upper seams of coal? A full description of which is as follows:—

1st.—The lease of a colliery, having 35 years to run from the time when the upper and lower seams of coal, iron ore, and clay are successively won, subject to a royalty of 3d. per ton upon all coal or other minerals raised from the mine. It is known from the surrounding collieries, that the roofs over the seams of coal are good, and that only a moderate quantity of water exists in the strata to be passed through. It is estimated that the first seams of coal will yield 60,000 tons per annum for the entire term, and that the cost of developing this portion of the mine, including plant, is £12,500. The time occupied in performing the work is estimated at 2 years, and the rate of interest allowed to a present purchaser is at the rate of 16 per cent. per annum, and to redeem the capital at the rate of 2½ per cent., and the annuity estimated to be realized during the entire period amounts to £6750.

2nd.—The lower seams of coal, won by the same shafts, are capable of yielding 120,000 tons per annum for a longer period than that of the lease, and assessing the profits at a moderately low rate per ton, it is estimated that an annuity of £15,000 may confidently be expected to be realized. This portion of the mine will require a further period of 2 years to develop it, at an extra cost of £15,000, including plant. Interest to be allowed to a present purchaser at the rate of 20 per cent. per annum, and redeeming the capital at the rate of 2½ per cent. per annum.

3rd.—By continuing the same shafts a short distance below

the lower coal seams, it is estimated that an output of 60,000 tons per annum of excellent Hematite iron ore may be secured at a further outlay of £8,000 at the expiration of 6 years from the time of commencing operations. The ore contains 50 per cent. of metallic iron ; it is, therefore, estimated that under the favourable circumstances by which this property is secured, an annuity of £12,000 may be realized. Interest to a present purchaser to be allowed at the rate of 22 per cent. per annum, and redeeming the capital at the rate of 2½ per cent. per annum.

4th.—Fire clay in beds, of an excellent quality, is also known to exist over the entire area included in the lease, and that 45,000 tons may be annually extracted during the entire period. There is a ready market for its disposal at a constant price per ton; it is therefore estimated that an annuity of £5875 will be realized. The additional cost for the development and plant will be about £7500, expended at the same time that the iron ore is won. Interest to be allowed to a present purchaser at the rate of 14 per cent. per annum, and redeeming the capital at the rate of 2½ per cent. per annum.

5th.—Wayleave of certain private branch Railways, and other accommodations, charged at the rate of 1d. per ton upon all minerals conveyed over them, amounting to an annuity of £250 per annum after 2 years. Also an annuity of £500 after a period of 4 years, and further an annuity of £437 10s. after 6 years. Interest to be allowed at the rate of 10 per cent. per annum, and to redeem the capital at the rate of 2½ per cent. per annum.

6th.—Ground rents derived from houses and other buildings, amounting to £140 per annum. Interest to be allowed to a present purchaser at the rate of 8 per cent. per annum, and to redeem the capital at the rate of 2½ per cent. per annum.

7th. — Royalties to the lessor, amounting to £750 per annum after 2 years, and to run 10 years afterwards. Interest at 10 per cent., and to redeem at 2½ per cent.

8th.—Royalty to lessor, amounting to £1500 per annum after 4 years, and to run 10 years afterwards. Interest as in last preceding case.

9th. — Royalty to lessor, derived from iron ore and clay,

amounting to £1312 10s. per annum after 6 years, and to run 10 years afterwards. Interest as before.

10th.—The lessor will not consent to sell the royalty of 3d. per ton on the estimated annual output from the mine, until after the expiration of 10 years from the time estimated for winning the upper and lower seams of coal, iron ore, and clay, in succession. The minerals contained in this property will not be exhausted in 35 years; the royalty is, consequently, worth more than 3d. per ton; the lessor will, however, convey it on the assumption that it may be exhausted in that time. Interest to be allowed at the rate of 10 per cent. per annum to a present purchaser, and to redeem the capital at 2½ per cent. per annum. The lessor consents to accept any *bonâ fide* incoming tenant introduced by the lessee; and the latter may sell his interest in the property at any time.

1st Valuation.

Annuity from 60,000 tons of coal per annum = £6750.
Interest allowed at 16 per cent. per annum.
Redemption of capital at 2½ ,, ,,

The present value of £1 per annum for 35 years, so as to allow a purchaser 16 per cent. per annum upon his purchase money, and to redeem the capital at 2½ per cent. per annum = 5·61149650. As the annuity is deferred 2 years, from Table IV we have v^2, or the present value of £1 due 2 years hence, at ·16 per cent. = ·74316290.

Then

$$P_{35+2} = P_{35}v^2 = 5\cdot61149650 \times \cdot74316290 = 4\cdot170256019,$$

years' purchase, or present value deferred.

Then, to prove the accuracy of the operation, we also have

$$4\cdot170256019 \times 1\cdot3456 = 5\cdot61149650,$$

the value immediate, as before.

Then . . . . 4·170256019
£6750
                                         _____
Present gross value . . . = £28149·228128250

2nd Valuation.

> Annuity from 120,000 tons of coal per annum = £15000.
> Interest allowed at   20 per cent. per annum.
> Redemption of capital at $2\frac{1}{2}$,,   ,,

The present value of £1 per annum for 35 years, so as to allow a purchaser 20 per cent. per annum upon his purchase money, and redeem the capital at $2\frac{1}{2}$ per cent. per annum = 4·58283418, and as the annuity is deferred 4 years, from Table IV we have v^4, or the present value of £1 due 4 years hence, at ·20 per cent. = ·48225309.

Then

$$P_{35+4} = P_{35}v^4 = 4 \cdot 58283418 \times \cdot 48225309 = 2 \cdot 21008594,$$

the present value deferred.

For proof, we have $2 \cdot 210085927 \times 2 \cdot 0736 = 4 \cdot 58283418$,

the present value immediate.

Then   .   .   .   .   2·210085927

£15000

Present gross value   .   .   .   = £33151·288905000

3rd Valuation.

> Annuity from 60,000 tons of iron ore per annum
> = £12,000.
> Interest allowed at   22 per cent. per annum.
> Redemption of capital at $2\frac{1}{2}$,,   ,,

The present value of £1 per annum for 35 years, so as to allow a purchaser 22 per cent. per annum upon his purchase money, and to redeem the capital at $2\frac{1}{2}$ per cent. per annum = 4·19805443; and as the annuity is deferred 6 years, we have

$$P_{35+6} = P_{35}v^6 = 4 \cdot 19805443 \times \cdot 3032780757 = 1 \cdot 273177869,$$

the present value deferred.

Then, for proof, we have

$$1 \cdot 273177869 \times 3 \cdot 297303989 = 4 \cdot 19805443,$$

the present value immediate.

And . . . . 1·273177869

£12000
_____

Present gross value . . . = £15278·134428000

4th Valuation.

Annuity from 45,000 tons of clay per annum = £5875.
Interest allowed at   14 per cent. per annum.
Redemption of capital at 2½   „   „

The present value of £1 per annum for 35 years, so as to allow a purchaser 14 per cent. per annum upon his purchase money, and to redeem the capital at 2½ per cent. per annum = 6·32088948; but, as it is also deferred 6 years, we have

$$P_{35+6} = P_{35} v^6.$$

Then by Table (VI) $P_{35} = 6·32088948$, and by Table (IV) $v^6 = ·45558655.$

Consequently $6·32088948 \times ·45558655 = 2·879712216$, the present value deferred.

Then, for proof of the accuracy of the mode of working, we have $2·879712216 \times 2·1949726239 = 6·32088948,$

the value immediate.

Then . . . . 2·879712216

£5875
_____

Present gross value . . . = £16918·309269000

5th Valuation, part 1.

Annuity from wayleave of 60,000 tons per annum = £250.
Interest allowed at   10 per cent. per annum.
Redemption of capital at 2½   „   „

The present value of £1 per annum for 35 years, so as to allow a purchaser 10 per cent. per annum upon his purchase money, and to redeem the capital at 2½ per cent. per annum = 8·45983736; but, as this annuity is deferred 2 years, we have

$$P_{35+2} = P_{35} v^2.$$

By Table (VI) $P_{35} = 8.45983736$, and by Table (IV) $v^2 = .82644628$.

Then $8.45983736 \times .82644628 = 6.991601118$,

the present value deferred.

For proof, we also have

$$6.991601118 \times 1.21 = 8.45983736,$$

the years' purchase immediate.

Then 6.991601118
£250

Present net value . . . . = £1747.900279500

5th Valuation, part 2.

Annuity from wayleave of 120,000 tons per annum = £500.

Interest to purchaser and for redemption, same as in part 1.

The present value of £1 being also 8.45983736, and as the annuity is deferred 4 years, we have the following expression :—

$$P_{35+4} = P_{35} v^4.$$

Here $P_{35} = 8.45983736$, and $v^4 = .68301346$.

Then $8.45983736 \times .68301346 = 5.778182744$,

the present value deferred.

Then, for proof, we have

$$5.778182744 \times 1.4641 = 8.45983736,$$

value immediate.

Then 5.778182744
£500

Present net value . . . . = £2889.091372000

5th Valuation, part 3.

Annuity from wayleave of 60,000 tons of iron ore, and 45,000 tons of clay = £437 10s.

Interest to purchaser and for redemption as in parts
1 and 2.

The present value of £1 per annum for 35 years, also, as
before = 8·45983736, but it is deferred 6 years, therefore we
have

$$P_{35} = 8·45983736, \text{ and } v^6 = ·56447393.$$

Then 8·45983736 × ·56447393 = 4·77535764,
the present value deferred.

Also 4·77535764 × 1·771561 = 8·45983736,
the present value immediate.

And . . . . 4·77535764
£437½
                                    ─────────────
Present net value . . . . = £2089·218967500

6th Valuation.

Annuity from freehold ground rents = £140.
Interest allowed at   8 per cent. per annum.
Redemption of capital at 2½   „   „

The present value of £1 per annum for
35 years, so as to allow a purchaser 8 per
cent. per annum upon his purchase money,
and to redeem the capital at the rate of
2½ per cent. per annum . . . =   10·18272055
£140
                                    ─────────────
Present net value . . . . . = £1425·58087700

7th Valuation.

Annuity from royalty to lessor = £750.
Interest allowed at   10 per cent. per annum.
Redemption of capital at 2½   „   „

The present value of £1 per annum for 10 years, so as to
allow a purchaser 10 per cent. per annum upon his purchase
money, and to redeem the capital at 2½ per cent. per annum
= 5·28377119, but, as the annuity is deferred 2 years, we have

$$P_{10} = 5·28377119, \text{ and } v^2 = ·82644628.$$

Then, $5\cdot28377119 \times \cdot82644628 = 4\cdot366753048$, the present value deferred.

The proof is $4\cdot366753048 \times 1\cdot21 = 5\cdot28377119$, value immediate.

And . . . . $4\cdot366753048$

$£750$

Present net value . . . . . $= £3275\cdot064786000$

8th Valuation.

Annuity from royalty to lessor $= £1500$.
Interest to purchaser and for redemption as before.

The present value of £1 per annum for 10 years, as previously given $= 5\cdot28377119$, and as the annuity is deferred 4 years, we have

$$P_{10} = 5\cdot28377119, \text{ and } v^4 = \cdot68301346.$$

Then $5\cdot28377119 \times \cdot68301346 = 3\cdot608886817$, the present value deferred.

For proof we also have $3\cdot608886817 \times 1\cdot4641 = 5\cdot28377119$, present value immediate.

And 3608886817

1500

Present net value . . . . . $= £5413\cdot330225500$

9th Valuation.

Annuity from royalty to lessor on iron ore and clay $= £1312$ 10s.
Interest to purchaser and for redemption the same as in 7th and 8th Valuations, and the present value of £1 per annum for 10 years, as before $= 5\cdot28377119$; and as the annuity is deferred 6 years, we have

$$P_{10} = 5\cdot28377119, \text{ and } v^6 = \cdot56447393.$$

Then $5\cdot28377119 \times \cdot56447393 = 2\cdot982551103$, the present value deferred.

The proof is $2·982551103 \times 1·771561 = 5·28377119$, value immediate.

Then . . . . $2·982551103$
$£1312\frac{1}{2}$

Present net value . . . . $= £3914·5983226875$

10th Valuation, part 1.

Annuity from lessor's interest in royalty, to be sold at the expiration of 10 years from the time of winning upper seams $= £750$.
Interest allowed at   10 per cent. per annum.
Redemption of capital at $2\frac{1}{2}$ „   „

Here, as the lessor receives royalty for 10 years from the time of winning the upper seams, and as the winning occupies 2 years, the lessor's interest, in this case, can only be realised or purchased after a period of 12 years.   We therefore have

P_{25} by Table (VI) $= 7·73539259$, and v^{12} by Table (IV) $= ·31863082$.

Then $7·73539259 \times ·31863082 = 2·464734463$, the present value deferred.

For $2·464734463 \times 3·13842838 = 7·73539259$, value immediate.

Then . . . . $2·464734463$
$£750$

Present net value . . . . $= £1848·550847250$

10th Valuation, part 2.

Annuity from lessor's interest in royalty, to be sold at the expiration of 10 years from the time of winning the middle seams . . . $£1500$
Interest allowed at   10 per cent. per annum.
Redemption of capital at $2\frac{1}{2}$ „   „

Again, the lessor receives royalty for 10 years from the time of winning the middle seams, which are won after 4 years; his interest for the remaining 25 years can therefore only be purchased after 14 years' deferrence; we have therefore,

P_{25} by Table (VI) = 7·73539259, and v^{14} by Table (IV) = ·26333125,

Then 7·73539259 × ·26333125 = 2·036970630,

the present value deferred.

And 2·036970630 × 3·79749834 = 7·73539259,

value immediate.

Then . . . . . . 2·036970630
£1 500
_____

Present net value . . . . = £3055·455945000

10th Valuation, part 3.

Annuity from lessor's interest in royalty, to be sold after the expiration of 10 years from the time of winning the lower seams of iron ore and clay . . . £1312 10 0
Interest on capital and for redemption as in last preceding case.

The lessor receives royalty from these seams for 10 years also, and sells his interest at the expiration of that time, but, as it is deferred 16 years, we have

P_{25} by Table (VI) = 7·73539259, and v^{16} by Table (IV) = ·21762914.

Then 7·73539259 × ·21762914 = 1·683446802,

the present value deferred.

And 1·683446802 × 4·59497299 = 7·73539259,

value immediate.

Then . . . . . . 1·683446102
£1312½
_____

Present net value . . . . = £2209·5239276250

REDUCTION FROM GROSS TO NET VALUES.

1st Valuation of upper coal seams   . = £28149·228128250

From which must be
deducted the cost of
developement =   12500·000000000

Customary interest
thereon at 5 per
cent. for 2 years =   1250·000000000

5th valuation, part 1,
of wayleaves. =   1747·900279500

7th valuation of
royalty to lessor =   3275·064786000   18772·965065500

Total present net value of the first or
upper seams of coal . . . = £9376·263062750

2nd Valuation of middle or lower coal
seams . . . . . . = £33151·288905000

From which must be
deducted the cost
of developement =   15000·000000000

Customary interest
thereon during the
time occupied in
winning the seams,
at 5 per cent. for
2 years . =   1500·000000000

5th valuation, part 2,
of wayleaves. =   2889·091372000

8th valuation of
royalty to lessor =   5413·330225500   24802·421597500

Total present net value of the middle
seams of coal . . . . = £8348·867307500

3rd Valuation. Lower seams of iron
ore . . . . . . = 15278·1344280000
4th Valuation. Lower seams of clay . = 16918·3092690000

Total gross value of iron ore and clay = £32196·4436970000

From which must be
deducted the cost
of developement of
iron ore . = 8000·0000000000

Customary interest
thereon at 5 per
cent. for 2 years = 800·0000000000

Also for the developement of the clay = 7500·0000000000

And customary interest at 5 per cent.
for 2 years . = 750·0000000000

5th valuation, part 3,
of wayleaves of iron
ore and clay . = 2089·2189675000

9th valuation of
royalty to lessor on
iron ore and clay = 3914·5983226875   23053·8172901875

Total present net value of iron ore and
clay . . . . . . = £9142·6264068125

SUMMARY OF VALUES.

Total net value of upper seams . . = 9376·2630627500
Total net value of middle seams . . = 8348·8673075000
Total net value of the lower seams of
iron ore and clay . . . . = 9142·6264068125
Total net value of ground rents, 6th
valuation . . . . . = 1425·5808770000

28293·3376540625

From which must be deducted the present value of the lessor's interest,
which is to be sold after 10 years,
10th valuation, parts 1, 2, and 3,
together equal . . . . 7113·5307198750

Total present net value of mineral property . . . . . . = £21179·8069341875

Under the peculiar conditions of the lease, it was deemed advisable, either for the purpose of a real or hypothetical sale, to enter upon a series of valuations, in order to arrive at the present net value of the mineral property on behalf of the lessee, who is responsible for the developement of the property, but who may nevertheless sell it now or hereafter.

The present interest held by different parties under existing circumstances, which enters into and affects the question, was to be fully set forth before the works were commenced.

After two, four, and six years, the deferred periods for winning each successive series of seams, royalty has to be paid to the lessor, extending to ten years' duration in each case; the property is therefore of less value *now* by the amount or present value of the estimated or prospective annuity to be paid to the lessor, which in the valuation is treated as a minus quantity. This is evident, as an incoming purchaser must be held to be responsible to the lessor for the payment of the annuity accruing on account of royalty. The same remark also applies to the lessor's interest, which can only be purchased after the expiration of ten years.

Presuming, however, that the seams were won, and it was proposed to sell the property at that time, the case would be very different, for then the party in possession would have a current going concern, and the present value from the annuity that has at that time accrued must be taken as immediate.

The lessor receives royalty for four and two years respectively, upon the output from the upper and middle series of seams, at the time the other minerals are won, and if taken as an immediate annuity it would then have six, eight, and ten years, respectively, to run.

Now, assuming that the time the royalty has to be paid to the lessor has elapsed, and for the remainder of the term of the lease, it has to be purchased, or the property cleared from such charge, the property would at that time assume a greater value, equivalent to the present value of the amount of the annuity derived from the royalty, but which will now merge into that due to the profits of the mine.

The party in possession could then fairly charge it to another purchaser, who would then, in point of fact, be in possession of a freehold property as far as the minerals are concerned.

Taking the time of the lease of the mineral property in the last preceding case at 21 years from the commencement of the works, instead of 35 years from the time the seams are won, all other conditions being the same, the comparative value will appear from the following deductions.

Here, the term of the lease being 21 years from the time of commencing the works, and considering the deferred periods for winning, the time to run afterwards will be represented by $21 - 2 = 19$ years, $21 - 4 = 17$ years, and $21 - 6 = 15$ years, respectively.

1st Valuation.

The present value of £1 per annum for 19 years after 2 years, so as to allow a purchaser 16 per cent. per annum upon his purchase money, and redeem the capital at 2½ per cent. per annum . = 3·683389351

Annuity . . . . . . = £6750

Present gross value . . . . = £24862·878119250

2nd Valuation.

The present value of £1 per annum for 17 years after 4 years, so as to allow a purchaser 20 per cent. per annum upon his purchase money, and redeem the capital at 2½ per cent. per annum . = 1·945135417

Annuity . . . . . . = £15000

Present gross value . . . . = £29177·031255000

3rd Valuation.

The present value of £1 per annum for 15 years after 6 years, so as to allow a purchaser 22 per cent. per annum upon his purchase money, and redeem the capital at 2½ per cent. per annum . = 1·099764199

Annuity . . . . . . = £12000

Present gross value . . . . = £13197·170388000

4th Valuation.

The present value of £1 per annum for
15 years after 6 years, so as to allow a
purchaser 14 per cent. per annum upon
his purchase money, and redeem the
capital at 2½ per cent. per annum   .=   2·327194131

Annuity   .   .   .   .   .   . =   £5875

Present gross value   .   .   . .=£13672·265519625

5th Valuation, part 1.

The present value of £1 per annum for
19 years after 2 years, so as to allow a
purchaser 10 per cent. per annum upon
his purchase money, and redeem the
capital at 2½ per cent. per annum   .=   5·829872283

Annuity   .   .   .   .   . =   £250

Present net value .   .   .   . .= £1457·468070750

5th Valuation, part 2.

The present value of £1 per annum for
17 years after 4 years, so as to allow a
purchaser 10 per cent. per annum upon
his purchase money, and redeem the
capital at 2½ per cent. per annum   .=   4·617209168

Annuity   .   .   .   .   .   . =   £500

Present net value .   .   .   . .= £2308·604584000

5th Valuation, part 3.

The present value of £1 per annum for
15 years after 6 years, so as to allow a
purchaser 10 per cent. per annum upon
his purchase money, and redeem the
capital at 2½ per cent. per annum   .=   3·623847804

Annuity   .   .   .   .   .   . =   £437½

Present net value .   .   .   . .= £1585·433414250

6th Valuation.

The present value of £1 per annum for
21 years, so as to allow a purchaser 8
per cent. per annum upon his purchase
money, and redeem the capital at 2½
per cent. per annum . . . . =   8·56257287

Annuity . . . . . . =   £140
_____

Present net value . . . . .   £1198·76020180

Here the present value of the royalty to lessor for 10 years
will be the same as in the last preceding cases, viz. :

7th Valuation .   .   .   .   £3275·0647860000
8th Valuation   .   .   .   £5413·3302255000
9th Valuation .   .   .   .   £3914·5983226875
_____

Total present net value   = £12602·9933341875

10th Valuation, part 1.

The present value of £1 per annum for
9 years after 12 years, so as to allow a
purchaser 10 per cent. per annum upon
his purchase money, and redeem the
capital at the rate of 2½ per cent. per
annum . . . . . . =   1·589522903

Annuity . . . . . . =   £750
_____

Present net value . . . . . = £1192·142177250

10th Valuation, part 2.

The present value of £1 per annum for
7 years after 14 years, so as to allow a
purchaser 10 per cent. per annum upon
his purchase money, and redeem the
capital at the rate of 2½ per cent. per
annum is . . . . . . =   1·132629809

Annuity . . . . . . =   £1500
_____

Present net value . . . . . = £1698·944713500

10th Valuation, part 3.

The present value of £1 per annum for 5
   years after 16 years, so as to allow a purchaser 10 per cent. per annum upon his
   purchase money, and redeem the capital
   at the rate of 2½ per cent. per annum =   0·749807026

Annuity   .   .   .   .   .   .   =   £1312½

Present net value .   .   .   .   .= £984·121721625

REDUCTION FROM GROSS TO NET VALUES.
2nd Series.

1st Valuation of upper coal seams .   . = £24862·878119250
From which must be
   deducted the cost of
   developement as in
   1st series   .   . = £12500·00000000
Customary   interest
   thereon   at   5 per
   cent. for 2 years   . =   1250·00000000
5th valuation, part 1,
   of wayleaves .   . =   1457·46807075
7th valuation of roy-
   alty to lessor .   . =   3275·06478600   18482·532856750

Total present net value of the first or
   upper seams of coal .   .   .   . = £6380·345262500

2nd Valuation of middle or lower coal seams = £29177·031255000
From which must be
   deducted the cost of
   developement as in
   1st series   .   . = 15000·0000000
Customary   interest
   thereon during the
   time   occupied   in
   winning the seams at
   5 per cent. for 2 years =   1500·0000000
5th valuation, part 2,
   of wayleaves .   . =   2308·6045840
8th valuation of roy-
   alty to lessor .   . =   5413·3302255   24221·934809500

Total present net value of the middle
   seams of coal .   .   .   .   . = £4955·096445500

3rd Valuation of lower seams of iron ore $= £13197\cdot1703880000$
4th Valuation of lower seams of clay . $= 13672\cdot2655196250$

Total gross value of iron ore and clay . $= £26869\cdot4359076250$

From which must be
  deducted the cost
  of developement of
  iron ore as in 1st
  series . . . $= 8000\cdot000000000$

Customary interest
  thereon at 5 per
  cent. for 2 years . $= 800\cdot000000000$

Also for the develope-
  ment of the clay as
  in 1st series . . $= 7500\cdot000000000$

Customary interest
  thereon at 5 per
  cent. for 2 years . $= 750\cdot000000000$

5th valuation, part 3,
  of wayleaves . . $= 1585\cdot433414250$

9th valuation of roy-
  alty to lessor on iron
  ore and clay . . $= 3914\cdot5983226875$ $22550\cdot0317369375$

Total present net value of iron ore and
  clay . . . . . . $= £4319\cdot4041706875$

SUMMARY OF VALUES.
2nd Series.

Total net value of upper coal seams $=.$ $6380\cdot3452625000$

Total net value of middle or lower coal
  seams . . . . . $=$ $4955\cdot0964455000$

Total net value of the lower seams of
  iron ore and clay . . . $=$ $4319\cdot4041706875$

Total net value of ground rents, 6th
  valuation . . . . $=$ $1198\cdot7602018000$

                      $£16853\cdot6060804875$

From which must be deducted the
  present value of the lessor's interest,
  which is to be sold after 10 years,
  10th valuation, parts 1, 2, and 3,
  together . . . . $=$ $3875\cdot2086123750$

Total present net value of mineral
  property, 2nd series. . . $=$ $£12978\cdot3974681125$

I

If several seams of coal and other minerals exist at different depths (which is a case of common occurrence) in an area leased, it is highly desirable, and to the interest of the lessee, that ample time is granted for developement before the royalty becomes due, and that the time embraced in the lease is sufficiently long to justify the adventure, and expenditure connected therewith. In the 1st series of valuations the total net value is £21179·8069341875, the duration being 35 years after the seams are won; but, for the sake of comparison, if we confine the period of a lease to 21 years from the commencement of the works, the present net value is £12978·3974681125; and the difference is £8201·409466075. The present value of the property in the last case is consequently less by that amount, and due as a matter of course to the shorter period of time.

The cost of winning minerals at any defined depth is the same, no matter what time is fixed for the lease to run, but the comparison of values above referred to demonstrates that the present value is very much affected by the time. It is, therefore, to be inferred that the time any lease has to run for working minerals at great depths should much exceed that granted when the minerals are much nearer the surface.

On the whole I am inclined to the opinion that 21 years' lease of any mineral property is much too short a period, when the time of developement extends to three, four, five, and six years. When, however, a longer period cannot be granted, it should, if possible, be made compulsory on the part of the lessor, or his representatives, to extend the time a further period of 21 years upon the lapse of the former period, at a reasonable royalty, and not to be in excess of the rate per ton as previously determined, unless the profits of the mine are such as to justify it.

It is undoubtedly an error in judgment on the part of those who suppose that, by allowing short periods of time for developement, fixing the royalty or other dues above the normal or customary rate in a mining district, or in excess of what any particular mine will bear, the landlord or lessor's interest is thereby either permanently augmented or established. In point of fact the very reverse is the case, and great consideration should be exercised by the landlord or lessor towards the lessee, upon whom devolves the risk of the adventure. It is a question of no small importance to the ultimate success of a mine; and I venture to assert that the high dues demanded have frequently operated to discourage and frighten away those who would otherwise have spent their capital in developing such mines.

It should be remembered that, in the majority of cases, landlords are not disposed or in a position to expend large sums of money in order to develop the mineral resources of their estates. While, therefore, such minerals lie dormant, the owner is in exactly the same position as he would be if the minerals did not exist at all. The interest of the landlord or lessor is therefore intimately bound up, if not exactly identical, with that of the lessee, and upon the degree of success of the latter entirely depends the income to be received by the former.

An equitable state of things should therefore exist between the parties, and every facility be offered for the encouragement of *bonâ fide* undertakings; and instead of raising the royalty or other dues, it is necessary in very many cases that these should be reduced, so as to enable capitalists to develop mineral properties with profit to themselves as well as to the landlord.

To attempt to raise the landlord's dues when making a new grant or assigning a lease, simply on account of temporary good trade, having the appearance of producing extra profits, is as unwise as it is unjust, operating as an impediment to future progress in opening up those mines coming under such restrictions.

There is also another point intimately connected with this question, and that is, the area included or described in the grant or lease. At first sight this would appear too simple a matter to require special notice; but, in reality, it is necessary that it should be treated as systematically as any other matter of importance connected with mining engineering; for, if taken at random, there may be no proper relation whatever existing between the area granted and the time the lease has to run. It may be in excess, to the injury of the landlord, or in defect of the proper quantity which should have been assigned, and consequently to the detriment of the lessee.

In assigning the area, due regard should be had to the increasing depth of the minerals, as compared to landworks, estimated cost of winning, annual output, time fixed for the.

grant or lease, probable profit per ton to be derived, and a proper and accurate valuation made before ultimately fixing the area, which should always be such as to justify the outlay to be incurred in the developement. In cases where parties are entitled to have grants of mineral tracts made to them from the lord of the manor, by virtue of some right, as in the Forest of Dean, two distinct interests exist, *i.e.*, that of the Crown, as lord of the manor, and that of persons called ' free miners,' who are entitled under existing law to have grants of mineral tracts made to them. The Crown exercises the right to make such grants conditionally upon certain payments being made by the grantee, such as dead rent and royalty dues, which are intended to represent *one-fifth* of the profits derived, or to be derived, as the share or interest of the Crown. The galee or grantee nominally undertakes to develop, or procure to be developed, the grant in question ; but, as those who are so entitled are not competent—by reason of their being working men—to attempt to open up any of the deeper mines, it is necessary that another party should be introduced to effect this for them.

Here, then, the representative of the galee is not only obliged to purchase the grant or interest of the galee, but is called upon to expend a sufficient sum in the developement of the mineral tract or grant, and also to pay a dead or certain rent, after a certain determined period, if the mine is undeveloped, or royalty or tonnage dues, when the mine is opened.

If, therefore, the dead rent and royalty are unusually heavy, a *double burden* has to be borne by those engaged in opening up the mines. Under such circumstances the difficulty of procuring capitalists willing to enter into such undertakings is all the greater.

The result is that, at the expiration of the fixed period when the dead rent becomes due, if no one is forthcoming to take the matter up, the grant or grants must lapse to the Crown, but subject to be re-granted to other persons over and over again, to the manifest injury of the Crown, the free miner, and the district in general. There is no remedy apparently for this state of things, unless the Crown would make grants of the ungranted tracts of minerals, and then purchase back the interest of the galee in such grants.

43. **What is the Present Value** of the royalty of an iron mine producing an income of £600 per annum, during the life of a person A, aged 52? Interest to be allowed to the purchaser at the rate of 10 per cent. per annum. Capital to be redeemed by effecting an insurance upon A's life at the office rate of, say, £4 10s. 4d. per £100.

Here the annuity of £1 is to be purchased on a life aged x, to yield r' per cent. on the purchase money P, and to redeem it at the determination of the contingency, by effecting a policy in an insurance office at the rate of r per pound; but while the annuity (a) is due at the end of the year, the premium must be paid at its commencement. To prevent, therefore, the possibility of the loss of a year's income in case the annuitant should die before the completion of a year, the sum to be insured will be represented by $P+a$; and v being the present value of £1 due one year hence, we have,

$$(20).— \qquad P = \frac{1}{(1-v)+r} - 1.$$

Substituting the numerical values, we also have

$$P = \frac{1}{(1-\cdot90909091)+\cdot04516667} - 1 = 6\cdot348847436$$

years' purchase.

Then, $\qquad 6\cdot348847436 \times £600 = £3809\cdot3084616,$

the present value of A's interest; but, by the conditions of the problem,

$$£3809\cdot3084616 + £600 = £4409\cdot3084616,$$

the total sum to be insured. The premium necessary to insure this sum on the death of A will be represented by

$$(P+a) \times r'.$$

Therefore $£4409\cdot3084616 \times \cdot04516667 \qquad . \qquad = £199\cdot15378$
And $(£3809\cdot3084616 + 199\cdot1537802) \times 10$ p.cent. $= \quad 400\cdot84622$

Together equal to the annuity $\quad . \qquad . \qquad . \qquad £600\cdot00000$

44. **What is the Present Value** of the royalty of a mine producing an income of £500 per annum during the life of a person *A* aged 47? The annuity has 60 years to run, and on the death of *A* reverts to his successor; whose interest is to be sold at the present time. Interest allowed to a purchaser on the capital at the rate of 10 per cent. per annum, and to be redeemed by effecting an insurance on *A*'s life at the office rate of £3 18s. 1d. per £100. The value of the successor's interest to be redeemed at the rate of 3 per cent. per annum.

The present value of £1 per annum for 60 years, allowing a purchaser 10 per cent. per annum upon his purchase money, and redeeming the same at 3 per cent. per annum = 9·42214381.

And 9·42214381 × £500 = £4711·071905,

the value of the annuity for the total period of 60 years; and, as in the last preceding case, the value of *A*'s interest is

$$P = \frac{1}{(1 - ·90909091) + ·03905208} - 1 = 6·694606011$$

years' purchase.

Then 6·694606011 × £500 = £3347·3030055,

the present value of *A*'s interest; and, by condition,

£3347·3030055 + £500 = £3847·3030055,

the total sum to be insured. The premium to insure this sum at the death of *A* is

£3847·3030055 × ·03905208 . . . = £150·24518
And,

(£3347·3030055 + 150·24518476) × 10 per cent. = 349·75482

Together equal to the annuity . . . = £500·00000

The total value of the annuity for 60 years = £4711·0719050
Value of *A*'s interest . . . . = 3347·3030055

Value of *A*'s successor's interest . . = £1363·7688995

I have devoted a considerable amount of time and thought to the construction of other problems, involving some of the more general cases of lives with immediate and deferred annuities. Originally it was intended to take up the whole range of such cases, but after entering fully into the solution of some of the more difficult deferred cases, I concluded that they were not of that class likely to be of any great value or service to the Civil and Mining Engineer, Colliery Proprietor, Colliery Viewer, or General Manager. It is true, however, that the cases devised were both curious and difficult ; although probably of more use to professional Actuaries and Assurance Offices, than for those for whom this work is more particularly intended.

Being fully aware that ample rules and examples illustrating the treatment of such cases are to be found in works already in existence, it would have been entirely out of place on my part to have gone over the same ground. The subject of lives, however, is one of great interest, and I confess it was with very considerable reluctance that I finally determined not to introduce anything further of that nature in this work.

The cases given in the preceding pages referring to the Valuation of Mines are those usually occurring in practice, but it is impossible to provide for all the modifications which it may be necessary to introduce, suitable for all the varying circumstances that may arise. Such modifications will be best applied to any such cases by those to whom they may occur.

It will be observed that throughout the problems where the condition was introduced that a certain sum was necessary to be expended upon open or unopened mines, with a view to obtain an estimated yield of minerals, and constant profit extending over a definite future period, the ordinary or customary mode of allowing 5 per cent. upon any such sum has been followed. It was considered advisable that this mode of solution should be fully exhibited, as it is believed to be good practice by some of the profession.

Others, however, entertain an opposite opinion, the nature of which will be best understood by putting a case. For this purpose, therefore, let us assume that a colliery is yielding a nett income of £8,000 per annum, and that after careful consideration, a valuer has estimated that to place the colliery in a position to yield a constant quantity of minerals extending over a period of 21 years, so that in all probability the income will be uniform for that period, the sum of £12,000 must be expended upon the works, during a period of 3 years, in equal sums of £4,000 each year. The interest to be allowed to a purchaser is 20 per cent. per annum, and the capital is to be redeemed at 3 per cent. per annum.

Under such conditions the present value
of the colliery would be . . . = £34,061·13800

The redemption fund to replace this gross
value of £34,061·1380 . . . = 1,187·77239
And interest on the gross value of
£34,061·1380 × 20 per cent. per annum = 6,812·22761

The proposed annuity . . = £8,000·00000

Then, it is customary to say,
From the gross value of the colliery . = 34,061·1380
Must be deducted the estimated
cost of works . . . = £12,000
And also interest thereon at
the rate of 5 per cent. for 3
years . . . . = 1,800
——— = 13,800·0000

Nett present value of the colliery . = £20,261·1380

Now, it is held that the gross value of the colliery is made up of two parts, *i.e.* £22,061·1380 and £12,000; because these two sums together = £34,061·1380, or the gross value; also, that the purchaser, or party in possession, is receiving 20 per cent. per annum upon £22,061·1380, and upon £12,000, the latter sum being contained in and part of the gross value. Further, that the vendor receives a less sum for the colliery than the gross value, by the difference between that value and £12,000, or £20,261·1380; and, therefore, that the purchaser is not entitled to be allowed 5 per cent. for 3 years upon £12,000, nor indeed the full sum of £12,000, but only such a sum as would, if it were invested at 3 per cent., accumulate to £12,000 at the end of 3 years. According to this view, by

Table (XIII), the present value of £4,000 per annum for 3 years, allowing interest at 3 per cent. per annum = £11,314·445.

The present value of the colliery, as previously stated	=	£34,061·1380
From which must be deducted the present value of £4,000 per annum for 3 years at 3 per cent. per annum . . .	=	11,314·4450
Present nett value of the colliery according to the new mode	=	£22,746·6930
Present nett value of the colliery first deduced	=	20,261·1380
Difference in value . . .	=	£2,485·5550

The difference between the values as found by the two modes is not large, but it is apparent that if the time over which the expenditure was distributed amounted to 8 or 10 years, the difference would be very considerable.

It will be seen in Parts I. and II., especially on pp. 10–14, 19–20, and 58–60, what elements are necessary to be considered in arriving at a valuation; but after the valuer has exercised his best judgment in determining all the necessary elements, involving of course the rate per cent. to be allowed, and the probable annuity to be derived over any fixed period in the future; then opinion as to the deduction and mode of valuation ceases, or ought to cease altogether.

When we have no better means for determining any point involved in a question than that of opinion, undoubtedly it must be accepted; but where science will aid us in arriving at any conclusion, it must be taken as definite, and must not be displaced by mere opinion.

DEDUCING VALUES FROM TABLES OF MULTIPLES OF YEARS' PURCHASE, ETC.

For those who prefer to arrive at the value of either immediate or deferred annuities, derived from any property, simply by adding the quantities together, instead of performing a long multiplication of the years' purchase by the annuity, additional Tables for a few percentages may be prepared to effect this.

For each percentage so treated there must be 10 columns of figures; the annuity in each may be found at the top of each column, as £1, £2, £3, £4, &c., up to £10. These numbers may be conceived to have as many noughts attached to them as there are decimal places in each column of figures. Thus £1, £2, &c., may represent £1 to £100,000,000, £2 to £200,000,000, up to £10 or £1,000,000,000. It is therefore evident that the numbers in the column under £1 are the years' purchase or values of that annuity, and that those in the other 9 columns are simply multiples of it.

The numbers £1, £2, £3, &c., to £10 may be called £10 or £100, £20 or £200, £30 or £300, or any other number of tens up to the limits before assigned.

In order, therefore, to find the proper value of any proposed annuity, the decimal point must be removed as many places to the right of the position it at first occupied to unity, as there may have been tens or noughts attached to the annuity digit. This mode of pointing off so as to form each number into *whole pounds* and decimals of a *pound*, under or for any annuity, may be best illustrated by example.

Taking, therefore, the interest to be allowed to a present purchaser at 21 per cent. per annum, and to redeem the capital at 3 per cent. per annum, for 30 years' duration, with an annuity of £1, the present value would be equivalent to £4·328643434; this number, therefore, stands as it is found, without alteration, but, by calling the £1 annuity £10, the present value would be changed to £43·28643434, and by assuming the annuity to be still greater, or £100 and £1000, the present value would be changed to £432·8643434, and £4328·643434 respectively. We may, therefore, continue this process of adding noughts to the original number corresponding to unity until we get up to £100,000,000; and the equivalent, as present value, would be £432864343·4. If instead, however, of attaching noughts to the annuity of £1, they are prefixed, the value will be *decreased* in the same ratio as they were *increased* in the former case.

Calling, therefore, £1 annuity £·1, the present value or

years' purchase would be £·4328643434, and supposing it to be £·01, £·001, £·0001, and £·00000001 respectively, the present value would be £·04328643434, £·004328643434, £·0004328643434, and £·00000004328643434 respectively, which mode of working would hold good throughout.

Everything that is necessary to be obtained within the limits of the rates of interest per cent. the Tables should be calculated for, may be deduced from the first five columns, but with a ten-column Table the lines of figures to be taken out and added together are considerably diminished.

Supposing the annuity consisted of four figures, or say £8448, before applying the five-column Table it would be best to divide it into parts, as

$$£5000 + £3000 + £400 + £40 + £5 + £3 ;$$

but if we were to employ the ten-column Table, the figures would be broken up into sections thus—

$$£8000 + £400 + £40 + £8.$$

Thus, assuming the annuity derived from any property is £8448, to last for 20 years, interest to be 21 per cent. per annum, and to redeem the capital at 3 per cent. per annum, the work would stand thus (See Table XIV) :—

SPECIMEN TABLE No. I.

Years	Annuity £1, £10, £100, £1000 ; or £·1, £·01, £·001, &c.	Annuity £2, £20, £200, £2000 ; or £·2, £·02, £·002, &c.	Annuity £3, £30, £300, £3000 ; or £·3, £·03, £·003, &c.	Annuity £4, £40, £400, £4000 ; or £·4, £·04, £·004, &c.	Annuity £5, £50, £500, £5000 ; or £·5, £·05, £·005, &c.	Years
20	4·045050412	8·090100824	12·135151236	16·180201648	20·225252060	20

Years	Annuity £6, £60, £600, £6000 ; or £·6, £·06, £·006, &c.	Annuity £7, £70, £700, £7000 ; or £·7, £·07, £·007, &c.	Annuity £8, £80, £800, £8000 ; or £·8, £·08, £·008, &c.	Annuity £9, £90, £900, £9000 ; or £·9, £·09, £·009, &c.	Annuity £10, £100, £1000, £10000 ; or £·01, £·001, £·0001, &c.	Years
20	24·270302472	28·315352884	32·360403296	36·405453708	40·450504120	20

Annuity	From Specimen Table No 1
£5000	£20225·252060
3000	12135·151236
400	1618·0201648
40	161·80201648
5	20·22525206
3	12·135151236
£8448	£34172·585880576

Employing the above specimen Table, using ten columns instead of five, we shall obtain the value more readily.

Annuity	From Specimen Table
£8000	£32360·403296
400	1618·0201648
40	161·80201648
8	32·360403296
£8448	£34172·585880576

Although valuation may be performed by this simple mode, it would nevertheless render this work too bulky and cumbrous to compute additional tables for all the rates per cent. in a similar manner, that is, to form tables of every rate per cent. for which the present values have been calculated.

When the time of duration of a mine, the rate per cent., and present value are given, to find the annuity, it may be deduced from the same Table by the following process, and may be thus expressed :—

Rule.-- Find in the Tables in line with the number of years' duration, and at the given rate per cent., the nearest value to the one proposed, and take their difference; the nearest value to this difference must again be found in one of the columns in line with the same number of years, and deducted as before. This operation of seeking a value nearest to every new difference must be repeated until the required or corresponding annuity is obtained.

At each operation of finding such a value nearest to any difference, the corresponding annuity to it, as found at the head of the column of figures from whence each value was obtained, must be noted down in a tabular form, and made to occupy a proper position with reference to the preceding figures; then the sum of all the lines or parts will express the annuity.

This rule will appear more clear from the following example :—

Required the annuity, all the other elements being as in the last preceding case.

Corresponding Annuity	From Specimen Table No. 1	
	£34172·585880576	= given value.
£8000	32360·403296	nearest value.
	1812·182584576	= 1st difference.
400	1618·0201648	nearest value.
	194·162419776	= 2nd difference.
40	161·80201648	nearest value.
	32·360403296	= 3rd difference.
8	32·360403296	the value.
£8448		

The annuity required is therefore £8448.

If, however, an annuity composed of whole numbers, and decimals, for a period of 30 years, and rates per cent. as in the last preceding case, were required to be valued, it would present no greater difficulty than a simple number.

Assuming it, therefore, to be £24362·29463, it must be disposed of thus :—

Annuity	From Table (XIV)
£20000·00000	£86572·86868
4000·00000	17314·573736
300·00000	1298·5930302
60·00000	259·71860604
2·00000	8·657286868
·20000	·8657286868
·09000	·38957790906
·00400	·017314573736
·00060	·0025971860604
·00003	·00012985930302
£24362·29463	£105455·68668732295942

The proof is 4·328643434 × £24362·29463 = the above result, or £105455·68668732295942.

The annuity may also be found from the present value as illustrated above, thus :—

Corresponding Annuity	From Table (XIV)	
	105455·68668732295942 = given value.	
£20000·00000	86572·86868	nearest value.
	18882·81800732295942 =: 1st difference.	
4000·00000	17314·573736	nearest value.
	1568·24427132295942 = 2nd difference.	
300·00000	1298·5930302	nearest value.
	269·65124112295942 = 3rd difference.	
60·00000	259·71860604	nearest value.
	9·93263508295942 = 4th difference.	
2·00000	8·657286868	nearest value.
	1·27534821495942 = 5th difference.	
·20000	·8657286868	nearest value.
	·40961952815942 = 6th difference.	
·09000	·38957790906	nearest value.
	·02004161909942 = 7th difference.	
·00400	·017314573736	nearest value.
	·00272704536342 = 8th difference.	
·00060	·0025971860604	nearest value.
	·00012985930302 = 9th difference.	
·00003	·00012985930302	the value.

£24362·29463

The value of a deferred annuity may also be determined in a similar manner, but it would first be necessary to construct a table which should be *multiples* of the years' purchase *deferred* under each rate per cent., according to the following Specimen Table :—

Specimen Table No. 2.

Years	Annuity £1, £10, £100, £1000; or £·1, £·01, £·001, &c.	Annuity £2, £20, £200, £2000; or £·2, £·02, £·002, &c.	Annuity £3, £30, £300, £3000; or £·3, £·03, £·003, &c.	Annuity £4, £40, £400, £4000; or £·4, £·04, £·004, &c.	Annuity £5, £50, £500, £5000; or £·5, £·05, £·005, &c.	Years
30	2·18195053	4·36390106	6·54585159	8·72780212	10·90975265	30

Years	Annuity £6, £60, £600, £6000; or £·6, £·06, £·006, &c.	Annuity £7, £70, £700, £7000; or £·7, £·07, £·007, &c.	Annuity £8, £80, £800, £8000; or £·8, £·08, £·008, &c.	Annuity £9, £90, £900, £9000; or £·9, £·09, £·009, &c.	Annuity £10, £100, £1000, £10000; or £·01, £·001, £·0001, &c.	Years
30	13·09170318	15·27365371	17·45560424	19·63755477	21·81950530	30

Thus an annuity of £18,254, to continue 30 years after 4 years, allowing 20 per cent. per annum to a present purchaser, and to redeem the capital at the rate of 3 per cent. per annum, would be dealt with by the following process :—

Annuity	From Specimen Table No. 2
£10000	£21819·5053
8000	17455·60424
200	436·390106
50	109·0975265
4	8·72780212
£18254	£39829·32497462

The accuracy of this deduction may be proved thus: The present value of £1 per annum deferred 4 years

$$= £2·18195053 \times £18254 = £39829·32497462,$$

the present value as before.

When all the other elements are given except the annuity, and it is required to be found, it may be readily deduced by the converse operation, as previously illustrated for immediate annuities.

Deferred values may also be obtained directly from the table of the *Present Value Immediate*, with the assistance of the table of values due at a future period, or the present value of £1 due in n years

Thus, presuming it were required to find the present value of £1 per annum at 15 per cent. per annum, redeeming the capital at the rate of 3 per cent. per annum, and to continue 35 years after 10 years' deferrence, we should have,

By Table (IV) the present value of £1 due 10 years hence = ·24718471, and considering it as an annuity, the present value deferred may be deduced from a conversion of Table (VII) as in Specimen Table No. 3.

SPECIMEN TABLE No. 3.

Years	Assumed Annuity of £·1, £·01, £·001, &c.	Assumed Annuity of £·2, £·02, £·002, &c.	Assumed Annuity of £·3, £·03, £·003, &c.	Assumed Annuity of £·4, £·04, £·004, &c.	Assumed Annuity of £·5, £·05, £·005, &c.	Years
35	·600458901	1·200917802	1·801376703	2·401835604	3·002294505	35

Years	Assumed Annuity of £·6, £·06, £·006, &c.	Assumed Annuity of £·7, £·07, £·007, &c.	Assumed Annuity of £·8, £·08, £·008, &c.	Assumed Annuity of £·9, £·09, £·009, &c.	Assumed Annuity of £·01, £·001, £·0001, &c.	Years
35	3·602753406	4·203212307	4·803671208	5·404130109	6·004589010	35

Assumed Annuity	Present Value deferred 10 Years (From Specimen Table No. 3)
·2	£1·200917802
·04	·2401835604
·007	·04203212307
·0001	·000600458901
·00008	·0004803671208
·000004	·00002401835604
·0000007	·000004203212307
·00000001	·000000060045890I
·24718471	£1·484242593106037I

or the present value of £1 per annum for 35 years, deferred 10 years.

The immediate value of £1 per annum corresponding to the given elements in the last preceding case is £6·00458901, and the proof of the above conclusion is £6·00458901 × ·24718471

= £1·4842425931060371, the present value deferred as before. The converse of this will result by operating as previously explained.

But taking an example, and assuming it were required to find the present value of £1 due in 10 years, by having given the present value of £1 per annum deferred 10 years,

= £1·4842425931060371, the work must be arranged as follows:

Assumed Annuity	From Specimen Table No. 3	
	1·4842425931060371	= given value.
·2	1·200917802	= nearest value.
	·2833247911060371	= 1st difference.
·04	·2401835604	= nearest value.
	·0431412307060371	= 2nd difference.
·007	·04203212307	= nearest value.
	·0011091076360371	= 3rd difference.
·0001	·000600458901	= nearest value.
	·0005086487350371	= 4th difference.
·00008	·0004803671208	= nearest value.
	·0000282816142371	= 5th difference.
·000004	·00002401835604	= nearest value.
	·0000042632581971	= 6th difference.
·0000007	·000004203212307	= nearest value.
	·000000600458901	= 7th difference.
·00000001	·000000600458901	= the value.

·24718471 the value sought.

We may also determine by similar means the annuity which may be purchased for a given sum, at a certain rate per cent. and for a given time. Thus the present value of £1 per annum, allowing 20 per cent. interest, and redeeming at 3 per cent. per annum, for a period of 50 years = £4·78777025, and,

$$\frac{1}{4·78777025} = £·2088654943,$$

or the annuity which £1 will purchase.

A table may be then formed, having such numbers for a basis, according to the following specimen for the fiftieth year.

SPECIMEN TABLE No. 4.

Years	Annuity £1, £10, £100, £1000; or £·1, £·01, £·001, &c.	Annuity £2, £20, £200, £2000; or £·2, £·02, £·002, &c.	Annuity £3, £30, £300, £3000; or £·3, £·03, £·003, &c.	Annuity £4, £40, £400, £4000; or £·4, £·04, £·004, &c.	Annuity £5, £50, £500, £5000; or £·5, £·05, £·005, &c.	Years
50	·2088654943	·4177309886	·6265964829	·8354619772	1·0443274715	50

Years	Annuity £6, £60, £600, £6000; or £·6, £·06, £·006, &c.	Annuity £7, £70, £700, £7000; or £·7, £·07, £·007, &c.	Annuity £8, £80, £800, £8000; or £·8, £·08, £·008, &c.	Annuity £9, £90, £900, £9000; or £·9, £·09, £·009, &c.	Annuity £10, £100, £1000, £10000; or £·1, £·01, £·001, &c.	Years
50	1·2531929658	1·4620584601	1·6709239494	1·8797894487	2·0886549430	50

Required the annuity which may be purchased for the sum of £46,842, interest to be at the rate of 20 per cent. per annum, and to redeem the same at the rate of 3 per cent. per annum, to continue 50 years.

Sum to be Invested in Purchasing Annuity	Annuity to be Purchased			
£40000	£8354·619772	From Specimen Table No. 4.		
6000	1253·1929658	,,	,,	,,
800	167·09239544	,,	,,	,,
40	8·354619772	,,	,,	,,
2	·4177309886	,,	,,	,,
£46842	£9783·6774840006			

Again, as a proof, we have

$$£46842 \times ·2088654943 = £9783·6774840006,$$

the annuity as before.

By reversing the operation, the purchase sum may be deduced by employing the annuity.

Purchase Money

	9783·6774840006	= given annuity.
£40000	8354·619772	= nearest value.
	1429·0577120006	= 1st difference.
6000	1253·1929658	= nearest value.
	175·8647462006	= 2nd difference.
800	167·09239544	= nearest value.
	8·7723507606	= 3rd difference.
40	8·354619772	= nearest value.
	·4177309886	= 4th difference.
2	·4177309886	= the value.

£46842

For proof, we also have,

$$£4·78777025 \times 9783·6774840006 = £46842,$$

as before, and also

$$\frac{9783·6771480006}{·2088654943} = £46842.$$

The same rule may also be applied in determining the deferred annuity which £1 or any other sum will purchase at a certain rate and for a given period.

Thus the present value of £1 per annum, allowing 20 per cent. per annum, deferred 3 years, and to continue 50 years afterwards = £2·77070036 (see Table X).

Then $\dfrac{1}{2·77070036} = ·3609195763$, the annuity

which £1 will purchase after 3 years. Then, by forming a table with this number as one of its bases, we have the following results for the fiftieth year:—

SPECIMEN TABLE No. 5.

Years	Annuity £1, £10, £100, £1000, &c., &c., &c.	Annuity £2, £20, £200, £2000, &c., &c. &c.	Annuity £3, £30, £300, £3000, &c., &c., &c.	Annuity £4, £40, £400, £4000, &c., &c. &c.	Annuity £5, £50, £500, £5000, &c., &c., &c.	Years
50	·3609195763	·7218391526	1·0827587289	1·4436783052	1·8045978815	50

Years	Annuity £6, £60, £600, £6000, &c., &c., &c.	Annuity £7, £70, £700, £7000, &c., &c., &c.	Annuity £8, £80, £800, £8000, &c., &c., &c.	Annuity £9, £90, £900, £9000, &c., &c., &c.	Annuity £10, £100, £1000, £10000, &c., &c., &c.	Years
50	2·1655174578	2·5264370341	2·8873566104	3·2482761867	3·6091957630	50

Required the annuity that may be purchased, deferred 3 years, and to continue 50 years afterwards, allowing interest at 20 per cent. per annum, and to redeem the capital for 3 per cent. per annum for the sum of £64,242.

Purchase Money	Annuity Purchased			
£60000	£21655·174578	From Specimen Table No. 5.		
4000	1443·6783052	,,	,,	,,
200	72·18391526	,,	,,	,,
40	14·436783052	,,	,,	,,
2	·7218391526	,,	,,	,,
£64242	£23186·1954206646			

Then, the proof is

$$£64242 \times ·3609195763 = £23186·1954206646,$$

the annuity as before.

And, conversely, we have

Purchase Money
Required

	$23186\cdot1954206646$ = given annuity.
£60000	$21655\cdot174578$ = nearest value.
	$1531\cdot0208426646$ = 1st difference.
4000	$1443\cdot6783052$ = nearest value.
	$87\cdot3425374646$ = 2nd difference.
200	$72\cdot18391526$ = nearest value.
	$15\cdot1586222046$ = 3rd difference.
40	$14\cdot436783052$ = nearest value.
	$\cdot7218391526$ = 4th difference.
2	$\cdot7218391526$ = the value.
£64242	

For proof, we have

$$£2\cdot77070036 \times £23186\cdot1954206646 = £64242, \text{ as before};$$

$$\text{also} \quad \frac{23186\cdot1954206646}{\cdot3609195763} = £64242.$$

The redemption fund necessary to be set aside annually in order to redeem any capital sum, may also be determined by the same rule; but in this case also it would be necessary first to construct a table of redemption funds, which should be multiples of those corresponding to unity or £1, at different rates per cent. See Table (XV).

Thus, supposing it were necessary to redeem £38105·25 at 3 per cent. per annum, at the expiration of 30 years, we should thus proceed:—

SPECIMEN TABLE No. 6.

Years	Redemption Fund for £1, £10, £100, £1000; or £·1, £·01, £·001, &c.	Redemption Fund for £2, £20, £200, £2000; or £·2, £·02, £·002, &c.	Redemption Fund for £3, £30, £300, £3000; or £·3, £·03, £·003, &c.	Redemption Fund for £4, £40, £400, £4000; or £·4, £·04, £·004, &c.	Redemption Fund for £5, £50, £500, £5000; or £·5, £·05, £·005, &c.	Years
30	·0210192593	·0420385186	·0630577779	·0840770372	·1050962965	30

Years	Redemption Fund for £6, £60, £600, £6000; or £·6, £·06, £·006, &c.	Redemption Fund for £7, £70, £700, £7000; or £·7, £·07, £·007, &c.	Redemption Fund for £8, £80, £800, £8000; or £·8, £·08, £·008, &c.	Redemption Fund for £9, £90, £900, £9000; or £·9, £·09, £·009, &c.	Redemption Fund for £10, £100, £1000, £10000; or £·01, £·001, £·0001, &c.	Years
30	·1261155558	·1471348151	·1681540744	·1891733337	·2101925930	30

Capital to be redeemed	Redemption Fund from Specimen Table No. 6
£30000·00	£630·577779
8000·00	168·1540744
100·00	2·10192593
5·00	·1050962965
·20	·00420385186
·05	·001050962965
£38105·25	£800·944130441325

We may obtain a proof of this conclusion thus:—The redemption fund necessary to produce £1 in 30 years is £·0210192593, and £·0210192593 × £38105·25 = £800·944130441325, as before.

The capital sum which may be redeemed in any particular time at a certain rate per cent., and at a given redemption fund, may also be found by the converse operation to that given above; thus:

Required the capital sum which may be redeemed in 30 years, when the rate is 3 per cent. per annum, and the Redemption Fund = £800·944130441325.

Corresponding Capital	From Specimen Table No. 6	
£30000·00	£630·577779	= nearest value.
	170·366351441325	= 1st difference.
8000·00	168·1540744	= nearest value.
	2·212277041325	= 2nd difference.
100·00	2·10192593	= nearest value.
	·110351111325	= 3rd difference.
5·00	·105096296500	= nearest value.
	·005254814825	= 4th difference.
·20	·004203851860	= nearest value.
	·001050962965	= 5th difference.
·05	·001050962965	= the value.
£38105·25		

It may probably be considered by some of my readers that the scheme for the Specimen Tables of Multiples, and the examples worked by their means are needlessly diffuse; but I preferred allowing the decimals to run to the greatest number of places possible, in order to exhibit more fully the general arrangement of the mode and power of the Tables. When the plan of working is understood, the labour of writing down so many figures can be abbreviated; indeed, it would not be convenient or advantageous to retain in the work more than four or five places of decimals.

Tables of multiples of value may be arranged in a different manner to those previously given. Thus in Specimen Table No. 7, nine values will be found sufficient for each rate per cent. and number of years.

SPECIMEN TABLE NO. 7.

Interest on Capital **10** per cent. Redemption **3** per cent.

An-nuity	10 years	12 years	13 years	14 years	15 years	An-nuity
1	5·34100996	5·86640717	6·09646272	6·30810001	6·50336371	1
2	10·68201992	11·73281434	12·19292544	12·61620002	13·00672742	2
3	16·02302988	17·59922151	18·28938816	18·92430003	19·51009113	3
4	21·36403984	23·46562868	24·38585088	25·23240004	26·01345484	4
5	26·70504980	29·33203585	30·48231360	31·54050005	32·51681855	5
6	32·04605976	35·19844302	36·57877632	37·84860006	39·02018226	6
7	37·38706972	41·06485019	42·67523904	44·15670007	45·52354597	7
8	42·72807968	46·93125736	48·77170176	50·46480008	52·02690968	8
9	48·06908964	52·79766453	54·86816448	56·77290009	58·53027329	9

The annuity being £7428·375, to last 15 years, interest being at 10 per cent. per annum, and redeeming the capital at 3 per cent. per annum, we have the following deductions from Specimen Table No. 7.

Annuity	Value
7000	45523·546
400	2601·345
20	130·067
8	52·027
·3	1·951
·07	·455
·005	·033
7428·375	48309·424

This may be written in abbreviated form as follows :—

$$\text{Proposed annuity} = \pounds7428\cdot375$$

$$
\begin{aligned}
&45523\cdot546\\
&2601\cdot345\\
&130\cdot067\\
&52\cdot027\\
&1\cdot951\\
&\cdot455\\
&\cdot033
\end{aligned}
$$

$$\text{Equivalent present value} = \pounds48309\cdot42\,4$$

Specimen Table No. 8.

Present Value of £·1 per Annum. Redemption of Capital 3 per cent. per Annum.

Years	3½ per cent.	4 per cent.	4½ per cent.	5 per cent.	7 per cent.	Years
1	·096618357	·096153846	·095693780	·095238095	·093457943	1
2	·189533635	·187754347	·186008155	·184294144	·177742755	2
3	·278916405	·275080186	·271348062	·267715853	·254109998	3
4	·364927483	·358388198	·352079148	·345988383	·323596273	4
5	·447718618	·437915472	·428532423	·419543034	·387064953	5
6	·527433115	·513881215	·501008279	·488764524	·445240930	6
7	·604206411	·586488408	·569779941	·553997119	·498737312	7
8	·677766575	·655925283	·635096491	·615549815	·548076177	8
9	·749434980	·722366639	·697185463	·673700745	·593704863	9
10	·818126365	·785974980	·756255138	·728700946	·636008890	10

Years	10 per cent.	12 per cent.	15 per cent.	18 per cent.	20 per cent.	Years
1	·090909090	·089285714	·086956522	·084745762	·083333333	1
2	·168744805	·163235767	·155615178	·148674381	·144381223	2
3	·236110581	·225463707	·211179700	·198597756	·191010889	3
4	·294961719	·278530549	·257051537	·238648080	·227776400	4
5	·346795265	·324301986	·295547950	·271477559	·257496647	5
6	·392776833	·364169375	·328302103	·298866548	·282009884	6
7	·433827521	·399191472	·356498164	·322054601	·302566044	7
8	·470684834	·430188219	·381015682	·341931323	·320044664	8
9	·503946260	·457804488	·402521626	·359151725	·335082624	9
10	·534100996	·482554435	·421530947	·374208773	·348152434	10

All the numbers contained in the Tables in this work may be arranged to read as pure decimal numbers. Thus, in Table VII for 10 per cent. and for 10 years' duration by reading £1

as £·1, the corresponding present value would be changed from
5·34100996 to ·534100996, and by the same rule if we read £1
as £·01, the corresponding value of £1 per annum would be
changed to £·0534100996, etc.

Specimen Table No. 8 illustrates this principle.

The present value of an annuity of £3,000 for 8 years,
allowing interest at the rate of 10 per cent. per annum, and
redeeming the capital at 3 per cent. per annum, may be deduced
from the above table as follows:—

The present value of £·1 at 10 per cent. for 8 years
= ·470684834.

Then, ·470684834 × 3000 = £1412·054502000. We must
now remove the decimal point one place to the right (which is
equivalent to multiplying by 10), and we shall then have
£14120·54502000 for the present value.

This Table may also be arranged in another way, i.e. by
considering £1 to have a cypher affixed to it, so as to read £10,
the corresponding present value would be changed.

Thus in Table VII for 10 years' duration, and at 10 per
cent. per annum, the present value of £1 = 5·34100996, but for
£10 per annum for a similar period the value would be changed
to £53·4100996; if we were to affix two cyphers to £1 and
make it read £100, then the value would be changed from
5·34100996 to £534·100996. This method is shown in Specimen Tables No. 9 and 10.

SPECIMEN TABLE No. 9.

Present Value of £10 per Annum.   Redemption of Capital 3 per cent.
per Annum.

Years	3½ per cent.	4 per cent.	4½ per cent.	5 per cent.	7 per cent.	Years
1	9·6618357	9·6153846	9·5693780	9·5238095	9·3457943	1
2	18·9533635	18·7754347	18·6008155	18·4294144	17·7742755	2
3	27·8916405	27·5080186	27·1348062	26·7715853	25·4109998	3
4	36·4927483	35·8388198	35·2079148	34·5988383	32·3596273	4
5	44·7718618	43·7915472	42·8532423	41·9543034	38·7064953	5
6	52·7433115	51·3881215	50·1008279	48·8764524	44·5240930	6
7	60·4206411	58·6488408	56·9779941	55·3997119	49·8737312	7
8	67·7766574	65·5925283	63·5096491	61·5549815	54·8076177	8
9	74·9434980	72·2366639	69·7185463	67·3700745	59·3704863	9
10	81·8126365	78·5974980	75·6255138	72·8700946	63·6008890	10

SPECIMEN TABLE No. 10.

Present Value of £100 per Annum.   Redemption of Capital 3 per cent.
per Annum.

Years	10 per cent.	12 per cent.	15 per cent.	18 per cent.	20 per cent.	Years
1	90·909090	89·285714	86·956522	84·745762	83·333333	1
2	168·744805	163·235767	155·615178	148·674381	144·381223	2
3	236·110581	225·463707	211·179700	198·597756	191·010889	3
4	294·961719	278·530549	257·051537	238·648080	227·776400	4
5	346·795264	324·301986	295·547950	271·477559	257·496647	5
6	392·776833	364·169375	328·302103	298·866548	282·009884	6
7	433·827521	399·191472	356·498164	322·054601	302·566044	7
8	470·684834	430·188219	381·015682	341·931323	320·044664	8
9	503·946260	457·804488	402·521626	359·151725	335·082624	9
10	534·100996	482·554435	421·530947	374·208773	348·152434	10

The annuity being £3426, to continue 8 years, allowing
10 per cent. per annum, and redeeming the purchase money
at 3 per cent. per annum, the present value is deduced from
Specimen Table No. 10, as under.

The present value of £100 per annum under the conditions,
is £470·684834.

Then £470·684834 × £3426 = £1612566·241284.

By the decimal rule we have only cut off six decimal places,
which gives the result 100 times greater than it should be; we
must therefore remove the decimal point two places to the left
to obtain the required value, i.e. £16125·66241284.

In working with the Tables it may be sufficient for ordinary
purposes to employ a less number of decimal places than will
be found in the Tables.   This is of course effected by writing
down the required number of decimal places, not forgetting to
add an unit to the last place retained, when the first figure in
the portion cut off is 5 or more.   For instance, by Table VII
the present value of £1 for 8 years' continuance, at 5 per cent.
interest, and 3 per cent. redemption, is £6·15549815; but if we
employ only 4 decimal places, the value will read £6·1555.
Specimen Tables No. 11 and 12 give values to 3 and 4 places of
decimals respectively.

Specimen Table No. II.

Present Value of £1 per Annum. Redemption 3 per cent.

Years	3½ per cent.	4 per cent.	4½ per cent.	5 per cent.	7 per cent.	Years
1	0·966	0·962	0·957	0·952	0·935	1
2	1·895	1·878	1·860	1·843	1·777	2
3	2·789	2·751	2·713	2·677	2·541	3
4	3·649	3·584	3·521	3·460	3·236	4
5	4·777	4·379	4·285	4·195	3·870	5
6	5·274	5·139	5·010	4·888	4·452	6
7	6·042	5·865	5·698	5·540	4·987	7
8	6·778	6·559	6·351	6·155	5·480	8
9	7·494	7·224	6·972	6·737	5·937	9
10	8·181	7·860	7·563	7·287	6·360	10

Years	10 per cent.	12 per cent.	15 per cent.	18 per cent.	20 per cent.	Years
1	0·909	0·893	0·870	0·847	0·833	1
2	1·687	1·632	1·556	1·487	1·444	2
3	2·361	2·255	2·112	1·986	1·910	3
4	2·950	2·785	2·571	2·386	2·278	4
5	3·468	3·243	2·955	2·715	2·575	5
6	3·928	3·642	3·283	2·989	2·820	6
7	4·338	3·992	3·565	3·221	3·026	7
8	4·707	4·302	3·810	3·419	3·200	8
9	5·039	4·578	4·025	3·599	3·351	9
10	5·341	4·826	4·215	3·742	3·482	10

Specimen Table No. 12.

Present Value of £1 per Annum. Redemption 3 per cent.

Years	3½ per cent.	4 per cent.	4½ per cent.	5 per cent.	7 per cent.	Years
1	0·9662	0·9615	0·9569	0·9524	0·9346	1
2	1·8953	1·8775	1·8601	1·8429	1·7774	2
3	2·7891	2·7508	2·7135	2·6772	2·5411	3
4	3·6493	3·5839	3·5208	3·4599	3·2360	4
5	4·4772	4·3792	4·2853	4·1954	3·8706	5
6	5·2743	5·1388	5·0101	4·8876	4·4524	6
7	6·0421	5·8649	5·6978	5·5400	4·9874	7
8	6·7777	6·5593	6·3510	6·1555	5·4808	8
9	7·4943	7·2237	6·9719	6·7370	5·9370	9
10	8·1813	7·8597	7·5626	7·2870	6·3601	10

Years	10 per cent.	12 per cent.	15 per cent.	18 per cent.	20 per cent.	Years
1	0·9091	0·8929	0·8696	0·8475	0·8333	1
2	1·6874	1·6324	1·5562	1·4867	1·4438	2
3	2·3611	2·2546	2·1118	1·9860	1·9101	3
4	2·9496	2·7853	2·5705	2·3865	2·2778	4
5	3·4680	3·2430	2·9555	2·7148	2·5750	5
6	3·9278	3·6417	3·2830	2·9887	2·8201	6
7	4·3383	3·9919	3·5650	3·2205	3·0257	7
8	4·7068	4·3019	3·8102	3·4193	3·2004	8
9	5·0395	4·5780	4·0252	3·5915	3·3508	9
10	5·3410	4·8255	4·2153	3·7421	3·4815	10

I am not aware that any special advantage is obtained by arranging the Tables as in Specimen Tables No. 7, 8, 9, 10, 11, and 12, but they may have the effect of preventing persons pirating my Tables by adopting any of the modes of arrangement I have exhibited. Unpleasant reminiscences of having suffered by the dishonourable conduct of others in the past, have induced me to take this course, with a view of guarding as much as possible against such a contingency in the future.

REMARKS ON LOGARITHMIC CALCULATIONS.

In the last preceding division of this work I have given a few practical examples of the mode of solving some of the more difficult propositions in compound interest by Logarithms.

Had it been necessary, this mode of conducting computations could also have been applied to the solution of the cases in Valuation of Mines; but it was not desirable to encumber that portion of the work by introducing other rules, which, of necessity, would have been subject to frequent repetition.

The following rules are of very great importance in deducing results of an exceedingly accurate order, and may therefore be applied as an occasional test to the numbers composing the Tables.

The property of logarithms available for the facilitation of arithmetical operations is, that the sum of the logarithms of two or more numbers is equal to the logarithm of the product of those numbers. From this, it follows that,

1st. The difference of two logarithms is the logarithm of the quotient of the corresponding numbers. A particular case of this is, that the remainder arising from the subtraction of a logarithm from o, is the logarithm of the reciprocal of the corresponding number. For, the product of a number and its reciprocal being always 1, the logarithm of which is o, the sum of their logarithms is consequently o.

The indexes of the logarithms of two numbers reciprocal to each other are necessarily affected with contrary signs, the sign of the one being positive and that of the other negative; and it simplifies work to remember that, apart from their signs, the negative index always exceeds the positive by an unit.

2nd. Another consequence of the property above enunciated is, that n times the logarithm of a number is the logarithm of the nth power of the number. Thus $n \log a = \log a^n$; and in like manner one n^{th} part of the logarithm of a number is the logarithm of the nth root of that number :

$$\frac{\log a}{n} = \log \sqrt[n]{a.}$$

Tables of 7-figure logarithms, which suffice for most purposes, are very accessible. When results of more than seven figures are wanted, recourse may be had to Gray's *Tables of logarithms to twelve places*.*

I now give a few examples.

1. Required the amount of £1 at 3½ per cent. in 100 years.

The required amount here is R^{100}, where R is the amount at the specified rate, of £1 in one year, $= 1·035$.

Hence,

$$\log R^{100} = 100 \log R;$$

that is,

$$\log (1·035)^{100} = 100 \log 1·035 = 0·0149403 \times 100 = 1·4940300$$

and the number corresponding to this is, 31·1915, which is the amount required.

It is to be observed that two places in the logarithm being in effect lost in the multiplication by 100, the result cannot be depended on to more than five or six places in all.

---

* Published by C. and E. Layton, London.

Using the 12-figure logarithms we find,.

$$100 \log 1\cdot035 = 1\cdot494034979300,$$

the number corresponding to which, to 10 places, beyond which it is useless to go, is $31\cdot19140798$.

2. Required the present value of £1 per annum, the rate of interest on the purchase money being 16 per cent. per annum, and that for redemption being 3 per cent. per annum, for a duration of 40 years.

The formula for solution here is, p. 31,

$$P_n = \frac{1}{q^{\prime\prime} + s_n} ;$$

that is,

$$P_{40} = \frac{1}{\cdot16 + s_{40}}.$$

s_{40}, (Table V) $= \cdot01326238$

$\cdot16 = \cdot16$

———————

$\cdot17326238 \log 1\cdot2387043$

P_{40} .   . $= 5\cdot771593$,, $0\cdot7612957$ complement.

The complement (which is the logarithm of the reciprocal) is obtained by subtracting the logarithm here formed from 0; and the subtraction is most readily performed by deducting each figure in the decimal portion from 9, except the last, which is deducted from 10. And the index of the logarithm being 1, that of the complement is 0, in accordance with what has been said.

Using the 12-figure tables we get for the logarithm

$$0\cdot761295729529,$$

and for the corresponding number

$$5\cdot7715934 16075.$$

By Table VII the required value is $5\cdot77159342$, agreeing, as far as it goes, with that last found.

3. Required the present value of £1 per annum deferred 4 years, and to continue 40 years thereafter, allowing interest at 20 per cent. per annum, and for redemption 3 per cent. per annum.

The formula here is

$$P_{t+n} = P_n v^t; \text{ that is } P_{4+40} = P_{40} v^4.$$

P_{40}, (Table VII) . $= 4{\cdot}6890596 \log 0{\cdot}6710858$
v^4, (Table IV) . $= {\cdot}4822531$,, $\bar{1}{\cdot}6832750$

P_{4+40} . . $= 2{\cdot}261313$,, $0{\cdot}3543608$

By the 12-figure table we have,

$$P_{4+40} = 2{\cdot}261313463908;$$

the value by Table X being $2{\cdot}26131348$.

4. Required the redemption fund that will amount to £1 in 20 years at the rate of 3 per cent. per annum.

The formula here is, p. 29,

$$s_n = \frac{r}{R^n - 1}, \text{ that is,}$$

$$s_{20} = \frac{{\cdot}03}{(1{\cdot}03)^{20} - 1}.$$

$\qquad\qquad\qquad {\cdot}03 \qquad \log \bar{2}{\cdot}4771213$
$(1{\cdot}03)^{20} - 1,$ (Table I) $= {\cdot}8061112$,, $\bar{1}{\cdot}9063949$

s_{20} . . $= {\cdot}03721571$,, $\bar{2}{\cdot}5707264$

The value by Table V is ${\cdot}03721571$. The 12-figure process gives ${\cdot}0372157075953$.

5. Required the present value of £1 due 20 years hence.

The value here required is the reciprocal of the amount of £1 in the same time, being denoted by v^{20}. We therefore have

$$v^{20} = \frac{1}{R^{20}}, \text{ and } \log v^{20} = \text{co } \log R^{20}.$$

R^{20}, (Table I) $= 1{\cdot}806111 \quad \log 0{\cdot}2567444$
v^{20} . . $= {\cdot}5536759$,, $\bar{1}{\cdot}7432556$ complement.

The 12-figure process gives ${\cdot}553675754178$; and the value given by the tables is ${\cdot}55367575$.

6. Required the annuity which £1 will purchase, the elements being as in (2).

The annuity that £1 will purchase is, in all cases, the reciprocal of the present value of the same annuity. Hence the annuity here required will be found by forming the reciprocal of P_{40} found in example 2.

$$P_{40} \quad = 5\text{·}771593 \quad \log 0\text{·}7612957$$
$$\text{Annuity} = \text{·}1732624 \quad \text{,,} \quad \bar{1}\text{·}2380043 \text{ complement.}$$

By the 12-figure tables we have for the required value ·173262377895.

7. Required the annuity, deferred, which £1 will purchase, the elements being as in (3).

The value here sought is the reciprocal of that found in example 3.

By the 7-figure process it comes out ·4422208, and by the 12-figure process, ·4422208667492.

PART IV.

SOURCES FOR REDEMPTION OF CAPITAL

BY

REINVESTMENT OF SURPLUS ANNUITY;

WITH REMARKS ON THE ADVANTAGES OF HOME AND
FOREIGN MINING, ETC. ETC.

L

PART IV.

SOURCES FOR REDEMPTION OF CAPITAL BY REINVESTMENT OF SURPLUS ANNUITY.

WHEN a proper Valuation has been made, and a mineral property purchased upon its basis, with a view to its ultimate development, it is of the greatest consequence to be in a position to thoroughly examine into all available sources for the redemption of any capital sum so invested, and select that which under all the circumstances is most reliable and profitable.

This being settled, any values may be obtained from the Tables having corresponding rates of interest for redemption at $2\frac{1}{2}$, 3, $3\frac{1}{2}$, and 4 per cent. per annum.

For the sake of comparison, and to illustrate the difference in value at different rates of redemption, we may employ the years' purchase or present value of £1 per annum at a given constant rate of interest and time; thus, the present value of £1 per annum, or years' purchase, allowing 20 per cent. to a purchaser for a period of 20 years, and to redeem the capital at rates of interest of 4, $4\frac{1}{2}$, and 5 per cent., is 4·281156377, 4·312647180, and 4·343245149 respectively.

The difference between the first and second years' purchase is ·031490803 or $7\frac{1}{2}$ pence, that between the second and third, ·030597969 or $7\frac{1}{2}$ pence, and that between the first and third is ·062088772, or 15 pence. Presuming, therefore, the interest allowed on any purchase to be at the same rate, and for the same time of duration as stated, it appears that the present value of £1 per annum, or years' purchase, is augmented as the rate per cent. for redemption is increased. Thus, at 20 per cent. per annum upon a purchase, at $4\frac{1}{2}$ per cent. for redemption, and for 20 years' duration, every £1 annuity purchased would cost more by about $7\frac{1}{2}$ pence, than it would, presuming the redemption rate of interest had been fixed at 4 per cent. per annum. The difference between the years' purchase at redemptive rates of interest at 4 and 5 per cent. per annum, comes out more prominent, amounting to 15 pence more than would be paid for each £1 annuity in case the capital were redeemed at the rate of 4 per cent. per annum. High rates of interest for redemption, therefore, are against the interest of a purchaser, and in favour of that of a vendor. On the other hand, however, as the rate of interest for redemption increases, the redemption fund necessary to re-produce £1 in the given time, decreases; which is of course due to the increase at compound interest of £1 in any number of years.

Thus, $\dfrac{\cdot 04}{2 \cdot 19112314303 - 1} = \cdot 0335817503$, or s_{20},

and $\dfrac{\cdot 045}{2 \cdot 411714024 - 1} = \cdot 0318761443$, or s_{20};

also, $\dfrac{\cdot 05}{2 \cdot 653297785 - 1} = \cdot 0302425872$, or s_{20}.

The redemption fund being s_{20}, and corresponding to the rates of interest at which the years' purchase were computed.

If we pay down the present sums of £4·281156377, £4·312647180, and £4·343245149, and expect to realise 20 per cent. per annum for 20 years, and redeem such capital sums at rates of interest of 4, 4½, and 5 per cent., we shall have the available and capitalisation sums thus derived and represented:

£4·281156377 at 20 per cent. per annum = ·8562312755
£4·281156377 × ·0335817503　　　.　　. = ·1437687245

　　　　　Annuity purchased　　　 = £1·0000000000
　And,

£4·312647180 at 20 per cent. per annum = ·8625294362
£4·312647180 × ·0318761443　　　.　　. = ·1374705638

　　　　　Annuity purchased　　　 = £1·0000000000
　　Also,

£4·343245149 at 20 per cent. per annum = ·8686490298
£4·343245149 × ·033581750　　　.　　. = ·1313509702

　　　　　Annuity purchased　　　 = £1·0000000000

which, as previously mentioned, shows a decreasing redemption fund for reproducing each capital sum in twenty years.

Perhaps one of the most reliable sources for investments at low rates of interest is presented by the English Government Funds. The Consolidated 3 per Cent. Annuities are, however, subject to much uncertainty as to the price to be realised by them in the market. The highest rate of interest ever obtained, occurred I believe in 1797, when Consols were sold at 52, the rate per cent. on the purchase being $\frac{300}{52} = £5\ 15s.\ 4\frac{1}{2}d.$ The least rate realised upon purchase appears to have been in 1737, when Consols were sold as high as 106, the rate being $\frac{300}{106} = £2\ 16s.\ 7\frac{1}{8}d.$

One of the chief causes which operate to influence the price of stock is the limitation in the demand in proportion to the supply, and *vice versâ*. Consols, as well as most other kinds of stocks, are also affected by a variety of circumstances, such as the storing or withdrawal of gold from the Bank, political changes, apprehensions of the disturbance or restoration of peace, and many other causes influencing the condition of the money market, known best to stock brokers, their agents, and jobbers, who are adepts not only in understanding, but sometimes in producing certain fluctuations in the value of stock for special benefit. Indeed any excitement of public feelings, due either to real or imaginary causes, is sufficient to produce a temporary change at least in the value of funded property.

Those whose business it is to deal in stocks endeavour to make a profit by purchasing at low prices, and selling out at higher rates, at favourable opportunities.

The money market is also much influenced by the press; and it is curious to note with what anxiety and expectancy City and other business men will turn to the Money Article in the day's newspapers, and the eagerness they exhibit in exchanging comments and eliciting opinions upon it, evidently with a view to extend or curtail their financial operations, according as the general tone may seem favourable or otherwise.

At the time of writing this portion of the work the sale of stock in about 28 of the British railways was producing from 2 to 5½ per cent.; the best in the list being the Bristol and Exeter,

Furness, North-Eastern, Shropshire Union, and the Taff Vale, which were selling at rates of interest of $5\frac{3}{8}$, $5\frac{1}{2}$, $5\frac{1}{4}$, $5\frac{3}{8}$, and $5\frac{1}{4}$ per cent. respectively. Railway preference stocks were selling to realise from 4 to $5\frac{1}{4}$ per cent. per annum; railway debenture stock was also selling to realise from $3\frac{7}{8}$ to $5\frac{1}{2}$ per cent. per annum, and is considered to be a safe investment.

The stock of 13 Indian railways were selling to realise from $3\frac{3}{4}$ to $4\frac{3}{8}$ per cent. per annum, and are considered very safe; the interest on the issue being at the rates of $4\frac{1}{2}$ and 5 per cent., which, with the principal, is guaranteed by the Secretary of State for India in Council.

Of 43 colonial railways, 27 were selling stock which realises from 3 to 9 per cent. per annum. The highest percentage represents the European and North American 6 per cent. issue first mortgage bonds, redeemable at par in 1898; but it appears that the most reliable are the Melbourne and Hobson's Bay united 6 per cent. bonds, payable in 1880, and 5 per cent. bonds redeemable in 1895, the latter having 20 years to run. Also Tasmania Main Line, Limited, guarantees 5 per cent., the stocks of which were selling to realise $5\frac{1}{4}$, 5, and $6\frac{1}{4}$ per cent. per annum respectively.

Of 85 American railways, the shares in 44 of them were selling to realise from $4\frac{1}{2}$ to $9\frac{3}{8}$ per cent., the former rate representing that obtained by the sale of the shares in the Illinois Central redemption mortgage, payable in 1875, and the latter the Paris and Decanture.

The sale of the shares in Baltimore and Potomack Main Line first mortgage, and the Galveston and Harrisburgh first mortgage, were selling to realise $6\frac{1}{2}$ and $8\frac{1}{2}$ per cent. per annum, and redeemable in 1911, having 31 years to run.

Of foreign railway obligations, the bonds in the Central Argentine first issue 7 per cent. were selling to realise $5\frac{7}{8}$ per cent. per annum; and out of 20 others, the bonds of 16 of them were selling to realise from $4\frac{3}{8}$ to 7 per cent. per annum.

Out of 30 colonial government investments, the stock in 29 of them were sold at from 4 to $5\frac{3}{4}$ per cent. per annum.

The shares in 12 Insurance Companies were realising $6\frac{1}{2}$, 7, $6\frac{1}{4}$, $5\frac{1}{4}$, 6, $5\frac{7}{8}$, $5\frac{3}{8}$, $4\frac{3}{8}$, $4\frac{1}{2}$, 5, and $6\frac{3}{4}$ per cent. per annum respectively, the highest rate being realised by the 'Universal Marine' insurance company, limited, and the lowest by the

' Royal' insurance company, limited. The highest and lowest rate obtained by the sale of shares in 12 land companies was 7⅝ and 4¾ per cent.; in 7 dock companies 5¾ and 4 per cent.; and in 8 shipping companies it was 8½ and 6 per cent. per annum.

The shares in the Globe Telegraph and Trust Company were realising 8½ per cent. per annum. Those in 11 other telegraph companies were realising from 5¾ to 7⅝ per cent. per annum upon the market value of stock.

Of other industrial companies, the sale of shares in Hooper's Telegraph Works, Limited, realised as much as 13½ per cent. at the then market value. The shares in most other companies were selling to realise from 5 to 10 per cent. per annum.

There are a great variety of foreign stocks, loans, and bonds, which were realising rates in the market from 3¾ to 11⅜ per cent. per annum, such as Argentine, Columbian, and Costa Rica 6 per cents., Paraguay 8 per cents., and many others of an uncertain character belonging to the South American States. Indeed, mention may be made of many not far removed from an entire collapse.

Great care, and the exercise of sound judgment, are of necessity required on the part of an intending purchaser of stocks, if. they are to be regarded as a means of profitable investment.

Indian railways, Indian debenture bonds, colonial government investments, safe home railway debenture stock, and joint stock limited banks, &c., working on a safe basis, and possessing firm guarantees and good management, are very inviting, and would doubtless yield a good percentage upon the capital invested in such undertakings.

Surplus annuities derived from mining may therefore be employed to advantage in the purchase of stock or shares in such of these undertakings as may be considered to be absolutely safe, and so from time to time redeem any capital sum, or at least a portion of such sum invested elsewhere, at a higher rate of interest than could be realised by investing in 3 per cent. Consols. On the average 4, 4½, or probably 5 per cent. may be realised for limited periods. A considerable advantage is also connected in possessing property of this class, as it may always be turned into ready cash, at the market value.

REMARKS ON FOREIGN AND HOME MINING.

As to Foreign Mines on the whole—with some exceptions—I consider it to be a great mistake to invest in them indiscriminately, as the majority of them are of such uncertain character as to render it a very unsafe venture. Much, however, depends on the part of the world in which they exist, the surrounding circumstances, and other associations. Very valuable beds of coal of great thickness, and iron ore deposits in immense masses exist, and are of such frequent occurrence in the different mineral basins in the United States, as to entitle them to be considered as the future storehouses of untold wealth; but the isolated condition, and want of transit of many of them, renders it very improbable that they will receive much attention from English capitalists, at least for some years to come. Those, however, who can afford to lock up a considerable amount of capital in the purchase of large tracts of minerals in the States, for the benefit of their successors, might do a more unwise thing.

With regard to the silver, and gold mines of the far West, some of them are unworthy the attention of English capitalists, as it is not in the nature of things—considering their great distance from home, and all the surrounding circumstances —that any permanent profit can be realised from many of those offered in the market, even supposing that such mines are really in existence. Many of these mines are too much in the hands of a class of men whose chief aim in a general way is to interest English persons in their behalf with a view to carry on some illegitimate speculation. It is also a most surprising fact that persons of apparent respectability are to be found in London to co-operate in such barefaced schemes, some of which are now and then exposed and held up to the light in the law courts. If, instead of taking up with all the mines introduced in England from Colorado, Utah, California, and places similar to the late *salted* diamond fields, the English public were to turn their attention to portions of the Argentine Republic, they would find some of the silver mines of those regions far more worthy of their attention.

At the time I was appointed Engineer to survey, and draw

up a report, estimate, and valuation, for the purpose of carrying out an immense drainage scheme, projected for the Argentine Government, and intended to unwater the silver mines of the Cero-de-Pasco, high up in the Andes, I had ample opportunity of collecting information as to the great riches existing in that district, and from further evidence, since published by Major Rickard, it is ascertained beyond a doubt that many portions of the mountainous districts of the Argentine Republic are *replete* with rich silver ore deposits, containing a greater percentage of silver than can be obtained from similar mines in many other parts of the world.

The chief drawback to mining being carried on in this Republic, is the want of special and speedy means of transit from the mines to the Towns, and seaboard, but the great elevation, dangerous passes, and gorges, have hitherto prevented this, other than what may be performed by *pack-mules*, the load of each being from ¾ to one hundred weight. This difficulty may, however, be remedied to some extent, by extracting the metal at the mines upon an improved principle, and then conveying it to its destination by mules. If it is necessary to adventure capital in American mining at all, the Argentine Republic should undoubtedly have the preference, on account of the unusually high profits expected to be derived from the silver mines there. To obtain a large concession in this region, and colonise it with English people, would be a far more wise and profitable scheme than scores of those laid before the public from time to time.

My professional visit to the Brazils in 1851 did not strongly impress me with the idea that it would ever be likely to prove a legitimate and permanent mining field for the expenditure of English capital. Indeed, the small amount of labour to be obtained from the natives,—naturally an indolent race—under a burning tropical sun, is not such as to encourage English adventurers to speculate in such a place.

Australia has undoubtedly created and is still creating considerable interest as a mining colony, being rich in tin, copper, iron, coal, and the precious metal, gold ; but I am not of opinion that the quantity of the latter to be found there in the future will ever affect the value in the currency so as to cause a depreciation. I visited Australia professionally in 1853, and at that time there seemed to exist such a tendency, but it soon became manifest, that, even to obtain a moderate supply of gold, the search would have to be continued in a more regular mining way, involving of course more labour, and the expenditure of adequate capital in order to obtain fair or corresponding returns. I anticipate great things for Australia, from the future yield of her mineral fields.

India, and also New Zealand, are legitimate fields for mining enterprise; they are British Possessions so to speak, and therefore it is natural Englishmen should turn their attention in these directions. It is only a question of colonising these places with young English people, so that they may become acclimatised, and the more general introduction of railways, and then capitalists may fearlessly venture their cash in developing the mineral resources in conjunction with manufactories. On the whole, New Zealand, as a mineral field, is in some respects preferable to some of the other colonies, on account of its splendid climate.

Spain has from very ancient times been far famed for her mineral wealth, it is much nearer home than any of the places previously referred to. Many parts abound in different kinds of minerals, such as silver, copper, lead, quicksilver, sulphur, iron ore, and coal. The English obtained concessions in Spain and caused a great excitement there in 1825, but the speculators in a general way were so ill advised, that the mania soon subsided to its proper level. Since that time very valuable concessions have been obtained, and worked by English companies to a very great profit. The nation has hitherto been much crippled by its internal disorganisation, producing a re-action in speculation to a considerable extent. Mercantile relations are thus injured, and thrown out of proper order. Spain, however, obtains a large revenue from her mines, and should the new government secure permanent peace to the nation, the tide of English speculators would again flow in that direction, and the export of minerals from there to England would undoubtedly prove large. I am quite persuaded that by perseverance, there is ample opportunity to amass large fortunes from working Spanish mines.

There are, however, great profits to be gained from legitimate Home Mining, and it will no doubt be preferred by many to foreign adventure ; England, however, creates rich men, and in too many instances, foreigners reap the benefit of them.

However, it is the nature of Englishmen to desire to become rich, and as the population increases so enormously, and this passion will always exist, they will naturally seek that field of enterprise most likely to enable them to achieve their object ; hence, we shall always find English capitalists adventuring, some in one thing, and some in another, in different parts of the world.

There are many good mines in every mining district throughout the United Kingdom, and there are also a good many inferior ones ; but it is known to many persons that there are thousands and thousands of acres of virgin ground, containing iron ore, lead, tin, coal, and other minerals, not yet explored. When however the time comes for winning, a rich harvest will be yielded to those embarking in it. Coal, iron, and other minerals will always be in requisition for the purpose of carrying on the commerce of the world, no matter what may be urged against new undertakings, consequent upon fluctuations of trade. There are at all times capitalists who will purchase properties for the purpose of working the minerals, and the proper time to make these purchases is when trade is dull, and prices are low. It is only in a season of great commercial excitement and prosperity, that all sorts of persons join together in order to float bubble companies, endeavour to pocket the cash obtained, perhaps upon glaring and false representations, ruin the share holders, and bring legitimate mining into ill repute. Of all places for palming off shams, and hatching swindles, there is none equal to London.

It always has been, and will still be the case, that those capable of producing the best article in the greatest abundance, and at the cheapest rate, must win the day ; but whether it is a colliery, iron, silver, or other mine, a good one in the first instance must be possessed, that is, it must contain minerals in sufficient quantity, and quality, to justify the contemplated outlay, otherwise it will be so much capital lost. When a good property is obtained, it is only a question of time, and capital judiciously expended, in order to produce proportionate results as to profit. In these times, too, it is necessary to obtain a very large and constant annual output or yield from the mine, and with regard

to the expected profit per ton, or annual income, it is also a
question of importance to consider whether, on the whole, it
would not be the wisest plan at the commencement of a new
winning, to arrange all the works above and below ground, so
as to be in a position to raise from the mine two or three times
as much as is being raised from any single colliery or iron mine
in the same district, or indeed as may be required from such
new colliery by the demand of trade during the first few years
of its life. If 800 or 1000 tons per diem could be considered a
good output from any one of the several mines surrounding the
winning about to be developed, the machinery should be calculated to raise at least three times that quantity, or from 2400 to
3000 tons per day, and arrange all the surface and underground
works to correspond.

The position would then be this, that if at 800 or 1000 tons
per diem, and say 1s. 6d. or 2s. per ton profit, the smaller
collieries in the district would but just or scarcely pay at
the rate of profit named or that actually realised, the larger
colliery would live, and flourish at a less profit per ton than
the smaller ones, because a large colliery properly laid out
would cost less in working expenses in proportion, than those
of less capacity. Another advantage would also accrue
from laying out a new winning according to this plan, and it
is this: in a time of great demand, and with high rates of profit,
the larger quantity could be obtained, supposing the colliery
had not been worked up to its full capacity, consequently a
greater annual revenue would result. I believe it is a great
mistake, in laying out a new winning of ample area, to copy even
the best example in the same neighbourhood, its capacity being
only equal to that of an ordinary going concern.

Provisions should be always made in surface arrangements
for future underground extension, at any time it may be required
to double or triple the output, but if the machinery is under
power this cannot be accomplished without making additions,
which are always more expensive, and incomplete, besides producing vexation from delay.

As to the underground works, if there are good seams of coal
to work upon, provision can always be made in times of moderate
demand, to have a reserve ready, to be extracted as an additional
supply, at the least notice. I am of opinion that if all new

collieries were laid out as suggested, it would tend to equalise trade, and place the proprietors in a position to meet a downward market : I think this is clear, as 1000 tons per diem at 1s. per ton profit would produce the same annuity as 500 tons per diem at 2s. per ton profit, and although more capital would of necessity be employed, nevertheless a colliery so circumstanced would pay a dividend, when those of more limited capacity would be struggling for life, or perhaps would be closed altogether.

In the case where proprietors require a certain definite and invariable income from their mines, no matter whether times are good or bad, I think they would be more likely to succeed by adopting the plan proposed.

The unprecedented struggle of labour against capital experienced of late throughout the United Kingdom, has created, and laid severe infliction upon all the parties involved. Whether English proprietors will, with neighbouring countries, be content to accept in the future less profit than heretofore, is not easy to judge ; but as labour can be commanded on the Continent at a much cheaper rate than at home, enabling the firms established there to produce the raw and manufactured article at a cheaper price than we can, under present circumstances, necessity seems to force itself upon us, to adopt some measures for preventing a destructive competition, and to retain and increase the trade of this country. The only way to this which seems to commend itself, is, reasonable wages for men, such indeed as the demand of trade seems to warrant, and smaller profits to the employers.

I have no doubt the plan of working I have suggested, that is, to produce large vends and receive small profits, would meet with considerable opposition from proprietors of limited means, but the policy is undoubtedly a sound one. I am also of opinion that gentlemen of limited means are not the proper persons to embark in such adventures ; in point of fact, it is only those private persons who can afford to risk from £50,000 to £150,000 if necessary, in mining, that should enter into it, and then if success crowned the efforts put forth, it would be very pleasing, and highly satisfactory ; on the contrary, the loss of the capital would not prove an entirely ruinous matter. It is a great mistake for gentlemen who do not understand mining to embark their *all* in it, especially as is too commonly the case, when their capital is limited to a few thousands, and they enter into the undertaking for the sake of a *managing salary*, and a *living*. The issue generally is, that the development is carried on by feeble attempts; gets crippled, and drags on to a premature death; but if by any chance it survives, it only yields respectable poverty to all connected with it: whereas, if sufficient capital had been put into it, under more favourable auspices, it would have proved a lucrative undertaking.

One great drawback to the success of a mine is that gentlemen proprietors, some of whom are only half educated in mining matters, are continually interfering in the management, leading to neglect and disappointment on the part of those in charge, and disastrous government.

Persons in the possession of limited means, are, therefore, fit subjects for joining a Limited Liability Company, where their dominant spirit can only be exercised in direct ratio to the amount of capital they are able to throw into the concern.

When foreign mines are found to compare favourably in productiveness and estimated worth with our best home mines of a similar class, more risk is undoubtedly incurred in the purchase of the former than of the latter. A purchaser is therefore entitled to be allowed a greater rate per cent. per annum upon the Capital he may invest in foreign than in home mines. Extra contingencies, which are impossible to be foreseen, arising out of the policy or internal disorganisation of foreign Governments, may frequently operate to prejudicially affect the value of the interest held by home capitalists in foreign mines. This must be evident, especially when it is necessary to ship the minerals to this country. If, then, 20 or 25 per cent. per annum may be considered to be the maximum rate allowed upon the purchase of home metalliferous mines, foreign mines should be purchased at least from 10 to 15 per cent. per annum higher. Since printing Tables I to XVIII, I have been called upon to consider this question; and, after mature deliberation, have arrived at the above conclusion.

A small Table (C) of Immediate Values for 30, 35, 40, and 45 per cent., and also a Table (D) of Deferred Values for the same rates per cent., Deferred 1, 2, 3,4, and 5 years, have been prepared, and will be found of special use to purchasers of foreign mines at high rates of interest; but it is probable that such mines would find little favour with the English public when the time for development and consequent delay of dividends extended to or beyond 3 or 4 years. Table A was calculated for the purpose of constructing Table B, and the latter was employed in deferring the values in Table C.

In Table C, at 45 per cent. per annum for 40 years, the years' purchase is 2·1586, and for the same rate and time, but deferred 3 years, Table D gives a years' purchase of ·7081 ; and for 4 or 5 years' deferrence it is ·4883 and ·3368 years' purchase respectively. These values are apparently small, but when it is considered that gold, silver, and other foreign mines are frequently purchased upon representations which assume that the annual income will be very large, it is clear that the value would amount to a very considerable sum. For instance, assuming that a gold mine were to be offered which, according to representation, would yield an Annuity of £60,000, but to prepare for this 3 years must be spent in development. Then, at 45 per cent., the Value would be ·7081 × £60,000 = £42,486 = the Gross Value; but if deferred 4 years, it would be ·4883 × £60,000 = £29,298 = the Gross Value; if deferred 5 years, ·3368 × £60,000 = £20,208 = the Gross Value. Of course the cost of opening the mines must be deducted from the Gross Value.

As a constant yield from gold, silver, and some other metalliferous mines is very uncertain for any length of time, it is highly desirable that they should be purchased upon the most advantageous terms.

TABLE A.

Amount of £1 in n years, up to 5 years, at the following rates per cent. :—

n Years	30 per cent.	35 per cent.	40 per cent.	45 per cent.	n Years
1	1·30000	1·35000	1·40000	1·45000	1
2	1·69000	1·82250	1·96000	2·10250	2
3	2·19700	2·46038	2·74400	3·04863	3
4	2·85610	3·32153	3·84160	4·42051	4
5	3·71293	4·48407	5·37824	6·40974	5

TABLE B.

Present Value of £1 in n years, up to 5 years, at the following rates per cent.:—

n Years	30 per cent.	35 per cent.	40 per cent.	45 per cent.	n Years
1	·76923	·75074	·71429	·68965	1
2	·59172	·54869	·51020	·47562	2
3	·45517	·40644	·36443	·32802	3
4	·35013	·30107	·26031	·22622	4
5	·26933	·22302	·18593	·15602	5

TABLE C.

Present Value of £1 per Annum in n years. Redemption of Capital being at 3 per cent., with Interest allowed to a Purchaser at the following rates per cent.:—

n Years	30 per cent.	35 per cent.	40 per cent.	45 per cent.	n Years
1	·7692	·7407	·7143	·6897	1
3	1·6038	1·4847	1·3821	1·2928	3
5	2·0477	1·8575	1·6997	1·5665	5
6	2·1997	1·9818	1·8031	1·6540	6
8	2·4245	2·1624	1·9514	1·7779	8
10	2·5826	2·2871	2·0524	1·8614	10
12	2·6993	2·3783	2·1256	1·9214	12
15	2·8267	2·4767	2·2038	1·9850	15
20	2·9655	2·5825	2·2877	2·0525	20
22	3·0053	2·6127	2·3108	2·0715	22
25	3·0541	2·6495	2·3396	2·0946	25
27	3·0811	2·6698	2·3554	2·1072	27
30	3·1151	2·6953	2·3752	2·1231	30
32	3·1343	2·7097	2·3864	2·1320	32
35	3·1592	2·7282	2·3915	2·1434	35
37	3·1735	2·7389	2·4090	2·1500	37
40	3·1922	2·7528	2·4198	2·1586	40
42	3·2032	2·7610	2·4261	2·1636	42
45	3·2177	2·7702	2·4344	2·1702	45
47	3·2262	2·7781	2·4393	2·1741	47
50	3·2377	2·7866	2·4458	2·1793	50
52	3·2445	2·7916	2·4497	2·1824	52
55	3·2536	2·7984	2·4549	2·1865	55
57	3·2591	2·8024	2·4580	2·1890	57
60	3·2666	2·8079	2·4622	2·1923	60
65	3·2771	2·8157	2·4682	2·1971	65
70	3·2858	2·8222	2·4732	2·2010	70
75	3·2931	2·8275	2·4773	2·2043	75
80	3·2991	2·8320	2·4807	2·2070	80
90	3·3085	2·8388	2·4866	2·2111	90
100	3·3151	2·8438	2·4898	2·2141	100

TABLE D.

Present Value (or Years' Purchase) of £1 per Annum in
n years, after t years' Deferrence. Redemption of Capital being at
3 per cent., with Interest allowed to a Purchaser at 30 per cent.

n Years	Deferred 1 Year	Deferred 2 Years	Deferred 3 Years	Deferred 4 Years	Deferred 5 Years	n Years
1	·5917	·4552	·3501	·2693	·2072	1
3	1·2337	·9490	·7305	·5615	·4320	3
5	1·5752	1·2117	·9321	·7205	·5515	5
6	1·6921	1·3016	1·2374	·7702	·5924	6
8	1·8650	1·4346	1·1036	·8489	·6530	8
10	1·9866	1·5282	1·1755	·9042	·6956	10
12	2·0764	1·5972	1·2290	·9451	·7270	12
15	2·1744	1·6726	1·2866	·9897	·7613	15
20	2·2812	1·7547	1·3500	1·0383	·7987	20
22	2·3118	1·7783	1·3679	1·0522	·8094	22
25	2·3493	1·8072	1·3901	1·0693	·8226	25
27	2·3701	1·8231	1·4024	1·0788	·8298	27
30	2·3962	1·8433	1·4179	1·0907	·8390	30
32	2·4110	1·8546	1·4266	1·0974	·8442	32
35	2·4302	1·8694	1·4380	1·1061	·8509	35
37	2·4412	1·8778	1·4445	1·1111	·8547	37
40	2·4555	1·8889	1·4530	1·1177	·8598	40
42	2·4639	1·8953	1·4580	1·1215	·8627	42
45	2·4752	1·9040	1·4646	1·1266	·8666	45
47	2·4817	1·9659	1·5122	1·1632	·8689	47
50	2·4905	1·9666	1·5128	1·1637	·8720	50
52	2·4958	1·9670	1·5131	1·1639	·8738	52
55	2·5028	1·9676	1·5135	1·1643	·8763	55
57	2·5070	1·9679	1·5138	1·1645	·8778	57
60	2·5128	1·9684	1·5141	1·1647	·8798	60
65	2·5214	1·9690	1·5146	1·1651	·8825	65
70	2·5275	1·9695	1·5150	1·1654	·8850	70
75	2·5332	1·9700	1·5154	1·1657	·8869	75
80	2·5378	1·9704	1·5157	1·1659	·8885	80
90	2·5450	1·9709	1·5161	1·1662	·8911	90
100	2·5501	1·9713	1·5146	1·1665	·8929	100

TABLE D.

Present Value (or Years' Purchase) of £1 per Annum in
n years, after **t** years' Deferrence. Redemption of Capital being at
3 per cent., with Interest allowed to a Purchaser at **35** per cent.

n Years	Deferred 1 Year	Deferred 2 Years	Deferred 3 Years	Deferred 4 Years	Deferred 5 Years	n Years
1	·5561	·4064	·3011	·2230	·1652	1
3	1·1146	·8146	·6034	·4470	·3311	3
5	1·3945	1·0192	·7550	·5592	·4142	5
6	1·4878	1·0874	·8055	·5967	·4420	6
8	1·6234	1·1865	·8789	·6510	·4822	8
10	1·7170	1·2549	·9296	·6886	·5100	10
12	1·7855	1·3049	·9666	·7160	·5504	12
15	1·8594	1·3589	1·0066	·7457	·5523	15
20	1·9388	1·4170	1·0496	·7775	·5759	20
22	1·9615	1·4336	1·0619	·7886	·5827	22
25	1·9891	1·4538	1·0769	·7977	·5909	25
27	2·0043	1·4610	1·0822	·8017	·5938	27
30	2·0235	1·4789	1·0955	·8115	·6011	30
32	2·0343	1·4868	1·1013	·8158	·6043	32
35	2·0482	1·4969	1·1088	·8214	·6084	35
37	2·0562	1·5028	1·1132	·8246	·6108	37
40	2·0666	1·5104	1·1188	·8288	·6139	40
42	2.0728	1·5149	1·1222	·8313	·6157	42
45	2·0797	1·5200	1·1259	·8340	·6178	45
47	2·0856	1·5243	1·1291	·8364	·6196	47
50	2·0920	1·5290	1·1326	·8390	·6214	50
52	2·0958	1·5317	1·1346	·8407	·6226	52
55	2·1009	1·5355	1·1374	·8425	·6241	55
57	2·1039	1·5376	1·1390	·8436	·6250	57
60	2·1080	1·5407	1·1412	·8454	·6262	60
65	2·1139	1·5449	1·1444	·8477	·6279	65
70	2·1187	1·5485	1·1471	·8497	·6294	70
75	2·1227	1·5514	1·1492	·8513	·6306	75
80	2·1261	1·5539	1·1511	·8527	·6316	80
90	2·1312	1·5576	1·1538	·8547	·6331	90
100	2·1350	1·5604	1·1558	·8562	·6342	100

TABLE D.

Present Value (or Years' Purchase) of £1 per Annum in n years, after t years' Deferrence. Redemption of Capital being at 3 per cent., with Interest allowed to a Purchaser at 40 per cent.

n Years	Deferred 1 Year	Deferred 2 Years	Deferred 3 Years	Deferred 4 Years	Deferred 5 Years	n Years
1	·5102	·3644	·2603	·1859	·1328	1
3	·9872	·7052	·5037	·3598	·2570	3
5	1·2141	·8672	·6194	·4424	·3160	5
6	1·2880	·9200	·6571	·4694	·3353	6
8	1·3939	·9956	·7111	·5080	·3628	8
10	1·4660	1·0470	·7480	·5343	·3818	10
12	1·5183	1·0845	·7746	·5533	·3952	12
15	1·5747	1·1244	·8031	·5737	·4098	15
20	1·6344	1·1672	·8337	·5953	·4254	20
22	1·6508	1·1790	·8421	·6015	·4296	22
25	1·6712	1·1937	·8526	·6090	·4350	25
27	1·6824	1·2016	·8584	·6131	·4379	27
30	1·6986	1·2119	·8656	·6183	·4416	30
32	1·7046	1·2177	·8697	·6212	·4439	32
35	1·7082	1·2202	·8715	·6225	·4447	35
37	1·7207	1·2291	·8779	·6271	·4479	37
40	1·7284	1·2346	·8818	·6300	·4499	40
42	1·7329	1·2378	·8841	·6315	·4511	42
45	1·7389	1·2421	·8872	·6337	·4526	45
47	1·7424	1·2446	·8890	·6350	·4535	47
50	1·7470	1·2479	·8913	·6367	·4547	50
52	1·7498	1·2499	·8927	·6384	·4555	52
55	1·7535	1·2525	·8946	·6390	·4564	55
57	1·7557	1·2541	·8958	·6398	·4570	57
60	1·7587	1·2562	·8973	·6409	·4578	60
65	1·7630	1·2593	·8995	·6425	·4589	65
70	1·7666	1·2619	·9013	·6438	·4598	70
75	1·7695	1·2639	·9028	·6449	·4606	75
80	1·7719	1·2656	·9040	·6458	·4612	80
90	1·7757	1·2684	·9060	·6471	·4622	90
100	1·7784	1·2703	·9074	·6481	·4629	100

TABLE D.

Present Value (or Years' Purchase) of £1 per Annum in
n years, after **t** years' Deferrence. Redemption of Capital being at
3 per cent., with Interest allowed to a Purchaser at **45** per cent.

n Years	Deferred 1 Year	Deferred 2 Years	Deferred 3 Years	Deferred 4 Years	Deferred 5 Years	n Years
1	·4757	·3280	·2262	·1560	·1084	1
3	·8916	·6149	·4241	·2925	·2017	3
5	1·0803	·7451	·5138	·3544	·2444	5
6	1·1407	·7867	·5425	·3742	·2581	6
8	1·2261	·8456	·5832	·4024	·2774	8
10	1·2837	·8853	·6106	·4211	·2904	10
12	1·3251	·9139	·6303	·4347	·2998	12
15	1·3690	·9441	·6511	·4490	·3097	15
20	1·4155	·9762	·6733	·4643	·3202	20
22	1·4286	·9852	·6793	·4686	·3232	22
25	1·4445	·9962	·6871	·4738	·3268	25
27	1·4532	1·0022	·6912	·4767	·3288	27
30	1·4642	1·0098	·6964	·4803	·3312	30
32	1·4703	1·0140	·6993	·4823	·3326	32
35	1·4782	1·0194	·7031	·4849	·3344	35
37	1·4827	1·0226	·7052	·4864	·3354	37
40	1·4887	1·0267	·7081	·4883	·3368	40
42	1·4927	1·0291	·7097	·4894	·3376	42
45	1·4967	1·0322	·7119	·4909	·3386	45
47	1·4994	1·0340	·7131	·4918	·3392	47
50	1·5030	1·0365	·7149	·4930	·3400	50
52	1·5051	1·0380	·7159	·4937	·3405	52
55	1·5079	1·0399	·7172	·4946	·3411	55
57	1·5096	1·0416	·7180	·4952	·3415	57
60	1·5119	1·0427	·7191	·4961	·3420	60
65	1·5152	1·0450	·7207	·4970	·3428	65
70	1·5172	1·0468	·7220	·4979	·3434	70
75	1·5202	1·0484	·7231	·4987	·3439	75
80	1·5221	1·0497	·7239	·5003	·3443	80
90	1·5249	1·0516	·7253	·5008	·3450	90
100	1·5270	1·0531	·7263	·5009	·3454	100

NOTATION.

r' = Interest on £1 (or any other monetary unit, say on 1) for one year.

r = Rate of interest for redemption.

n = An integral number of years

R = Amount of £1 in one year, $\Big\} = 1 + r$

R^n = Amount of £1 in n years (Compound interest), $\Big\} = (1 + r)^n$

$v^n = \begin{cases} \text{Present value of £1 due } n \\ \text{years hence,} \end{cases} = \dfrac{1}{R^n}.$

$\qquad \begin{cases} \text{If interest can be realised } m \\ \text{times in a year,} \end{cases} = \dfrac{1}{\left(1 + \dfrac{r}{m}\right)^{mn}}$

V_n = Present value of a perpetuity of £1, payable once in every nth year, $\Big\} = \dfrac{v^n}{(1 - v^n)}, \text{ or } \dfrac{1}{R^n - 1}.$

D_n = Present value of a perpetuity of £1 deferred n years, $\Big\} = \dfrac{v^n}{r}.$

e = Base of Napierian Logarithms = $2\cdot7182818$, and its Logarithm = $\cdot43429448$

e^r = Amount of £1 in one year, the rate being $\dfrac{r}{m}$ per moment.

e^{rn} = Amount of £1 in n years at that rate.

$$M_n = \begin{cases} \text{Amount of £1 per annum in } n \text{ years,} & = \dfrac{R^n - 1}{r}. \\[2em] \begin{array}{l}\text{Amount of £1 per annum if}\\ \text{interest can be converted}\\ m \text{ times in a year at the}\\ \text{rate } \dfrac{r}{m},\end{array} & = \dfrac{\left(1 + \dfrac{r}{m}\right)^{mn} - 1}{\dfrac{r}{m}}. \end{cases}$$

$$s_n = \begin{cases} \begin{array}{l}\text{Redemption Fund to be in-}\\ \text{vested at the end of the}\\ \text{year to realise £1,}\end{array} & = \dfrac{r}{R^n - 1}, \text{ or } \dfrac{1}{M_n}. \\[2em] \begin{array}{l}\text{If interest can be realised } m\\ \text{times in a year,}\end{array} & = \dfrac{r}{\left(1 + \dfrac{r}{m}\right)^{mn} - 1}. \end{cases}$$

$$p_n = \begin{cases} \begin{array}{l}\text{Present value of £1 per}\\ \text{annum, calculated at one}\\ \text{rate of interest,}\end{array} & = \dfrac{R^n - 1}{R^n r}, \text{ or } \dfrac{1 - v^n}{r}. \end{cases}$$

$$P_n = \begin{cases} \begin{array}{l}\text{Present value of £1 per}\\ \text{annum in } n \text{ years, interest}\\ \text{on capital being at one}\\ \text{rate, } r', \text{ and for redemption}\\ \text{another rate, } r, \text{ per cent.}\end{array} & = \dfrac{1}{\dfrac{r}{R^n - 1} + r'}, \text{ or } \dfrac{1}{r' + s_n}. \\[3em] \begin{array}{l}\text{Present value, if interest can}\\ \text{be realised } m \text{ times in a}\\ \text{year at the rate } \dfrac{r}{m},\end{array} & = \dfrac{1}{\dfrac{r}{\left(1 + \dfrac{r}{m}\right)^{mn} - 1} + r'}. \end{cases}$$

$$P_{t+n} = \begin{cases} \begin{array}{l}\\ \\ \text{Present value of £1 per}\\ \text{annum for a duration of}\\ n \text{ years after } t \text{ years,}\\ \text{allowing interest on capital}\\ \text{at one rate, } r', \text{ and for re-}\\ \text{demption another rate, } r,\\ \text{per cent.,}\end{array} & \begin{array}{l} = \dfrac{M_n}{(1 + r')^t (1 + r' M_n)}, \\[2em] = \dfrac{1}{\dfrac{(1 + r')^t - 1}{P_n} + r' + s_n}, \\[2em] = \dfrac{1}{\dfrac{r}{R^n - 1} + r'} v^n, \\[2em] \text{or } \dfrac{1}{s_n + r'} v^n. \end{array} \end{cases}$$

$$A = \begin{cases} \text{Annuity immediate which} \\ \text{£1 will purchase} \\ \text{Annuity deferred which £1} \\ \text{will purchase} \end{cases}$$

$$= \frac{1}{P_n}, \text{ or } s_n + r'.$$

$$= \frac{1}{P_{t+n}}.$$

$$P = \begin{cases} \text{Present value of £1 per} \\ \text{annum on a single life,} \\ \text{allowing interest on capi-} \\ \text{tal at one rate, } r', \text{ per cent,} \\ \text{and to redeem the capital} \\ \text{by an assurance on the life} \\ \text{at the office rate, } r, \text{ per cent.,} \end{cases} = \frac{1}{(1-v)+r} - 1.$$

TABLE I.

The sum to which £1 will amount in n *years up to one hundred at the rates of* ½, ¾, 1, 1¼, 1½, 1¾, 2, 2¼, 2½, 2¾, 3, 3½, 4, 4½, 5, 5½, 6, 7, 8, 9, 10, 11, 12, 13, 14, 15, 16, 17, 18, 19, 20, 21, 22, 23, 24, *and* 25 *per cent.*

Calculated to 10 *places of decimals for each percentage to* 9 *per cent.; to* 6 *places to* 15 *per cent., and to* 5 *places to* 25 *per cent.*

TABLE I. iii

Amount of £1 in n years at the following rates per cent.

Years	½ per cent.	Years	¼ per cent.	Years	¾ per cent.	Years	¾ per cent.
1	1·005	51	1·2896419440	1	1·0075	51	1·4638541068
2	1·010025	52	1·2960901537	2	1·01505625	52	1·4748330126
3	1·015075125	53	1·3025706045	3	1·0226691719	53	1·4858942602
4	1·0201505006	54	1·3090834575	4	1·0303391907	54	1·4970384672
5	1·0252512531	55	1·3156288748	5	1·0380667346	55	1·5082662557
6	1·0303775094	56	1·3222070192	6	1·0458522351	56	1·5195782526
7	1·0355293969	57	1·3288180543	7	1·0536961269	57	1·5309750895
8	1·0407070439	58	1·3354621446	8	1·0615988478	58	1·5424574027
9	1·0459105791	59	1·3421394553	9	1·0695608392	59	1·5540258332
10	1·0511401320	60	1·3488501526	10	1·0775825455	60	1·5656810269
11	1·0563958327	61	1·3555944033	11	1·0856644146	61	1·5774236346
12	1·0616778119	62	1·3623723753	12	1·0938068977	62	1·5892543119
13	1·0669862009	63	1·3691842372	13	1·1020104494	63	1·6011737192
14	1·0723211319	64	1·3760301584	14	1·1102755278	64	1·6131825221
15	1·0776827376	65	1·3829103092	15	1·1186025942	65	1·6252813911
16	1·0830711513	66	1·3898248607	16	1·1269921137	66	1·6374710015
17	1·0884865070	67	1·3967739850	17	1·1354445545	67	1·6497520340
18	1·0939289396	68	1·4037578550	18	1·1439603887	68	1·6621251743
19	1·0993985843	69	1·4107766442	19	1·1525400916	69	1·6745911131
20	1·1048955772	70	1·4178305275	20	1·1611841423	70	1·6871505464
21	1·1104200551	71	1·4249196801	21	1·1698930234	71	1·6998041755
22	1·1159721553	72	1·4320442785	22	1·1786672210	72	1·7125527068
23	1·1215520161	73	1·4392044999	23	1·1875072252	73	1·7253968521
24	1·1271597762	74	1·4464005224	24	1·1964135294	74	1·7383373285
25	1·1327955751	75	1·4536325250	25	1·2053866309	75	1·7513748585
26	1·1384595530	76	1·4609006876	26	1·2144270306	76	1·7645101699
27	1·1441518507	77	1·4682051911	27	1·2235352333	77	1·7777439962
28	1·1498726100	78	1·4755462170	28	1·2327117476	78	1·7910770762
29	1·1556219730	79	1·4829239481	29	1·2419570857	79	1·8045101542
30	1·1614000829	80	1·4903385678	30	1·2512717638	80	1·8180439804
31	1·1672070833	81	1·4977902607	31	1·2606563021	81	1·8316793102
32	1·1730431187	82	1·5052792120	32	1·2701112243	82	1·8454169051
33	1·1789083343	83	1·5128056080	33	1·2796370585	83	1·8592575319
34	1·1848028760	84	1·5203696361	34	1·2892343364	84	1·8732019633
35	1·1907268904	85	1·5279714843	35	1·2989035940	85	1·8872509781
36	1·1966805248	86	1·5356113417	36	1·3086453709	86	1·9014053604
37	1·2026639275	87	1·5432893984	37	1·3184602112	87	1·9156659006
38	1·2086772471	88	1·5510058454	38	1·3283486628	88	1·9300333949
39	1·2147206333	89	1·5587608746	39	1·3383112778	89	1·9445086453
40	1·2207942365	90	1·5665546790	40	1·3483486123	90	1·9590924602
41	1·2268982077	91	1·5743874524	41	1·3584612269	91	1·9737856536
42	1·2330326987	92	1·5822593896	42	1·3686496861	92	1·9885890460
43	1·2391978622	93	1·5901706866	43	1·3789145588	93	2·0035034639
44	1·2453938515	94	1·5981215400	44	1·3892564180	94	2·0185297398
45	1·2516208208	95	1·6061121477	45	1·3996758411	95	2·0336687129
46	1·2578789249	96	1·6141427085	46	1·4101734099	96	2·0489212282
47	1·2641683195	97	1·6222134220	47	1·4207497105	97	2·0642881375
48	1·2704891611	98	1·6303244891	48	1·4314053333	98	2·0797702985
49	1·2768416069	99	1·6384761116	49	1·4421408733	99	2·0953685757
50	1·2832258149	100	1·6466684921	50	1·4529569299	100	2·1110838400

Amount of £1 in n years at the following rates per cent.

Years	1 per cent.	Years	1 per cent.	Years	1¼ per cent.	Years	1¼ per cent.
1	1·01	51	1·6610781401	1	1·0125	51	1·8842851532
2	1·0201	52	1·6776889215	2	1·02515625	52	1·9078387177
3	1·030301	53	1·6944658107	3	1·0379707031	53	1·9316867016
4	1·04060401	54	1·7114104688	4	1·0509453369	54	1·9558327854
5	1·0510100501	55	1·7285245735	5	1·0640821536	55	1·9802806952
6	1·0615201506	56	1·7458098192	6	1·0773831805	56	2·0050342039
7	1·0721353321	57	1·7632679174	7	1·0908504703	57	2·0300971315
8	1·0828567056	58	1·7809005966	8	1·1044861012	58	2·0554733456
9	1·0936852727	59	1·7987096025	9	1·1182921774	59	2·0811667624
10	1·1046221254	60	1·8166966986	10	1·1322708297	60	2·1071813470
11	1·1156683467	61	1·8348636655	11	1·1464242150	61	2·1335211138
12	1·1268250301	62	1·8532123022	12	1·1607545177	62	2·1601901277
13	1·1380932804	63	1·8717444252	13	1·1752639492	63	2·1871925043
14	1·1494742132	64	1·8904618695	14	1·1899547486	64	2·2145324106
15	1·1609689554	65	1·9093664882	15	1·2048291829	65	2·2422140657
16	1·1725786449	66	1·9284601531	16	1·2198895477	66	2·2702417416
17	1·1843044314	67	1·9477447546	17	1·2351381670	67	2·2986197633
18	1·1961474757	68	1·9672222021	18	1·2505773941	68	2·3273525104
19	1·2081089504	69	1·9868944242	19	1·2662096116	69	2·3564444168
20	1·2201900399	70	2·0067633684	20	1·2820372317	70	2·3858999720
21	1.2323919403	71	2·0268310021	21	1·2980626971	71	2·4157237216
22	1·2447158597	72	2·0470993121	22	1·3142884808	72	2·4459202681
23	1·2571630183	73	2·0675703052	23	1·3307170868	73	2·4764942715
24	1·2697346485	74	2·0882460083	24	1·3473510504	74	2·5074504499
25	1·2824319950	75	2·1091284684	25	1·3641929385	75	2·5387935805
26	1.2952563150	76	2·1302197530	26	1·3812453503	76	2·5705285003
27	1·3082088781	77	2·1515219506	27	1·3985109172	77	2·6026601065
28	1·3212909669	78	2·1730371701	28	1·4159923036	78	2·6351933578
29	1·3345038766	79	2·1947675418	29	1·4336922074	79	2·6681332748
30	1·3478489153	80	2·2167152172	30	1·4516133600	80	2·7014849408
31	1·3613274045	81	2·2388823694	31	1·4697585270	81	2·7352535025
32	1·3749406785	82	2·2612711931	32	1·4881305086	82	2·7694441713
33	1·3886900853	83	2·2838839050	33	1·5067321400	83	2·8040622234
34	1·4025769862	84	2·3067227440	34	1·5255662917	84	2·8391130012
35	1·4166027560	85	2·3297899715	35	1·5446358703	85	2·8746019137
36	1·4307687836	86	2·3530878712	36	1·5639438187	86	2·9105344377
37	1·4450764714	87	2·3766187499	37	1·5834931165	87	2·9469161181
38	1·4595272361	88	2·4003849374	38	1·6032867804	88	2·9837525696
39	1·4741225085	89	2·4243887868	39	1·6233278652	89	3·0210494767
40	1·4888637336	90	2·4486326746	40	1·6436194635	90	3·0588125952
41	1·5037523709	91	2·4731190014	41	1·6641647068	91	3·0970477526
42	1·5187898946	92	2·4978501914	42	1·6849667656	92	3·1357608495
43	1·5339777936	93	2·5228286933	43	1·7060288502	93	3·1749578602
44	1·5493175715	94	2·5480569803	44	1·7273542108	94	3·2146448334
45	1·5648107472	95	2·5735375501	45	1·7489461383	95	3·2548278938
46	1·5804588547	96	2·5992729256	46	1·7708079652	96	3·2955132425
47	1·5962634432	97	2·6252656548	47	1·7929430647	97	3·3367071580
48	1·6122260777	98	2·6515183114	48	1·8153548531	98	3·3784159975
49	1·6283483385	99	2·6780334945	49	1·8380467887	99	3·4206461975
50	1·6446318218	100	2·7048138294	50	1·8610223736	100	3·4634042749

TABLE I.   V

Amount of £1 in n years at the following rates per cent.

Years	1½ per cent.	Years	1½ per cent.	Years	1¾ per cent.	Years	1¾ per cent.
1	1·015	51	2·1368210569	1	1·0175	51	2·4224527382
2	1·030225	52	2·1688733728	2	1·03530625	52	2·4648456611
3	1·045678375	53	2·2014064734	3	1·0534241094	53	2·5079804602
4	1·0613635506	54	2·2344275705	4	1·0718590313	54	2·5518701182
5	1·0772840039	55	2·2679439840	5	1·0906165643	55	2·5965278453
6	1·0934432639	56	2·3019631438	6	1·1097023542	56	2·6419670826
7	1·1098449129	57	2·3364925909	7	1·1291221454	57	2·6882015065
8	1·1264925866	58	2·3715399798	8	1·1488817830	58	2·7352450329
9	1·1433899754	59	2·4071130795	9	1·1689872142	59	2·7831118210
10	1·1605408250	60	2·4432197757	10	1·1894444904	60	2·8318162778
11	1·1779489374	61	2·4798680723	11	1·2102597690	61	2·8813730627
12	1·1956181715	62	2·5170660934	12	1·2314393149	62	2·9317970913
13	1·2135524440	63	2·5548220848	13	1·2529895030	63	2·9831035404
14	1·2317557307	64	2·5931444161	14	1·2749168193	64	3·0353078523
15	1·2502320667	65	2·6320415823	15	1·2972278636	65	3·0884257398
16	1·2689855477	66	2·6715222061	16	1·3199293512	66	3·1424731902
17	1·2880023309	67	2·7115950391	17	1·3430281149	67	3·1974664710
18	1·3073406358	68	2·7522689647	18	1·3665311069	68	3·2534221343
19	1·3269507454	69	2·7935529992	19	1·3904454012	69	3·3103570216
20	1·3468550065	70	2·8354562942	20	1·4147781958	70	3·3682882695
21	1·3670578316	71	2·8779881386	21	1·4395368142	71	3·4272333142
22	1·3875636991	72	2·9211579607	22	1·4647287084	72	3·4872098972
23	1·4083771546	73	2·9649753301	23	1·4903614608	73	3·5482360704
24	1·4295028119	74	3·0094499600	24	1·5164427864	74	3·6103302017
25	1·4509453541	75	3·0545917094	25	1·5429805352	75	3·6735109802
26	1·4727095344	76	3·1004105851	26	1·5699826945	76	3·7377974223
27	1·4948001774	77	3·1469167439	27	1·5974573917	77	3·8032088772
28	1·5172221801	78	3·1941204950	28	1·6254128960	78	3·8697650326
29	1·5399805128	79	3·2420323025	29	1·6538576217	79	3·9374859206
30	1·5630802205	80	3·2906627870	30	1·6828001301	80	4·0063919243
31	1·5865264238	81	3·3400227288	31	1·7122491324	81	4·0765037829
32	1·6103243202	82	3·3901230697	32	1·7422134922	82	4·1478425991
33	1·6344791850	83	3·4409749158	33	1·7727022283	83	4·2204298446
34	1·6589963727	84	3·4925895395	34	1·8037245173	84	4·2942873369
35	1·6838813183	85	3·5449783826	35	1·8352896963	85	4·3694373958
36	1·7091395381	86	3·5981530583	36	1·8674072660	86	4·4459025503
37	1·7347766312	87	3·6521253542	37	1·9000868932	87	4·5237058449
38	1·7607982806	88	3·7069072345	38	1·9333384138	88	4·6028706972
39	1·7872102548	89	3·7625108430	39	1·9671718361	89	4·6834209344
40	1·8140184087	90	3·8189485057	40	2·0015973432	90	4·7653808007
41	1·8412286848	91	3·8762327333	41	2·0366252967	91	4·8487749647
42	1·8688471151	92	3·9343762243	42	2·0722662394	92	4·9336285266
43	1·8968798218	93	3·9933918676	43	2·1085308986	93	5·0199670258
44	1·9253330191	94	4·0532927457	44	2·1454301893	94	5·1078164488
45	1·9542130144	95	4·1140921368	45	2·1829752176	95	5·1972032366
46	1·9835262096	96	4·1758035189	46	2·2211772839	96	5·2881542933
47	2·0132791028	97	4·2384405717	47	2·2600478864	97	5·3806969934
48	2·0434782893	98	4·3020171803	48	2·2995987244	98	5·4748591908
49	2·0741304637	99	4·3665474380	49	2·3398417021	99	5·5706692266
50	2·1052424206	100	4·4320456495	50	2·3807889319	100	5·6681559381

Amount of £1 in n years at the following rates per cent.

Years	2 per cent.	Years	2 per cent.	Years	2¼ per cent.	Years	2¼ per cent.
1	1·02	51	2·7454197897	1	1·0225	51	3·1104924437
2	1·0404	52	2·8003281855	2	1·04550625	52	3·1804785237
3	1·061208	53	2·8563347492	3	1·0690301406	53	3·2520392904
4	1·08243216	54	2·9134614441	4	1·0930833188	54	3·3252101745
5	1·1040808032	55	2·9717306730	5	1·1176776935	55	3·4000274034
6	1·1261624193	56	3·0311652865	6	1·1428254416	56	3·4765280200
7	1·1486856676	57	3·0917885922	7	1·1685390140	57	3·5547499004
8	1·1716593810	58	3·1536243641	8	1·1948311418	58	3·6347317732
9	1·1950925686	59	3·2166968513	9	1·2217148425	59	3·7165132381
10	1·2189944200	60	3·2810307884	10	1·2492034265	60	3·8001347859
11	1·2433743084	61	3·3466514041	11	1·2773105036	61	3·8856378186
12	1·2682417946	62	3·4135844322	12	1·3060499899	62	3·9730646695
13	1·2936066305	63	3·4818561209	13	1·3354361147	63	4·0624586246
14	1·3194787631	64	3·5514932433	14	1·3654834272	64	4·1538639437
15	1·3458683383	65	3·6225231081	15	1·3962068044	65	4·2473258824
16	1·3727857051	66	3·6949735703	16	1·4276214575	66	4·3428907148
17	1·4002414192	67	3·7688730417	17	1·4597429402	67	4·4406057558
18	1·4282462476	68	3·8442505025	18	1·4925871564	68	4·5405193853
19	1·4568111725	69	3·9211355126	19	1·5261703674	69	4·6426810715
20	1·4859473960	70	3·9995582229	20	1·5605092007	70	4·7471413956
21	1·5156663439	71	4·0795493873	21	1·5956206577	71	4·8539520770
22	1·5459796708	72	4·1611403751	22	1·6315221225	72	4·9631659988
23	1·5768992642	73	4·2443631826	23	1·6682313703	73	5·0748372337
24	1·6084372495	74	4·3292504462	24	1·7057665761	74	5·1890210715
25	1·6406059945	75	4·4158354551	25	1·7441463240	75	5·3057740456
26	1·6734181144	76	4·5041521642	26	1·7833896163	76	5·4251539616
27	1·7068864766	77	4·5942352075	27	1·8235158827	77	5·5472199258
28	1·7410242062	78	4·6861199117	28	1·8645449901	78	5·6720323741
29	1·7758446903	79	4·7798423099	29	1·9064972523	79	5·7996531025
30	1·8113615841	80	4·8754391561	30	1·9493934405	80	5·9301452973
31	1·8475888158	81	4·9729479392	31	1·9932547929	81	6·0635735665
32	1·8845405921	82	5·0724068980	32	2·0381030258	82	6·2000039717
33	1·9222314039	83	5·1738550360	33	2·0839603439	83	6·3395040611
34	1·9606760320	84	5·2773321367	34	2·1308494516	84	6·4821429025
35	1·9998895527	85	5·3828787794	35	2·1787935042	85	6·6279911178
36	2·0398873437	86	5·4905363550	36	2·2278164194	86	6·7771209179
37	2·0806850906	87	5·6003470821	37	2·2779422889	87	6·9296061386
38	2·1222987924	88	5·7123540237	38	2·3291959904	88	7·0855222767
39	2·1647447683	89	5·8266011042	39	2·3816029002	89	7·2449465279
40	2·2080396636	90	5·9431331263	40	2·4351889654	90	7·4079578248
41	2·2522004569	91	6·0619957888	41	2·4899807171	91	7·5746368759
42	2·2972444660	92	6·1832357046	42	2·5460052833	92	7·7450662056
43	2·3431893553	93	6·3069004187	43	2·6032904022	93	7·9193301952
44	2·3900531425	94	6·4330384271	44	2·6618644362	94	8·0975151246
45	2·4378542053	95	6·5616991956	45	2·7217563860	95	8·2797092149
46	2·4866112894	96	6·6929331795	46	2·7829955069	96	8·4660026722
47	2·5363435152	97	6·8267918431	47	2·8456133126	97	8·6564877324
48	2·5870703855	98	6·9633276800	48	2·9096396121	98	8·8512587063
49	2·6388117932	99	7·1025942336	49	2·9751065934	99	9·0504120272
50	2·6915880291	100	7·2446461183	50	3·0420463997	100	9·2540462979

TABLE I.   vii

Amount of £1 in n years at the following rates per cent.

Years	2¼ per cent.	Years	2½ per cent.	Years	2¾ per cent.	Years	2¾ per cent.
1	1·025	51	3·5230364377	1	1·0275	51	3·9890856203
2	1·050625	52	3·6111123486	2	1·05575625	52	4·0987854749
3	1·076890625	53	3·7013901574	3	1·0847895469	53	4·2115020754
4	1·1038128906	54	3·7939249113	4	1·1146212594	54	4·3273183825
5	1·1314082129	55	3·8887730341	5	1·1452733440	55	4·4463196380
6	1·1596934182	56	3·9859923599	6	1·1767683610	56	4·5685934281
7	1·1886857537	57	4·0856421689	7	1·2091294909	57	4·6942297474
8	1·2184028975	58	4·1877832231	8	1·2423805519	58	4·8233210654
9	1·2488629699	59	4·2924778037	9	1·2765460171	59	4·9559623947
10	1·2800845442	60	4·3997897488	10	1·3116510326	60	5·0922513606
11	1·3120866578	61	4·5097844925	11	1·3477214360	61	5·2322882730
12	1·3448888242	62	4·6225291048	12	1·3847837755	62	5·3761762005
13	1·3785110449	63	4·7380923325	13	1·4228653293	63	5·5240210460
14	1·4129738210	64	4·8565446408	14	1·4619941259	64	5·6759316248
15	1·4482981665	65	4·9779582568	15	1·5021989643	65	5·8320197444
16	1·4845056207	66	5·1024072132	16	1·5435094358	66	5·9924002874
17	1·5216182612	67	5·2299673936	17	1·5859559453	67	6·1571912953
18	1·5596587177	68	5·3607165784	18	1·6295697338	68	6·3265140559
19	1·5986501856	69	5·4947344929	19	1·6743829015	69	6·5004931925
20	1·6386164403	70	5·6321028552	20	1·7204284313	70	6·6792567553
21	1·6795818513	71	5·7729054266	21	1·7677402131	71	6·8629363160
22	1·7215713976	72	5·9172280622	22	1·8163530690	72	7·0516670647
23	1·7646106825	73	6·0651587638	23	1·8663027784	73	7·2455879090
24	1·8087259496	74	6·2167877329	24	1·9176261048	74	7·4448415765
25	1·8539440983	75	6·3722074262	25	1·9703608227	75	7·6495747199
26	1·9002927008	76	6·5315126118	26	2·0245457453	76	7·8599380247
27	1·9478000183	77	6·6948004271	27	2·0802207533	77	8·0760863203
28	1·9964950188	78	6·8621704378	28	2·1374268240	78	8·2981786942
29	2·0464073942	79	7·0332246988	29	2·1962060617	79	8·5263786082
30	2·0975675791	80	7·2095678162	30	2·2566017284	80	8·7608540200
31	2·1500067686	81	7·3898070116	31	2·3186582759	81	9·0017775055
32	2·2037569378	82	7·5745521869	32	2·3824213785	82	9·2493263869
33	2·2588508612	83	7·7639159916	33	2·4479379664	83	9·5036828626
34	2·3153221327	84	7·9580138914	34	2·5152562605	84	9·7650341413
35	2·3732051861	85	8·1569642387	35	2·5844258077	85	10·0335725802
36	2·4325353157	86	8·3608883446	36	2·6554975174	86	10·3094958261
37	2·4933486986	87	8·5699105533	37	2·7285236991	87	10·5930069613
38	2·5556824161	88	8·7841583171	38	2·8035581008	88	10·8843146528
39	2·6195744765	89	9·0037622750	39	2·8806559486	89	11·1836333057
40	2·6850638384	90	9·2288563319	40	2·9598739872	90	11·4911832216
41	2·7521904343	91	9·4595777402	41	3·0412705218	91	11·8071907602
42	2·8209951952	92	9·6960671837	42	3·1249054612	92	12·1318885061
43	2·8915200751	93	9·9384688633	43	3·2108403614	93	12·4655154401
44	2·9638080770	94	10·1869305849	44	3·2991384713	94	12·8083171147
45	3·0379032789	95	10·4416308495	45	3·3898647793	95	13·1605458353
46	3·1138508609	96	10·7026439457	46	3·4830860607	96	13·5224608458
47	3·1916971324	97	10·9702100444	47	3·5788709274	97	13·8943285190
48	3·2714895607	98	11·2444652955	48	3·6772898779	98	14·2764225533
49	3·3532767997	99	11·5255769279	49	3·7784153495	99	14·6690241735
50	3·4371087197	100	11·8137163511	50	3·8823217716	100	15·0724223383

Amount of £1 in n years at the following rates per cent.

Years	3 per cent.	Years	3 per cent.	Years	3¼ per cent.	Years	3½ per cent.
1	·1·03	51	4·5154231993	1	1·035	51	5·7803992956
2	1·0609	52	4·6508858952	2	1·071225	52	5·9827132709
3	1·092727	53	4·7904124721	3	1·108717875	53	6·1921082354
4	·1·12550881	54	4·9341248463	4	1·1475230006	54	6·4088320237
5	1·1592740743	55	5·0821485917	5	1·1876863056	55	6·6331411445
6	1·1940522965	56	5·2346130494	6	1·2292553263	56	6·8653010846.
7	·1·2298738654	57	5·3916514409	7	1·2722792628	57	7·1055866225
8	1·2667700814	58	5·5534009841	8	1·3168090369	58	7·3542821543
9	1·3047731838	59	5·7200030136	9	1·3628973533	59	7·6116820297
10	·1·3439163793	60	5·8916031040	10	1·4105987606	60	7·8780909008
11	·1·3842338707	61	6·0683511972	11	1·4599697172	61	8·1538240823
12	1·4257608868	62	6·2504017331	12	1·5110686573	62	8·4392079252
13	1·4685337135	63	6·4379137851	13	1·5639560604	63	8·7345802026
14	·1·5125897249	64	6·6310511986	14	1·6186945225	64	9·0402905096
15	1·5579674166	65	6·8299827346	15	1·6753488308	65	9·3567006775
16	1·6047064391	66	7·0348822166	16	1·7339860398	66	9·6841852012
17	1·6528476323	67	7·2459286831	17	1·7946755512	67	10·0231316832
18	1·7024330612	68	7·4633065436	18	1·8574891955	68	10·3739412921
19	1·7535060531	69	7·6872057399	19	1·9225013174	69	10·7370292374
20	1·8061112347	70	7·9178219121	20	1·9897888635	70	11·1128252607
21	1·8602945717	71	8·1553565695	21	2·0594314737	71	11·5017741448
22	1·9161034089	72	8·4000172666·	22	2·1315115753	72	11·9043362399
23	1·9735865111	73	8·6520177846	23	2·2061144804	73	12·3209880083
24	2·0327941065	74	8·9115783181	24	2·2833284872	74	12·7522225886
25	2·0937779297	75	9·1789256676	25	2·3632449843	75	13·1985503792
26	2·1565912675	76	9·4542934377	26	2·4459585587	76	13·6604996424
27	2·2212890056	77	9·7379222408	27	2·5315671083	77	14·1386171299
28	2·2879276757	78	10·0300599080	28	2·6201719571	78	14·6334687295
29	2·3565655060	79	10·3309617053	29	2·7118779756	79	15·1456401350
30	2·4272624712	80	10·6408905564	30	2·8067937047	80	15·6757375397
31	2·5000803453	81	10·9601172731	31	2·9050314844	81	16·2243883536
32	2·5750827557	82	11·2889207913	32	3·0067075863	82	16·7922419460
33	2·6523352384	83	11·6275884151	33	3·1119423518	83	17·3799704141
34	2·7319052955	84	11·9764160675	34	3·2208603342	84	17·9882693786
35	2·8138624544	85	12·3357085465	35	3·3335904459	85	18·6178588068
36	2·8982783280	86	12·7057798060	36	3·4502661115	86	19·2694838651
37	2·9852266778	87	13·0869532002	37	3·5710254254	87	19·9439158004
38	3·0747834782	88	13·4795617962	38	3·6960113152	88	20·6419528534
39	3·1670269825	89	13·8839486501	39	3·8253717113	89	21·3644212032
40	3·2620377920	90	14·3004671096	40	3·9592597212	90	22·1121759453
41	3·3598989258	91	14·7294811229	41	4·0978338114	91	22·8861021034
42	3·4606958935	92	15·1713655566	42	4·2412579948	92	23·6871156771
43	3·5645167703	93	15·6265065233	43	4·3897020246	93	24·5161647258
44	3·6714522734	94	16·0953017190	44	4·5433415955	94	25·3742304912
45	3·7815958417	95	16·5781607705	45	4·7023585513	95	26·2623285583
46	4·8950437169	96	17·0755055936	46	4·8669411006	96	27·1815100579
47	4·0118950284	97	17·5877707615	47	5·0372840392	97	28·1328629099
48	4·1322518793	98	18·1154038843	48	5·2135889805	98	29·1175131118
49	4·2562194356	99	18·6588660008	49	5·3960645949	99	30·1366260707
50	4·3839060187	100	19·2186319809	50	5·5849268557	100	31·1914079831

Amount of £1 in n years at the following rates per cent.

Years	4 per cent.	Years	4 per cent.	Years	4½ per cent.	Years	4½ per cent.
1	1·04	51	7·3909506801	1	1·045	51	9·4391049048
2	1·0816	52	7·6865887073	2	1·092025	52	9·8638646255
3	1·124864	53	7·9940522556	3	1·141166125	53	10·3077385337
4	1·16985856	54	8·3138143454	4	1·1925186006	54	10·7715867677
5	1·2166529024	55	8·6463669197	5	1·2461819377	55	11·2563081722
6	1·2653190185	56	8·9922215965	6	1·3022601248	56	11·7628420400
7	1·3159317792	57	9·3519104603	7	1·3608618305	57	12·2921699318
8	1·3685690504	58	9·7259868787	8	1·4221006128	58	12·8453175787
9	1·4233118124	59	10·1150263539	9	1·4860951404	59	13·4233568698
10	1·4802442849	60	10·5196274080	10	1·5529694217	60	14·0274079289
11	1·5394540563	61	10·9404125044	11	1·6228530457	61	14·6586412857
12	1·6010322186	62	11·3780290045	12	1·6958814328	62	15·3182801435
13	1·6650735073	63	11·8331501647	13	1·7721960972	63	16·0076027500
14	1·7316764476	64	12·3064761713	14	1·8519449216	64	16·7279448738
15	1·8009435055	65	12·7987352182	15	1·9352824431	65	17·4807023931
16	1·8729812457	66	13·3106846269	16	2·0223701530	66	18·2673340008
17	1·9479004956	67	13·8431120120	17	2·1133768099	67	19·0893640308
18	2·0258165154	68	14·3968364925	18	2·2084787664	68	19·9483854122
19	2·1068491760	69	14·9727099521	19	2·3078603108	69	20·8460627557
20	2·191231430	70	15·5716183502	20	2·4117140248	70	21·7841355797
21	2·2787680688	71	16·1944830842	21	2·5202411560	71	22·7644216808
22	2·3699187915	72	16·8422624076	22	2·6336520080	72	23·7888206565
23	2·4647155432	73	17·5159529039	23	2·7521663483	73	24·8593175860
24	2·5633041649	74	18·2165910201	24	2·8760138340	74	25·9779868774
25	2·6658363315	75	18·9452546609	25	3·0054344565	75	27·1469962869
26	2·7724697847	76	19·7030648473	26	3·1406790071	76	28·3686111198
27	2·8833685761	77	20·4911874412	27	3·2820095624	77	29·6451986202
28	2·9987033192	78	21·3108349389	28	3·4296999927	78	30·9792325581
29	3·1186514519	79	22·1632683364	29	3·5840364924	79	32·3732980232
30	3·2433975100	80	23·0497990699	30	3·7453181345	80	33·8300964342
31	3·3731334104	81	23·9717910327	31	3·9138574506	81	35·3524507738
32	3·5080587468	82	24·9306626740	32	4·0899810359	82	36·9433110586
33	3·6483810967	83	25·9278891809	33	4·2740301825	83	38·6057600562
34	3·7943163406	84	26·9650047482	34	4·4663615407	84	40·3430192587
35	3·9460889942	85	28·0436049381	35	4·6673478100	85	42·1584551254
36	4·1039325540	86	29·1653491356	36	4·8773784615	86	44·0555856060
37	4·2680898561	87	30·3319631010	37	5·0968604922	87	46·0380869583
38	4·4388134504	88	31·5452416251	38	5·3262192144	88	48·1098008714
39	4·6163659884	89	32·8070512901	39	5·5658990790	89	50·2747419106
40	4·8010206279	90	34·1193333417	40	5·8163645376	90	52·5371052966
41	4·9930614531	91	35·4841066753	41	6·0781009418	91	54·9012750350
42	5·1927839112	92	36·9034709424	42	6·3516154842	92	57·3718324115
43	5·4004952676	93	38·3796097801	43	6·6374381810	93	59·9535648701
44	5·6165150783	94	39·9147941713	44	6·9361228991	94	62·6514752892
45	5·8411756815	95	41·5113859381	45	7·2482484296	95	65·4707916772
46	6·0748227087	96	43·1718413756	46	7·5744196089	96	68·4169773027
47	6·3178156171	97	44·8987150307	47	7·9152684913	97	71·4957412813
48	6·5705282417	98	46·6946636319	48	8·2714555734	98	74·7130496390
49	6·8333493714	99	48·5624501772	49	8·6436710742	99	78·0751368727
50	7·1066833463	100	50·5049481842	50	9·0326362725	100	81·5885180320

b

Amount of £1 in n years at the following rates per cent.

Years	5 per cent.	Years	5 per cent.	Years	5½ per cent.	Years	5½ per cent.
1	1·05	51	12·0407697750	1	1·055	51	15·3417690708
2	1·1025	52	12·6428082638	2	1·113025	52	16·1855663697
3	1·157625	53	13·2749486769	3	1·174241375	53	17·0757725200
4	1·21550625	54	13·9386961108	4	1·2388246506	54	18·0149400086
5	1·2762815625	55	14·6356309164	5	1·3069600064	55	19·0057617091
6	1·3400956406	56	15·3674124622	6	1·3788428068	56	20·0510786031
7	1·4071004227	57	16·1357830853	7	1·4546791611	57	21·1538879262
8	1·4774554438	58	16·9425722396	8	1·5346865150	58	22·3173517622
9	1·5513282160	59	17·7897008516	9	1·6190942733	59	23·5448061091
10	1·6288946268	60	18·6791858941	10	1·7081444584	60	24·8397704451
11	1·7103393581	61	19·6131451888	11	1·8020924036	61	26·2059578196
12	1·7958563260	62	20·5938024483	12	1·9012074858	62	27·6472854996
13	1·8856491423	63	21·6234925707	13	2·0057738975	63	29·1678862021
14	1·9799315994	64	22·7046671992	14	2·1160914618	64	30·7721199432
15	2·0789281794	65	23·8399005592	15	2·2324764922	65	32·4645865401
16	2·1828745884	66	25·0318955872	16	2·3552626993	66	34·2501387998
17	2·2920183178	67	26·2834903665	17	2·4848021478	67	36·1338964338
18	2·4066192337	68	27·5976648848	18	2·6214662659	68	38·1212607377
19	2·5269501954	69	28·9775481291	19	2·7656469105	69	40·2179300782
20	2·6532977051	70	30·4264255355	20	2·9177574906	70	42·4299162325
21	2·7859625904	71	31·9477468123	21	3·0782341526	71	44·7635616253
22	2·9252607199	72	33·5451341529	22	3·2475370310	72	47·2255575147
23	3·0715237559	73	35·2223908606	23	3·4261515677	73	49·8229631780
24	3·2250999437	74	36·9835104036	24	3·6145899039	74	52·5632261528
25	3·3863549409	75	38·8326859238	25	3·8133923486	75	55·4542035912
26	3·5556726879	76	40·7743202199	26	4·0231289278	76	58·5041847888
27	3·7334563223	77	42·8130362310	27	4·2444010188	77	61·7219149521
28	3·9201291385	78	44·9536880425	28	4·4778430749	78	65·1166202745
29	4·1161355954	79	47·2013724446	29	4·7241244440	79	68·6980343896
30	4·3219423752	80	49·5614410669	30	4·9839512884	80	72·4764262810
31	4·5380394939	81	52·0395131202	31	5·2580686093	81	76·4626297265
32	4·7649414686	82	54·6414887762	32	5·5472623828	82	80·6680743614
33	5·0031885420	83	57·3735632150	33	5·8523618138	83	85·1048184513
34	5·2533479691	84	60·2422413758	34	6·1742417136	84	89·7855834661
35	5·5160153676	85	63·2543534446	35	6·5138250078	85	94·7237905568
36	5·7918161360	86	66·4170711168	36	6·8720853833	86	99·9335990374
37	6·0814069428	87	69·7379246726	37	7·2500500793	87	105·4209469845
38	6·3854772899	88	73·2248209063	38	7·6488028337	88	111·2285940686
39	6·7047511544	89	76·8860619516	39	8·0694869895	89	117·3461667424
40	7·0399887121	90	80·7303650492	40	8·5133087740	90	123·8002059132
41	7·3919881477	91	84·7668833016	41	8·9815407565	91	130·6092172384
42	7·7615875551	92	89·0052274667	42	9·4755254982	92	137·7927241865
43	8·1496669329	93	93·4554888400	43	9·9966794005	93	145·3713240168
44	8·5571502795	94	98·1282632820	44	10·5464967676	94	153·3667469377
45	8·9850077935	95	103·0346764461	45	11·1265540898	95	161·8019179138
46	9·4342581832	96	108·1864102685	46	11·7385145647	96	170·7010233991
47	9·9059710923	97	113·5957307819	47	12·3841328658	97	180·0895796860
48	10·4012696469	98	119·2755173210	48	13·0662601734	98	189·9945065687
49	10·9213331293	99	125·2392931870	49	13·7838494830	99	200·4442044300
50	11·4673997858	100	131·5012578464	50	14·5419612045	100	211·4686356737

Amount of £1 in n years at the following rates per cent.

Years	6 per cent.	Years	6 per cent.	Years	7 per cent.	Years	7 per cent.
1	1·06	51	19·5253635315	1	1·07	51	31·5190168175
2	1·1236	52	20·6968853434	2	1·1449	52	33·7253479947
3	1·191016	53	21·9386984640	3	1·225043	53	36·0861223543
4	1·26247696	54	23·2550203718	4	1·31079601	54	38·6121509191
5	1·3382255776	55	24·6503215941	5	1·4025517307	55	41·3150014834
6	1·4185191123	56	26·1293408898	6	1·5007303518	56	44·2070515873
7	1·5036305290	57	27·6971013432	7	1·6057814765	57	47·3015451984
8	1·5938480745	58	29·3589274238	8	1·7181861798	58	50·6126533623
9	1·6894789590	59	31·1204630692	9	1·8384592124	59	54·1555390976
10	1·7908476965	60	32·9876908533	10	1·9671513573	60	57·9464268345
11	1·8982985883	61	34·9669523045	11	2·1048519523	61	62·0026767129
12	2·0121964718	62	37·0649694428	12	2·2521915890	62	66·3428640828
13	2·1329282601	63	39·2888676094	13	2·4098450002	63	70·9868645686
14	2·2609039558	64	41·6461996659	14	2·5785341502	64	75·9559450884
15	2·3965581931	65	44·1449716459	15	2·7590315407	65	81·2728612446
16	2·5403516847	66	46·7936699447	16	2·9521637486	66	86·9619615317
17	2·6927727858	67	49·6012901413	17	3·1588152110	67	93·0492988389
18	2·8543391529	68	52·5773675498	18	3·3799322757	68	99·5627497577
19	3·0255995021	69	55·7320096028	19	3·6165275350	69	106·5321422407
20	3·2071354722	70	59·0759301790	20	3·8696844625	70	113·9893921975
21	3·3995636005	71	62·6204859897	21	4·1405623749	71	121·9686496514
22	3·6035374166	72	66·3777151491	22	4·4304017411	72	130·5064551270
23	3·8197496616	73	70·3603780580	23	4·7405298630	73	139·6419069858
24	4·0489346413	74	74·5820007415	24	5·0723669534	74	149·4168404749
25	4·2918707197	75	79·0569207860	25	5·4274326401	75	159·8760193081
26	4·5493829629	76	83·8003360332	26	5·8073529249	76	171·0673406597
27	4·8223459407	77	88·8283561952	27	6·2138676297	77	183·0420545058
28	5·1116866971	78	94·1580575669	28	6·6488383637	78	195·8549983212
29	5·4183878990	79	99·8075410209	29	7·1142570492	79	209·5648482037
30	5·7434911729	80	105·7959934821	30	7·6122550427	80	224·2343874780
31	6·0881006433	81	112·1437530911	31	8·1451128956	81	239·9307947085
32	6·4533866819	82	118·8723782765	32	8·7152707983	82	256·7259503380
33	6·8405898828	83	126·0047209731	33	9·3253397542	83	274·6967668617
34	7·2510252758	84	133·5650042315	34	9·9781135370	84	293·9255405420
35	7·6860867923	85	141·5790244854	35	10·6765814846	85	314·5003283799
36	8·1472519999	86	150·0736387545	36	11·4239421885	86	336·5153513666
37	8·6360871198	87	159·0780570798	37	12·2236181417	87	360·0714259622
38	9·1542523470	88	168·6227405046	38	13·0792714116	88	385·2764257796
39	9·7035074879	89	178·7401049348	39	13·9948204105	89	412·2457755842
40	10·2857179371	90	189·4645112309	40	14·9744578392	90	441·1029798750
41	10·9028610134	91	200·8323819048	41	16·0226698879	91	471·9801884663
42	11·5570326742	92	212·8823248191	42	17·1442567801	92	505·0188016589
43	12·2504546346	93	225·6552643082	43	18·3443547547	93	540·3701177751
44	12·9854819127	94	239·1945801667	44	19·6284595875	94	578·1960260193
45	13·7646108274	95	253·5462549767	45	21·0024517587	95	618·6697478407
46	14·5904874771	96	268·7590302753	46	22·4726233818	96	661·9766301895
47	15·4659167257	97	284·8845720918	47	24·0457070185	97	708·3149943028
48	16·3938717293	98	301·9776464174	48	25·7289065098	98	757·8970439040
49	17·3775040330	99	320·0963052024	49	27·5299299655	99	810·9498369773
50	18·4210542750	100	339·3020835145	50	29·4570250630	100	867·7163255657

Amount of £1 in n years at the following rates per cent.

Years	8 per cent.	Years	8 per cent.	Years	9 per cent.	Years	9 per cent.
1	1·08	51	50·6537415143	1	1·09	51	81·0496968827
2	1·1664	52	54·7060408355	2	1·1881	52	88·3441696021
3	1·259712	53	59·0825241023	3	1·295029	53	96·2951448663
4	1·36048896	54	63·8091260305	4	1·41158161	54	104·9617079043
5	1·4693280768	55	68·9138561129	5	1·5386239549	55	114·4082616157
6	1·5868743229	56	74·4269646020	6	1·6771001108	56	124·7050051611
7	1·7138242688	57	80·3811217701	7	1·8280391208	57	135·9284556256
8	1·8509302103	58	86·8116115117	8	1·9925626417	58	148·1620165319
9	1·9990046271	59	93·7565404327	9	2·1718932794	59	161·4965981287
10	2·1589249973	60	101·2570636673	10	2·3673636746	60	176·0312919603
11	2·3316389971	61	109·3576287606	11	2·5804264053	61	191·8741082367
12	2·5181701168	62	118·1062390615	12	2·8126647818	62	209·1427779780
13	2·7196237262	63	127·5547381864	13	3·0658046121	63	227·9656279961
14	2·9371936243	64	137·7599172413	14	3·3417270272	64	248·4825345157
15	3·1721691142	65	148·7798466206	15	3·6424824597	65	270·8459626221
16	3·4259426433	66	160·6822343503	16	3·9703058811	66	295·2220992581
17	3·7000180548	67	173·5368130983	17	4·3276334104	67	321·7920881913
18	3·9960194992	68	187·4197581462	18	4·7171204173	68	350·7533761286
19	4·3157010591	69	202·4133387979	19	5·1416612548	69	382·3211799801
20	4·6609571439	70	218·6064059017	20	5·6044107678	70	416·7300861784
21	5·0338337154	71	236·0949183738	21	6·1088077369	71	454·2357939344
22	5·4365404126	72	254·9825118437	22	6·6586004332	72	495·1170153885
23	5·8714636456	73	275·3811127912	23	7·2578744722	73	539·6775467735
24	6·3411807372	74	297·4116018145	24	7·9110831747	74	588·2485259831
25	6·8484751962	75	321·2045299597	25	8·6230806604	75	641·1908933216
26	7·3963532119	76	346·9008923565	26	9·3991579198	76	698·8980737205
27	7·9880614689	77	374·6529637450	27	10·2450821326	77	761·7989003553
28	8·6271063864	78	404·6252008446	28	11·1617395246	78	830·3608013873
29	9·3172748973	79	436·9952169122	29	12·1721820818	79	905·0932735122
30	10·0626568891	80	471·9548342651	30	13·2676784691	80	986·5516681283
31	10·8676694402	81	509·7112210063	31	14·4617695314	81	1075·3413182598
32	11·7370829954	82	550·4881186869	32	15·7633287892	82	1172·1220369032
33	12·6760496351	83	594·5271681818	33	17·1820283802	83	1277·6130202245
34	13·6901336059	84	642·0893416363	34	18·7284109344	84	1392·5981920447
35	14·7853442943	85	693·4564889673	35	20·4139679185	85	1517·9320293287
36	15·9681718379	86	748·9330080846	36	22·2512250312	86	1654·3450119683
37	17·2456255849	87	808·8476487314	37	24·2538352840	87	1803·5450440455
38	18·6252756317	88	873·5554606299	38	26·4366804596	·88	1965·7659980095
39	20·1152976822	89	943·4398974803	39	28·8159817009	89	2142·6849378304
40	21·7245214968	90	1018·9150892787	40	31·4094200540	90	2335·5265822351
41	23·4624832166	91	1100·4282964210	41	34·2362678589	91	2545·7239746463
42	25·3394818739	92	1188·4625601347	42	37·3175319662	92	2774·8391323536
43	27·3666404238	93	1283·5395649455	43	40·6761098431	93	3024·5746542654
44	29·5559716577	94	1386·2227301411	44	44·3369597290	94	3296·7863731493
45	31·9204493903	95	1497·1205485524	45	48·3272861046	95	3593·4971467327
46	34·4740853415	96	1616·8901924366	46	52·6767418540	96	3916·9118899387
47	37·2320121629	97	1746·2414078316	47	57·4176486209	97	4269·4339600331
48	40·2105731424	98	1885·9407204581	48	62·5852369968	98	4653·6830164361
49	43·4274189938	99	2036·8159780947	49	68·2179083265	99	5072·5144879154
50	46·9016125133	100	2199·7612563423	50	74·3575200758	100	5529·0407918277

TABLE I.   xiii

Amount of £1 in n years at the following rates per cent.

Years	10 per cent.	Years	10 per cent.	Years	11 per cent.	Years	11 per cent.
1	1·10	51	129·129938	1	1·11	51	204·866958
2	1·21	52	142·042932	2	1·2321	52	227·402323
3	1·331	53	156·247225	3	1·367361	53	252·416579
4	1·4641	54	171·871948	4	1·518070	54	280·182402
5	1·61051	55	189·059142	5	1·685058	55	311·002466
6	1·771561	56	207·965057	6	1·870415	56	345·212738
7	1·948717	57	228·761562	7	2·076160	57	383·186139
8	2·143589	58	251·637719	8	2·304538	58	425·336614
9	2·357948	59	276·801490	9	2·558037	59	472·123642
10	2·593742	60	304·481640	10	2·839421	60	524·057242
11	2·853117	61	334·929803	11	3·151757	61	581·703539
12	3·138428	62	368·422784	12	3·498851	62	645·690928
13	3·452271	63	405·265062	13	3·883280	63	716·716930
14	3·797498	64	445·791568	14	4·310441	64	795·555793
15	4·177248	65	490·370725	15	4·784589	65	883·066930
16	4·594973	66	539·407798	16	5·310894	66	980·204292
17	5·054470	67	593·348578	17	5·895093	67	1088·026764
18	5·559917	68	652·683435	18	6·543553	68	1207·709708
19	6·115909	69	717·951779	19	7·263344	69	1340·557776
20	6·727500	70	789·746957	20	8·062312	70	1488·019132
21	7·400250	71	868·721652	21	8·949166	71	1651·701236
22	8·140275	72	955·593818	22	9·933574	72	1833·388372
23	8·954302	73	1051·153200	23	11·026267	73	2035·061093
24	9·849733	74	1156·268519	24	12·239157	74	2258·917813
25	10·834706	75	1271·895371	25	13·585464	75	2507·398773
26	11·918177	76	1399·084909	26	15·079865	76	2783·212638
27	13·109994	77	1538·993399	27	16·738650	77	3089·366028
28	14·420994	78	1692·892739	28	18·579901	78	3429·196291
29	15·863093	79	1862·182013	29	20·623691	79	3806·407883
30	17·449402	80	2048·400215	30	22·892297	80	4225·112750
31	19·194342	81	2253·240236	31	25·410449	81	4689·875153
32	21·113777	82	2478·564260	32	28·205599	82	5205·761420
33	23·225154	83	2726·420686	33	31·308214	83	5778·395176
34	25·547670	84	2999·062754	34	34·752118	84	6414·018645
35	28·102437	85	3298·969030	35	38·574851	85	7119·560696
36	30·912681	86	3628·865933	36	42·818085	86	7902·712373
37	34·003949	87	3991·752526	37	47·528074	87	8772·010734
38	37·404343	88	4390·927778	38	52·756162	88	9736·931915
39	41·144778	89	4830·020556	39	58·559340	89	10807·994425
40	45·259256	90	5313·022612	40	65·000867	90	11996·873812
41	49·785181	91	5844·324873	41	·72·150963	91	13316·529932
42	54·763699	92	6428·757360	42	80·087569	92	14781·348224
43	60·240069	93	7071·633096	43	88·897201	93	16407·296529
44	66·264076	94	7778·796406	44	98·675893	94	18212·099147
45	72·890484	95	8556·676047	45	109·530242	95	20215·430053
46	80·179532	96	9412·343651	46	121·578568	96	22439·127359
47	88·197485	97	10353·578016	47	134·952211	97	24907·431368
48	97·017234	98	11388·935818	48	149·796954	98	27647·248819
49	106·718957	99	12527·829400	49	166·274619	99	30688·446189
50	117·390853	100	13780·612340	50	184·564827	100	34064·175270

Amount of £1 in n years at the following rates per cent.

Years	12 per cent.	Years	12 per cent.	Years	13 per cent.	Years	13 per cent.
1	1·12	51	323·682453	1	1·13	51	509·331595
2	1·2544	52	362·524347	2	1·2769	52	575·544703
3	1·404928	53	406·027269	3	1·442897	53	650·365514
4	1·573519	54	454·750541	4	1·630474	54	734·913031
5	1·762342	55	509·320606	5	1·842435	55	830·451725
6	1·973823	56	570·439078	6	2·081952	56	938·410449
7	2·210681	57	638·891768	7	2·352605	57	1060·403808
8	2·475963	58	715·558780	8	2·658444	58	1198·256303
9	2·773079	59	801·425833	9	3·004042	59	1354·029622
10	3·105848	60	897·596933	10	3·394567	60	1530·053473
11	3·478550	61	1005·308566	11	3·835861	61	1728·960425
12	3·895976	62	1125·945593	12	4·334523	62	1953·725280
13	4·363493	63	1261·059065	13	4·898011	63	2207·709566
14	4·887112	64	1412·386152	14	5·534753	64	2494·711810
15	5·473566	65	1581·872491	15	6·254270	65	2819·024345
16	6·130394	66	1771·697189	16	7·067326	66	3185·497510
17	6·866041	67	1984·300852	17	7·986078	67	3599·612186
18	7·689966	68	2222·416954	18	9·024268	68	4067·561770
19	8·612762	69	2489·106989	19	10·197423	69	4596·344800
20	9·646293	70	2787·799828	20	11·523088	70	5193·869624
21	10·803848	71	3122·335807	21	13·021089	71	5869·072675
22	12·100310	72	3497·016104	22	14·713831	72	6632·052123
23	13·552347	73	3916·658036	23	16·626629	73	7494·218899
24	15·178629	74	4386·657001	24	18·788091	74	8468·467356
25	17·000064	75	4913·055841	25	21·230542	75	9569·368112
26	19·040072	76	5502·622542	26	23·990513	76	10813·385967
27	21·324881	77	6162·937247	27	27·109279	77	12219·126143
28	23·883866	78	6902·489716	28	30·633486	78	13807·612541
29	26·749930	79	7730·788482	29	34·615839	79	15602·602172
30	29·959922	80	8658·483100	30	39·115898	80	17630·940454
31	33·555113	81	9697·501072	31	44·200965	81	19922·962713
32	37·581726	82	10861·201201	32	49·947090	82	22512·947866
33	42·091533	83	12164·545345	33	56·440212	83	25439·631089
34	47·142517	84	13624·290786	34	63·777439	84	28746·783130
35	52·799620	85	15259·205681	35	72·068506	85	32483·864937
36	59·135574	86	17090·310362	36	81·437412	86	36706·767379
37	66·231843	87	19141·147606	37	92·024276	87	41478·647138
38	74·179664	88	21438·085318	38	104·987432	88	46870·871266
39	83·081224	89	24010·655557	39	117·505798	89	52964·084530
40	93·050970	90	26891·934223	40	132·781552	90	59849·415520
41	104·217087	91	30118·966330	41	150·043153	91	67629·839537
42	116·723137	92	33733·242290	42	169·548763	92	76421·718677
43	130·729914	93	37781·231365	43	191·590103	93	86356·542105
44	146·417503	94	42314·979128	44	216·496816	94	97582·892578
45	163·987604	95	47392·776624	45	244·641402	95	110268·668614
46	183·666116	96	53079·909819	46	276·444784	96	124603·595533
47	205·706050	97	59449·498997	47	312·382606	97	140802·062953
48	230·390776	98	66583·438876	48	352·992345	98	159106·331137
49	258·037669	99	74573·451542	49	398·881350	99	179790·154184
50	289·002190	100	83522·265727	50	450·735925	100	203162·874228

Amount of £1 in n years at the following rates per cent.

Years	14 per cent.	Years	14 per cent.	Years	15 per cent.	Years	15 per cent.
1	1·14	51	798·265607	1	1·15	51	1246·206058
2	1·2996	52	910·022792	2	1·3225	52	1433·136966
3	1·481544	53	1037·425983	3	1·520875	53	1648·107511
4	1·688960	54	1182·665620	4	1·749006	54	1895·323638
5	1·925415	55	1348·238807	5	2·011357	55	2179·622184
6	2·194973	56	1536·992240	6	2·313061	56	2506·565512
7	2·502269	57	1752·171154	7	2·660020	57	2882·550338
8	2·852586	58	1997·475115	8	3·059023	58	3314·932889
9	3·251949	59	2277·121631	9	3·517876	59	3812·172822
10	3·707221	60	2595·918660	10	4·045558	60	4383·998746
11	4·226232	61	2959·347272	11	4·652391	61	5041·598558
12	4·817905	62	3373·655890	12	5·350250	62	5797·838341
13	5·492411	63	3845·967715	13	6·152788	63	6667·514092
14	6·261349	64	4384·403195	14	7·075706	64	7667·641206
15	7·137938	65	4998·219642	15	8·137062	65	8817·787387
16	8·137249	66	5697·970392	16	9·357621	66	10140·455495
17	9·276464	67	6495·686247	17	10·761264	67	11661·523819
18	10·575169	68	7405·082321	18	12·375454	68	13410·752392
19	12·055693	69	8441·793846	19	14·231772	69	15422·365251
20	13·743490	70	9623·644985	20	16·366537	70	17735·720039
21	15·667578	71	10970·955283	21	18·821518	71	20396·078045
22	17·861039	72	12506·889022	22	21·644746	72	23455·489751
23	20·361585	73	14257·853485	23	24·891458	73	26973·813214
24	23·212207	74	16253·952973	24	28·625176	74	31019·885196
25	26·461916	75	18529·506390	25	32·918953	75	35672·867976
26	30·166584	76	21123·637284	26	37.856796	76	41023·798172
27	34·389906	77	24080·946504	27	43·535315	77	47177·367898
28	39·204493	78	27452·279015	28	50·065612	78	54253·973082
29	44·693122	79	31295·598077	29	57·575454	79	62392·069045
30	50·950159	80	35676·981807	30	66·211772	80	71750·979401
31	58·083181	81	40671·759260	31	76·143538	81	82513·511312
32	66·214826	82	46365·805557	32	87·562068	82	94890·538008
33	75·484902	83	52857·018335	33	100·699829	83	109124·118710
34	86·052788	84	60257·000902	34	115·804803	84	125492·736516
35	98·100178	85	68692·981028	35	133·175523	85	144316·646994
36	111·834203	86	78309·998372	36	153·151852	86	165964·144043
37	127·490992	87	89273·398144	37	176·124630	87	190858·765649
38	145·339731	88	101771·673884	38	202·543324	88	219487·580496
39	165·687293	89	116019·708227	39	232·924823	89	252410·717571
40	188·883514	90	132262·467379	40	267·863546	90	290272·325206
41	215·327206	91	150779·212812	41	308·043078	91	333813·173987
42	245·473015	92	171888·302606	42	354·249540	92	383885·150085
43	279·839237	93	195952·664971	43	407·386971	93	441467·922598
44	319·016730	94	223386·038067	44	468·495017	94	507688·110988
45	363·679072	95	254660·083396	45	538·769269	95	583841·327636
46	414·594142	96	290312·495072	46	619·584659	96	671417·526781
47	472·637322	97	330956·244382	47	712·522358	97	772130·155799
48	538·806547	98	377290·118595	48	819·400712	98	887949·679168
49	614·239464	99	430110·735199	49	942·310819	99	1021142·131044
50	700·232988	100	490326·238126	50	1083·657442	100	1174313·450700

Amount of £1 in n years at the following rates per cent.

Years	16 per cent.	Years	16 per cent.	Years	17 per cent.	Years	17 per cent.
1	1·16	51	1938·01641	1	1·17	51	3002·47188
2	1·3456	52	2248·09904	2	1·3689	52	3512·89210
3	1·56090	53	2607·79488	3	1·60161	53	4110·08376
4	1·81064	54	3025·04207	4	1·87388	54	4808·79800
5	2·10034	55	3509·04880	5	2·19244	55	5626·29366
6	2·43640	56	4070·49660	6	2·56516	56	6582·76358
7	2·82621	57	4721·77606	7	3·00124	57	7701·83339
8	3·27841	58	5477·26023	8	3·51145	58	9011·14507
9	3·80296	59	6353·62187	9	4·10840	59	10543·03973
10	4·41144	60	7370·20137	10	4·80682	60	12335·35648
11	5·11726	61	8549·43358	11	5·62399	61	14432·36708
12	5·93603	62	9917·34296	12	6·58007	62	16885·86949
13	6·88579	63	11504·11783	13	7·69868	63	19756·46730
14	7·98752	64	13344·77668	14	9·00745	64	23115·06674
15	9·26552	65	15479·94095	15	10·53872	65	27044·62809
16	10·74800	66	17956·73150	16	12·33030	66	31642·21486
17	12·46768	67	20829·80855	17	14·42646	67	37021·39139
18	14·46251	68	24162·57791	18	16·87895	68	43315·02793
19	16·77652	69	28028·59038	19	19·74838	69	50678·58267
20	19·46076	70	32513·16484	20	23·10560	90	59293·94173
21	22·57448	71	37715·27121	21	27·03355	71	69373·91182
22	26·18640	72	43749·71461	22	31·62925	72	81167·47683
23	30·37622	73	50749·66895	23	37·00623	73	94965·94789
24	35·23642	74	58869·61598	24	43·29729	74	111110·15904
25	40·87424	75	68288·75453	25	50·65783	75	129998·88607
26	47·41412	76	79214·95526	26	59·26966	76	152098·69670
27	55·00038	77	91889·34810	27	69·34550	77	177955·47514
28	63·80044	78	106591·64379	28	81·13423	78	208207·90592
29	74·00851	79	123646·30680	29	94·92705	79	243603·24992
30	85·84988	80	143429·71589	30	111·06465	80	285015·80241
31	99·58586	81	166378·47043	31	129·94564	81	333468·48882
32	115·51959	82	192999·02570	32	152·03640	82	390158·13192
33	134·00273	83	223878·86981	33	177·88259	83	456485·01435
34	155·44317	84	259699·48898	34	208·12263	84	534087·46679
35	180·31407	85	301251·40722	35	243·50347	85	624882·33614
36	209·16432	86	349451·63238	36	284·89906	86	731112·33329
37	242·63062	87	405363·89356	37	333·33191	87	855401·42994
38	281·45151	88	470222·11653	38	389·99833	88	1000819·67303
39	326·48376	89	545457·65517	39	456·29805	89	1170959·01745
40	378·72116	90	632730·88000	40	533·86871	90	1370022·05042
41	439·31654	91	733967·82080	41	624·62639	91	1602925·79899
42	509·60719	92	851402·67213	42	730·81288	92	1875423·18482
43	591·14434	93	987627·09967	43	855·05107	93	2194245·22623
44	685·72744	94	1145647·43561	44	1000·40975	94	2567266·79769
45	795·44383	95	1328951·02531	45	1170·47941	95	3003702·15330
46	922·71484	96	1541583·18936	46	1369·46091	96	3514331·51936
47	1070·34921	97	1788236·49966	47	1602·26927	97	4111767·87766
48	1241·60509	98	2074354·33961	48	1874·65504	98	4810768·41686
49	1440·26190	99	2406251·03394	49	2193·34640	99	5628599·04772
50	1670·70380	100	2791251·19938	50	2566·21528	100	6585460·88584

Amount of £1 in n years at the following rates per cent.

Years	18 per cent.	Years	18 per cent.	Years	19 per cent.	Years	19 per cent.
1	1·18	51	4634·28109	1	1·19	51	7126·80754
2	1·3924	52	5468·45169	2	1·4161	52	8480·90098
3	1·64303	53	6452·77300	3	1·68516	53	10092·27216
4	1·93878	54	7614·27214	4	2·00534	54	12009·80387
5	2·28776	55	8984·84112	5	2·38635	55	14291·66661
6	2·69955	56	10602·11252	6	2·83976	56	17007·08327
7	3·18547	57	12510·49278	7	3·37932	57	20238·42909
8	3·75886	58	14762·38148	8	4·02139	58	24083·73061
9	4·43545	59	17419·61014	9	4·78545	59	28659·63943
10	5·23384	60	20555·13997	10	5·69468	60	34104·97092
11	6·17593	61	24255·06516	11	6·77667	61	40584·91539
12	7·28759	62	28620·97689	12	8·06424	62	48296·04932
13	8·59936	63	33772·75273	13	9·59645	63	57472·29869
14	10·14724	64	39851·84822	14	11·41977	64	68392·03544
15	11·97375	65	47025·18090	15	13·58953	65	81386·52217
16	14·12902	66	55489·71346	16	16·17154	66	96849·96139
17	16·67225	67	65477·86188	17	19·24413	67	115251·45405
18	19·67325	68	77263·87702	18	22·90052	68	137149·23032
19	23·21444	69	91171·37489	19	27·25162	69	163207·58408
20	27·39303	70	107582·22237	20	32·42942	70	194217·02506
21	32·32378	71	126947·02239	21	38·59101	71	231118·25982
22	38·14206	72	149797·48643	22	45·92331	72	275030·72918
23	45·00763	73	176761·03398	23	54·64873	73	327286·56773
24	53·10901	74	208578·02010	24	65·03199	74	389471·01560
25	62·66863	75	246122·06372	25	77·38807	75	463470·50856
26	73·94898	76	290424·03518	26	92·09181	76	551529·90518
27	87·25980	77	342700·36152	27	109·58925	77	656320·58717
28	102·96656	78	404386·42659	28	130·41121	78	781021·49873
29	121·50054	79	477175·98338	29	155·18934	79	929415·58349
30	143·37064	80	563067·66039	30	184·67531	80	1106004·54435
31	169·17735	81	664419·83926	31	219·76362	81	1316145·40778
32	199·62928	82	784015·41032	32	261·51871	82	1566213·03526
33	235·56255	83	925138·18418	33	311·20726	83	1863793·51196
34	277·96381	84	1091663·05733	34	370·33664	84	2217914·27923
35	327·99729	85	1288162·40765	35	440·70061	85	2639317·99228
36	387·03680	86	1520031·64103	36	524·43372	86	3140788·41082
37	456·70343	87	1793637·33641	37	624·07613	87	3737538·20887
38	538·91004	88	2116492·05697	38	742·65059	88	4447670·46856
39	635·91385	89	2497460·62722	39	883·75421	89	5292727·85759
40	750·37834	90	2947003·54012	40	1051·66751	90	6298346·15053
41	885·44645	91	3477464·17734	41	1251·48433	91	7495031·91913
42	1044·82681	92	4103407·72926	42	1489·26636	92	8919087·98376
43	1232·89563	93	4842021·12053	43	1772·22696	93	10613714·70068
44	1454·81685	94	5713584·92223	44	2108·95009	94	12630320·49381
45	1716·68388	95	6742030·20823	45	2509·65060	95	15030081·38763
46	2025·68698	96	7955595·64571	46	2986·48422	96	17885796·85128
47	2390·31063	97	9387602·86194	47	3553·91622	97	21284098·25302
48	2820·56655	98	11077371·37708	48	4229·16030	98	25328076·92110
49	3328·26853	99	13071298·22496	49	5032·70076	90	30140411·53611
50	3927·35686	100	15424131·90545	50	5988·91390	100	35867089·72797

c

Amount of £1 in n years at the following rates per cent.

Years	20 per cent.	Years	20 per cent.	Years	21 per cent.	Years	21 per cent.
1	1·20	51	10920·52578	1	1·21	51	16674·54093
2	1·440	52	13104·63094	2	1·4641	52	20176·19453
3	1·7280	53	15725·55712	3	1·77156	53	24413·19538
4	2·07360	54	18870·66855	4	2·14359	54	29539·96641
5	2·48832	55	22644·80226	5	2·59374	55	35743·35935
6	2·98598	56	27173·76271	6	3·13843	56	43249·46482
7	3·58318	57	32608·51525	7	3·79750	57	52331·85243
8	4·29982	58	39130·21830	8	4·59497	58	63321·54144
9	5·15978	59	46956·26196	9	5·55992	59	76619·06514
10	6·19174	60	56347·51435	10	6·72750	60	92709·06882
11	7·43008	61	67617·01722	11	8·14027	61	112177·97327
12	8·91610	62	81140·42067	12	9·84973	62	135735·34766
13	10·69932	63	97368·50480	13	11·91818	63	164239·77066
14	12·83918	64	116842·20576	14	14·42099	64	198730·12250
15	15·40702	65	140210·64692	15	17·44940	65	240463·44823
16	18·48843	66	168252·77630	16	21·11378	66	290960·77236
17	22·18611	67	201903·33156	17	25·54767	67	352062·53455
18	26·62333	68	242283·99787	18	30·91268	68	425995·66681
19	31·94800	69	290740·79744	19	37·40434	69	515454·75684
20	38·33760	70	348888·95693	20	45·25926	70	623700·25577
21	46·00512	71	418666·74832	21	54·76370	71	754677·30949
22	55·20614	72	502400·09798	22	66·26408	72	913159·54448
23	66·24737	73	602880·11758	23	80·17953	73	1104923·04882
24	79·49685	74	723456·14109	24	97·01723	74	1336956·88907
25	95·39622	75	868147·36931	25	117·39085	75	1617717·83578
26	114·47546	76	1041776·84318	26	142·04293	76	1957438·58129
27	137·37055	77	1250132·21181	27	171·87195	77	2368500·68336
28	164·84466	78	1500158·65417	28	207·96506	78	2865885·82686
29	197·81359	79	1800190·38501	29	251·63772	79	3467721·85051
30	237·37631	80	2160228·46201	30	304·48164	80	4195943·43911
31	284·85158	81	2592274·15441	31	368·42278	81	5077091·56133
32	341·82189	82	3110728·98529	32	445·79157	82	6143280·78921
33	410·18627	83	3732874·78235	33	539·40780	83	7433369·75494
34	492·22352	84	4479449·73882	34	652·68344	84	8994377·40348
35	590·66823	85	5375339·68659	35	789·74696	85	10883196·65821
36	708·80187	86	6450407·62391	36	955·59382	86	13168667·95643
37	850·56225	87	7740489·14869	37	1156·26852	87	15934088·22728
38	1020·67470	88	9288586·97843	38	1399·08491	88	19280246·75501
39	1224·80964	89	11146304·37411	39	1692·89274	89	23329098·57356
40	1469·77157	90	13375565·24893	40	2048·40021	90	28228209·27401
41	1763·72588	91	16050678·29872	41	2478·56426	91	34156133·22154
42	2116·47106	92	19260813·95847	42	2999·06275	92	41328921·19807
43	2539·76527	93	23112976·75016	43	3628·86593	93	50007994·64967
44	3047·71832	94	27735572·10019	44	4390·92778	94	60509673·52610
45	3657·26199	95	33282686·52023	45	5313·02261	95	73216704·96658
46	4388·71439	96	39939223·82427	46	6428·75736	96	88592213·00956
47	5266·45726	97	47927068·58913	47	7778·79641	97	107196577·74156
48	6319·74872	98	57512482·30695	48	9412·34365	98	129707859·06729
49	7583·69846	99	69014978·76834	49	11388·93582	99	156946509·47142
50	9100·43815	100	82817974·52201	50	13780·61234	100	189905276·46042

Amount of £1 in n years at the following rates per cent.

Years	22 per cent.	Years	22 per cent.	Years	23 per cent.	Years	23 per cent.
1	1·22	51	25371·80497	1	1·23	51	38473·41024
2	1·4884	52	30953·60207	2	1·5129	52	47322·29460
3	1·81585	53	37763·39452	3	1·86087	53	58206·42235
4	2·21553	54	46071·34132	4	2·28887	54	71593·89950
5	2·70271	55	56207·03641	5	2·81531	55	88060·49638
6	3·29730	56	68572·58441	6	3·46283	56	108314·41055
7	4·02271	57	83658·55299	7	4·25928	57	133226·72497
8	4·90771	58	102060·43464	8	5·23891	58	163868·87172
9	5·98740	59	124517·39026	9	6·44386	59	201558·71221
10	7·30463	60	151911·21612	10	7·92595	60	247917·21602
11	8·91165	61	185331·68367	11	9·74891	61	304938·17571
12	10·87221	62	226104·65408	12	11·99116	62	375073·95612
13	13·26410	63	275847·67797	13	14·74913	63	461340·96603
14	16·18220	64	336534·16713	14	18·14143	64	567449·38821
15	19·74229	65	410571·68369	15	22·31396	65	697962·74750
16	24·08559	66	500897·45435	16	27·44617	66	858494·17943
17	29·38442	67	611094·89431	17	33·75859	67	1055947·84070
18	35·84899	68	745535·77105	18	41·52331	68	1298815·84406
19	43·73577	69	909553·64069	19	51·07368	69	1597543·48819
20	53·35764	70	1109655·44164	20	62·82062	70	1964978·49048
21	65·09632	71	1353779·63880	21	77·26936	71	2416923·54329
22	79·41751	72	1651611·15933	22	95·04132	72	2972815·95824
23	96·88936	73	2014965·61439	23	116·90082	73	3656563·62864
24	118·20502	74	2458258·04955	24	143·78801	74	4497573·26322
25	144·21013	75	2999074·82045	25	176·85925	75	5532015·11376
26	175·93636	76	3658871·28095	26	217·53688	76	6804378·58993
27	214·64236	77	4463822·96276	27	267·57036	77	8369385·66561
28	261·86368	78	5445864·01457	28	329·11155	78	10294344·36870
29	319·47368	79	6643954·09778	29	404·80720	79	12662043·59351
30	389·75789	80	8105623·99929	30	497·91286	80	15574313·59541
31	475·50463	81	9888861·27913	31	612·43282	81	19156405·72236
32	580·11565	82	12064410·76054	32	753·29237	82	23562379·03850
33	707·74109	83	14718581·12786	33	926·54961	83	28981726·21735
34	803·44413	84	17956668·97599	34	1139·65602	84	35647523·24734
35	1053·40184	85	21907136·15071	35	1401·77690	85	43846453·59423
36	1285·15025	86	26726706·10386	36	1724·18559	86	53931137·92091
37	1567·88330	87	32606581·44671	37	2120·74828	87	66335299·64272
38	1912·81763	88	39780029·36499	38	2608·52038	88	81592418·56054
39	2333·63751	89	48531635·82529	39	3208·48007	89	100358674·82947
40	2847·03776	90	59208595·70685	40	3946·43049	90	123441170·04024
41	3473·38607	91	72234486·76236	41	4854·10950	91	151832639·14950
42	4237·53100	92	88126073·85007	42	5970·55469	92	186754146·15388
43	5169·78782	93	107513810·09709	43	7343·78226	93	229707599·76928
44	6307·14114	94	131166848·31845	44	9032·85218	94	282540347·71621
45	7694·71219	95	160023554·94851	45	11110·40819	95	347524627·69094
46	9387·54887	96	195228737·03718	46	13665·80207	96	427455292·05986
47	11452·80963	97	238179059·18536	47	16808·93654	97	525770009·23362
48	13972·42774	98	290578452·20614	48	20674·99195	98	646697111·35736
49	17046·36185	99	354505711·69149	49	25430·24010	99	795437446·96955
50	20796·56145	100	432496968·26362	50	31279·19532	100	978388059·77254

Amount of £1 in n years at the following rates per cent.

Years	24 per cent.	Years	24 per cent.	Years	25 per cent.	Years	25 per cent.
1	1·24	51	58144·13892	1	1·25	51	87581·15402
2	1·5376	52	72098·73226	2	1·5625	52	109476·44253
3	1·90662	53	89402·42801	3	1·95313	53	136845·55316
4	2·36421	54	110859·01073	4	2·44141	54	171056·94145
5	2·93163	55	137465·17330	5	3·05176	55	213821·17681
6	3·63522	56	170456·81489	6	3·81470	56	267276·47101
7	4·50767	57	211366·45047	7	4·76837	57	334095·58876
8	5·58951	58	262094·39858	8	5·96046	58	417619·48595
9	6·93099	59	324997·05424	9	7·45058	59	522024·35744
10	8·59443	60	402996·34726	10	9·31323	60	652530·44680
11	10·65709	61	499715·47060	11	11·64153	61	815663·05850
12	13·21479	62	619647·18355	12	14·55192	62	1019578·82312
13	16·38634	63	768362·50760	13	18·18989	63	1274473·52891
14	20·31906	64	952769·50942	14	22·73737	64	1593091·91113
15	25·19563	65	1181434·19168	15	28·42171	65	1991364·88892
16	31·24259	66	1464978·39769	16	35·52714	66	2489206·11114
17	38·74081	67	1816573·21313	17	44·40892	67	3111507·63893
18	48·03860	68	2252550·78428	18	55·51115	68	3889384·54866
19	59·56786	69	2793162·97251	19	69·38894	69	4861730·68583
20	73·86415	70	3463522·08591	20	86·73617	70	6077163·35729
21	91·59155	71	4294767·38653	21	108·42022	71	7596454·19661
22	113·57352	72	5325511·55930	22	135·52527	72	9495567·74576
23	140·83116	73	6603634·33353	23	169·40659	73	11869459·68220
24	174·63064	74	8188506·57358	24	211·75824	74	14836824·60275
25	216·54199	75	10153748·15124	25	264·69780	75	18546030·75344
26	268·51207	76	12590647·70754	26	230·87225	76	23182538·44180
27	332·95497	77	15612403·15735	27	413·59031	77	28978173·05225
28	412·86416	78	19359379·91512	28	516·98788	78	36222716·31531
29	511·95156	79	24005631·09474	29	646·23485	79	45278395·39414
30	634·81993	80	29766982·55748	30	807·79357	80	56597994·24267
31	787·17672	81	36911058·37128	31	1009·74196	81	70747492·80334
32	976·09913	82	45769712·38038	32	1262·17745	82	88434366·00418
33	1210·36292	83	56754443·35168	33	1577·72181	83	110542957·50522
34	1500·85002	84	70375509·75608	34	1972·15226	84	138178696·88152
35	1861·05403	85	87265632·09754	35	2465·19033	85	172723371·10191
36	2307·70699	86	108209383·80094	36	3081·48791	86	215904213·87738
37	2861·55667	87	134179635·91317	37	3851·85989	87	269880267·34673
38	3548·33027	88	166382748·53233	38	4814·82486	88	337350334·18341
39	4399·92954	89	206314608·18009	39	6018·53108	89	421687917·92926
40	5455·91262	90	255830114·14331	40	7523·16385	90	527109897·16158
41	6765·33165	91	317229341·53771	41	9403·95481	91	658888371·45197
42	8389·01125	92	393364383·50676	42	11754·94351	92	823609214·31497
43	10402·37395	93	487771835·54838	43	14693·67939	93	1029511517·89371
44	12898·94370	94	604837076·07999	44	18367·09923	94	1286889397·36713
45	15994·69019	95	749997974·33919	45	22958·87404	95	1608611746·70892
46	19833·41583	96	929997488·18060	46	28698·59255	96	2010764683·38615
47	24593·43563	97	1153196885·34394	47	35873·24069	97	2513455854·23268
48	30495·86018	98	1429964137·82648	48	44841·55086	98	3141819817·79085
49	37814·86662	99	1773155530·90484	49	56051·93857	99	3927274772·23857
50	46890·43461	100	2198712858·32200	50	70064·92322	100	4909093465·29821

TABLE II.

The sum to which £1 will amount in n *years up to one hundred, by half-yearly and quarterly payments, at the rate of* 3 *per cent., the half-yearly and quarterly ratio being* 0·015 *and* 0·0075 *respectively.*

Calculated to 10 *decimal places.*

TABLE II. xxiii

Amount of £1 in n years at the rate of 3 per cent. Payable by half-yearly and quarterly instalments.

Years	Amount of £1 in n years at 3 per cent. Payable half-yearly. Ratio = 0·015	Amount of £1 in n years at 3 per cent. Payable quarterly. Ratio = 0·0075	Years	Years	Amount of £1 in n years at 3 per cent. Payable half-yearly. Ratio = 0·015	Amount of £1 in n years at 3 per cent. Payable quarterly. Ratio = 0·0075	Years
0¼	...	1·0075000000	0¼	¼	...	1·4858942602	¼
½	1·0150000000	1·0150562500	½	½	1·4948001774	1·4970384672	½
¾	...	1·0226691719	¾	¾	...	1·5082662557	¾
1	1·0302250000	1·0303391907	1	14	1·5172221801	1·5195782526	14
1¼	...	1·0380667346	1¼	¼	...	1·5309750895	¼
½	1·0456783750	1·0458522351	½	½	1·5399805128	1·5424574027	½
¾	...	1·0536961269	¾	¾	...	1·5540258332	¾
2	1·0613635506	1·0615988478	2	15	1·5630802205	1·5656810269	15
2¼	...	1·0695608392	2¼	¼	...	1·5774236346	¼
½	1·0772840039	1·0775825455	½	½	1·5865264238	1·5892543119	½
¾	...	1·0856644146	¾	¾	...	1·6011737192	¾
3	1·0934432639	1·0938068977	3	16	1·6103243202	1·6131825221	16
3¼	...	1·1020104494	3¼	¼	...	1·6252813911	¼
½	1·1098449129	1·1102755278	½	½	1·6344791850	1·6374710015	½
¾	...	1·1186025942	¾	¾	...	1·6497520340	¾
4	1·1264925866	1·1269921137	4	17	1·6589963727	1·6621251743	17
4¼	...	1·1354445545	4¼	¼	...	1·6745911131	¼
½	1·1433899754	1·1439603887	½	½	1·6838813183	1·6871505464	½
¾	...	1·1525400916	¾	¾	...	1·6998041755	¾
5	1·1605408250	1·1611841423	5	18	1·7091395381	1·7125527068	18
5¼	...	1·1698930234	5¼	¼	...	1·7253968521	¼
½	1·1779489374	1·1786672210	½	½	1·7347766312	1·7383373285	½
¾	...	1·1875072252	¾	¾	...	1·7513748585	¾
6	1·1956181715	1·1964135294	6	19	1·7607982806	1·7645101699	19
6¼	...	1·2053866309	6¼	¼	...	1·7777439962	¼
½	1·2135524440	1·2144270306	½	½	1·7872102548	1·7910770762	½
¾	...	1·2235352333	¾	¾	...	1·8045101542	¾
7	1·2317557307	1·2327117476	7	20	1·8140184087	1·8180439804	20
7¼	...	1·2419570857	7¼	¼	...	1·8316793102	¼
½	1·2502320667	1·2512717638	½	½	1·8412286848	1·8454169051	½
¾	...	1·2606563021	¾	¾	...	1·8592575319	¾
8	1·2689855477	1·2701112243	8	21	1·8688471151	1·8732019633	21
8¼	...	1·2796370585	8¼	¼	...	1·8872509781	¼
½	1·2880203309	1·2892343364	½	½	1·8968798218	1·9014053604	½
¾	...	1·2989035940	¾	¾	...	1·9156659006	¾
9	1·3073406358	1·3086453709	9	22	1·9253330191	1·9300333949	22
9¼	...	1·3184602112	9¼	¼	...	1·9445086453	¼
½	1·3269507454	1·3283486628	½	½	1·9542130144	1·9590924602	½
¾	...	1·3383112778	¾	¾	...	1·9737856536	¾
10	1·3468550065	1·3483486123	10	23	1·9835262096	1·9885890460	23
10¼	...	1·3584612269	10¼	¼	...	2·0035034639	¼
½	1·3670578316	1·3686496861	½	½	2·0132791028	2·0185297398	½
¾	...	1·3789145588	¾	¾	...	2·0336687129	¾
11	1·3875636991	1·3892564180	11	24	2·0434782893	2·0489212282	24
11¼	...	1·3996758411	11¼	¼	...	2·0642881375	¼
½	1·4083771546	1·4101734099	½	½	2·0741304637	2·0797702985	½
¾	...	1·4207497105	¾	¾	...	2·0953685757	¾
12	1·4295028119	1·4314053333	12	25	2·1052424206	2·1110838400	25
12¼	...	1·4421408733	12¼	¼	...	2·1269169688	¼
½	1·4509453541	1·4529569299	½	½	2·1368210569	2·1428688461	½
¾	...	1·4638541068	¾	¾	...	2·1589403625	¾
13	1·4727095344	1·4748330126	13	26	2·1688733728	2·1751324152	26

Amount of £1 in n years at the rate of 3 per cent. Payable by half-yearly and quarterly instalments.

Years	Amount of £1 in n years at 3 per cent. Payable half-yearly. Ratio = 0·015	Amount of £1 in n years at 3 per cent. Payable quarterly. Ratio = 0·0075	Years	Years	Amount of £1 in n years at 3 per cent. Payable half-yearly. Ratio = 0·015	Amount of £1 in n years at 3 per cent. Payable quarterly. Ratio = 0·0075	Year
¼	... ...	2·1914459083	¼	¼	... ...	3·2320167709	¼
½	2·2014064734	2·2078817526	½	½	3·2420323025	3·2562568967	½
¾	... ...	2·2244408657	¾	¾	... ...	3·2806788235	¾
27	2·2344275705	2·2411241727	27	40	3·2906627870	3·3052839146	40
¼	... ...	2·2579326035	¼	¼	... ...	3·3300735440	¼
½	2·2679439840	2·2748670980	½	½	3·3400227288	3·3550490956	½
¾	... ...	2·2919286013	¾	¾	... ...	3·3802119638	¾
28	2·3019631438	2·3091180658	28	41	3·3901230697	3·4055635535	41
¼	... ...	2·3264364513	¼	¼	... ...	3·4311052802	¼
½	2·3364925909	2·3438847247	½	½	3·4409749158	3·4568385698	½
¾	... ...	2·3614638601	¾	¾	... ...	3·4827648590	¾
29	2·3715399798	2·3791748391	29	42	3·4925894395	3·5088855955	42
¼	... ...	2·3970186504	¼	¼	... ...	3·5352022375	¼
½	2·4071130795	2·4149962902	½	½	3·5449783826	3·5617162542	½
¾	... ...	2·4331087624	¾	¾	... ...	3·5884291261	¾
30	2·4432197757	2·4513570781	30	43	3·5981530583	3·6153423446	43
¼	... ...	2·4697422562	¼	¼	... ...	3·6424574122	¼
½	2·4798680723	2·4882653231	½	½	3·6521253542	3·6697758428	½
¾	... ...	2·5069273131	¾	¾	... ...	3·6972991616	¾
31	2·5170660934	2·5257292679	31	44	3·7069072345	3·7250289053	44
¼	... ...	2·5446722374	¼	¼	... ...	3·7529666221	¼
½	2·5548220848	2·5637572792	½	½	3·7625108430	3·7811138717	½
¾	... ...	2·5829854588	¾	¾	... ...	3·8094722258	¾
32	2·5931444161	2·6023578497	32	45	3·8189485057	3·8380432675	45
¼	... ...	2·6218755336	¼	¼	... ...	3·8668285920	¼
½	2·6320415823	2·6415396001	½	½	3·8762327333	3·8958298064	½
¾	... ...	2·6613511471	¾	¾	... ...	3·9250485300	¾
33	2·6715222061	2·6813112807	33	46	3·9343762243	3·9544863939	46
¼	... ...	2·7014211153	¼	¼	... ...	3·9841450419	¼
½	2·7115950391	2·7216817737	½	½	3·9933918676	4·0140261297	½
¾	... ...	2·7420943870	¾	¾	... ...	4·0441313257	¾
34	2·7522689647	2·7626600949	34	47	4·0532927457	4·0744623106	47
¼	... ...	2·7833800456	¼	¼	... ...	4·1050207780	¼
½	2·7935529992	2·8042553959	½	½	4·1140921368	4·1358084338	½
¾	... ...	2·8252873114	¾	¾	... ...	4·1668269971	¾
35	2·8354562942	2·8464769662	35	48	4·1758035189	4·1980781995	48
¼	... ...	2·8678255435	¼	¼	... ...	4·2295637860	¼
½	2·8779881386	2·8893342351	½	½	4·2384405717	4·2612855144	½
¾	... ...	2·9110042418	¾	¾	... ...	4·2932451558	¾
36	2·9211579607	2·9328367736	36	49	4·3020171803	4·3254444944	49
¼	... ...	2·9548330494	¼	¼	... ...	4·3578853282	¼
½	2·9649753301	2·9769942973	½	½	4·3665474380	4·3905694681	½
¾	... ...	2·9993217545	¾	¾	... ...	4·4234987391	¾
37	3·0094499600	3·0218166677	37	50	4·4320456495	4·4566749797	50
¼	... ...	3·0444802927	¼	¼	... ...	4·4901000420	¼
½	3·0545917094	3·0673138949	½	½	4·4985263343	4·5237757923	½
¾	... ...	3·0903187491	¾	¾	... ...	4·5577041108	¾
38	3·1004105851	3·1134961397	38	51	4·5660042293	4·5918868916	51
¼	... ...	3·1368473608	¼	¼	... ...	4·6263266433	¼
½	3·1469167439	3·1603737160	½	½	4·6344942927	4·6610234886	½
¾	... ...	3·1840765189	¾	¾	... ...	4·6959811648	¾
39	3·1941204950	3·2079570928	39	52	4·7040117071	4·7312010235	52

TABLE II.   XXV

Amount of £1 in n years at the rate of 3 per cent.   Payable by half-yearly and quarterly instalments.

Years	Amount of £1 in n years at 3 per cent. Payable half-yearly. Ratio = 0·015	Amount of £1 in n years at 3 per cent. Payable quarterly. Ratio = 0·0075	Years	Years	Amount of £1 in n years at 3 per cent. Payable half-yearly. Ratio = 0·015	Amount of £1 in n years at 3 per cent. Payable quarterly. Ratio = 0·0075	Years
¼	... ...	4·7666850312	¼	¼	... ...	6·8230583759	¼
½	4·7745718827	4·8024351689	½	½	6·8252639321	6·8742313137	½
¾	... ...	4·8384534327	¾	¾	... ...	6·9257880485	¾
53	4·8461904610	4·8747418334	53	65	6·9276428911	6·9777314589	65
¼	... ...	4·9113023972	¼	¼	... ...	7·0300644449	¼
½	4·9188833179	4·9481371652	½	½	7·0315575345	7·0827899282	½
¾	... ...	4·9852481939	¾	¾	... ...	7·1359108527	¾
54	4·9926665676	5·0226375554	54	66	7·1370308975	7·1894301840	66
¼	... ...	5·0603073370	¼	¼	... ...	7·2433509104	¼
½	5·0675565662	5·0982596421	½	½	7·2440863609	7·2976760423	½
¾	... ...	5·1364965894	¾	¾	... ...	7·3524086126	¾
55	5·1435699146	5·1750203138	55	67	7·3527476563	7·4075516772	67
¼	... ...	5·2138329661	¼	¼	... ...	7·4631083147	¼
½	5·2207234634	5·2529367134	½	½	7·4630388712	7·5190816271	½
¾	... ...	5·2923337387	¾	¾	... ...	7·5754747393	¾
56	5·2990343153	5·3320262418	56	68	7·5749844543	7·6322907999	68
¼	... ...	5·3720164386	¼	¼	... ...	7·6895329809	¼
½	5·3785198300	5·4123065619	½	½	7·6886092211	7·7472044782	½
¾	... ...	5·4528988611	¾	¾	... ...	7·8053085118	¾
57	5·4591976275	5·4937956026	57	69	7·8039383594	7·8638483256	69
¼	... ...	5·5349990696	¼	¼	... ...	7·9228271881	¼
½	5·5410855919	5·5765115626	½	½	7·9209974348	7·9822483920	½
¾	... ...	5·6183353993	¾	¾	... ...	8·0421152549	¾
58	5·6242018758	5·6604729148	58	70	8·0398123963	8·1024311193	70
¼	... ...	5·7029264617	¼	¼	... ...	8·1631993527	¼
½	5·7085649039	5·7456984101	½	½	8·1604095822	8·2244233479	½
¾	... ...	5·7887911482	¾	¾	... ...	8·2861065230	¾
59	5·7941933775	5·8322070818	59	71	8·2828157260	8·3482523219	71
¼	... ...	5·8759486349	¼	¼	... ...	8·4108642143	¼
½	5·8811062781	5·9200182497	½	½	8·4070579619	8·4739456959	½
¾	... ...	5·9644183866	¾	¾	... ...	8·5375002886	¾
60	5·9693228723	6·0091515245	60	72	8·5331638313	8·6015315408	72
¼	... ...	6·0542201609	¼	¼	... ...	8·6660430274	¼
½	6·0588627154	6·0996268121	½	½	8·6611612888	8·7310383501	½
¾	... ...	6·1453740132	¾	¾	... ...	8·7965211377	¾
61	6·1497456561	6·1914643183	61	73	8·7910787081	8·8624950462	73
¼	... ...	6·2379003007	¼	¼	... ...	8·9289637591	¼
½	6·2419918410	6·2846845529	½	½	8·9229448887	8·9959309873	½
¾	... ...	6·3318196871	¾	¾	... ...	9·0634004697	¾
62	6·3356217186	6·3793083347	62	74	9·0567890621	9·1313759732	74
¼	... ...	6·4271531473	¼	¼	... ...	9·1998612930	¼
½	6·4306560444	6·4753567959	½	½	9·1926408980	9·2688602527	½
¾	... ...	6·5239219718	¾	¾	... ...	9·3383767046	¾
63	6·5271158850	6·5728513866	63	75	9·3305305115	9·4084145299	75
¼	... ...	6·6221477720	¼	¼	... ...	9·4789776389	¼
½	6·6250226233	6·6718138803	½	½	9·4704884691	9·5500699711	½
¾	... ...	6·7218524844	¾	¾	... ...	9·6216954959	¾
64	6·7243979627	6·7722663780	64	76	9·6125457962	9·6938582121	76

Amount of £1 in n years at the rate of 3 per cent. Payable by half-yearly and quarterly instalments.

Years	Amount of £1 in n years at 3 per cent. Payable half-yearly. Ratio = 0·015	Amount of £1 in n years at 3 per cent. Payable quarterly. Ratio = 0·0075	Years	Years	Amount of £1 in n years at 3 per cent. Payable half-yearly. Ratio = 0·015	Amount of £1 in n years at 3 per cent. Payable quarterly. Ratio = 0·0075	Years
¼	...	9·7665621487	¼	¼	...	13·9799091478	¼
½	9·7567339831	9·8398113649	½	½	13·9472786641	14·0847584665	½
¾	...	9·9136099501	¾	¾	...	14·1903941550	¾
77	9·9030849929	9·9879620247	77	89	14·1564878441	14·2968221111	89
¼	...	10·0628717399	¼	¼	...	14·4040482769	¼
½	10·0516312677	10·1383432779	½	½	14·3688351617	14·5120786390	½
¾	...	10·2143808525	¾	¾	...	14·6209192288	¾
78	10·2024057368	10·2909887089	78	90	14·5843676891	14·7305761230	90
¼	...	10·3681711242	¼	¼	...	14·8410554440	¼
½	10·3554418228	10·4459324077	½	½	14·8031332045	14·9523633598	½
¾	...	10·5242769007	¾	¾	...	15·0645060850	¾
79	10·5107734502	10·6032089775	79	91	15·0251802025	15·1774898806	91
¼	...	10·6827330448	¼	¼	...	15·2913210547	¼
½	10·6684350519	10·7628535427	½	½	15·2505579056	15·4060059626	½
¾	...	10·8435749442	¾	¾	...	15·5215510074	¾
80	10·8284615777	10·9249017563	80	92	15·4793162742	15·6379626399	92
¼	...	11·0068385195	¼	¼	...	15·7552473597	¼
½	10·9908885013	11·0893898084	½	½	15·7115060183	15·8734117149	½
¾	...	11·1725602319	¾	¾	...	15·9924623028	¾
81	11·1557518289	11·2563544337	81	93	15·9471786086	16·1124057700	93
¼	...	11·3407770919	¼	¼	...	16·2332488133	¼
½	11·3230881063	11·4258329201	½	½	16·1863862877	16·3549981794	½
¾	...	11·5115266670	¾	¾	...	16·4776606658	¾
82	11·4929344279	11·5978631170	82	94	16·4291820820	16·6012431208	94
¼	...	11·6848470904	¼	¼	...	16·7257524442	¼
½	11·6653284443	11·7724834436	½	½	16·6756198132	16·8511955875	½
¾	...	11·8607770694	¾	¾	...	16·9775795544	¾
83	11·8403083710	11·9497328974	83	95	16·9257541104	17·1049114011	95
¼	...	12·0393558942	¼	¼	...	17·2331982366	¼
½	12·0179129965	12·1296510634	½	½	17·1796404221	17·3624472233	½
¾	...	12·2206234463	¾	¾	...	17·4926655775	¾
84	12·1981816915	12·3122781222	84	96	17·4373350284	17·6238605693	96
¼	...	12·4046202081	¼	¼	...	17·7560395236	¼
½	12·3811544169	12·4976548597	½	½	17·6988950538	17·8892098200	½
¾	...	12·5913872711	¾	¾	...	18·0233788937	¾
85	12·5668717331	12·6858226757	85	97	17·9643784796	18·1585542354	97
¼	...	12·7809663457	¼	¼	...	18·2947433922	¼
½	12·7553748091	12·8768235933	½	½	18·2338441568	18·4319539676	½
¾	...	12·9733997703	¾	¾	...	18·5701936224	¾
86	12·9467054313	13·0707002685	86	98	18·5073518192	18·7094700745	98
¼	...	13·1687395206	¼	¼	...	18·8497911001	¼
½	13·1409060127	13·2674959995	½	½	18·7849620965	18·9911645333	½
¾	...	13·3670022195	¾	¾	...	19·1335982673	¾
87	13·3380196029	13·4672547361	87	99	19·0667365279	19·2771002543	99
¼	...	13·5682591466	¼	¼	...	19·4216785063	¼
½	13·5380898970	13·6700210902	½	½	19·3527375758	19·5673410950	½
¾	...	13·7725462484	¾	¾	...	19·7140961533	¾
88	13·7411612454	13·8758403453	88	100	19·6430286395	19·8619518744	100

TABLE III.

The sum to which £1 per annum will amount in n years up to **100**, *at the rates of* $\frac{1}{2}$, $\frac{3}{4}$, 1, 1$\frac{1}{4}$, 1$\frac{1}{2}$, 1$\frac{3}{4}$, 2, 2$\frac{1}{4}$, 2$\frac{1}{2}$, 2$\frac{3}{4}$, 3, 3$\frac{1}{2}$, 4, 4$\frac{1}{2}$, 5, 5$\frac{1}{2}$, 6, 7, 8, 9, *and* 10 *per cent.*

Calculated to **10** *places of decimals to* 7 *per cent., and to* **6** *places to* 10 *per cent.*

TABLE III. xxix

Amount of £1 per annum in **n** years at the following rates per cent.

Years	½ per cent.	Years	½ per cent.	Years	¾ per cent.	Years	¾ per cent.
1	1·	51	57·9283888021	1	1·	51	61·8472142443
2	2·005	52	59·2180307461	2	2·0075	52	63·3110683512
3	3·015025	53	60·5141208998	3	3·02255625	53	64·7859013638
4	4·030100125	54	61·8166915043	4	4·0452254219	54	66·2717956241
5	5·0502506256	55	63·1257749619	5	5·0755646125	55	67·7688340912
6	6·0755018788	56	64·4414038367	6	6·1136313471	56	69·2771003469
7	7·1058793881	57	65·7636108559	7	7·1594835822	57	70·7966785995
8	8·1414087851	58	67·0924289101	8	8·2131797091	58	72·3276536890
9	9·1821158290	59	68·4278910547	9	9·2747785569	59	73·8701110917
10	10·2280264082	60	69·7700305100	10	10·3443393961	60	75·4241369249
11	11·2791665402	61	71·1188806625	11	11·4219219416	61	76·9898179518
12	12·3355623729	62	72·4744750658	12	12·5075863561	62	78·5672415865
13	13·3972401848	63	73·8368474412	13	13·6013932538	63	80·1564958984
14	14·4642263857	64	75·2060316784	14	14·7034037032	64	81·7576696176
15	15·5365475176	65	76·5820618368	15	15·8136792310	65	83·3708521398
16	16·6142302552	66	77·9649721460	16	16·9322818252	66	84·9961335308
17	17·6973014065	67	79·3547970067	17	18·0592739389	67	86·6336045323
18	18·7857879135	68	80·7515709917	18	19·1947184934	68	88·2833565663
19	19·8797168531	69	82·1553288467	19	20·3386788821	69	89·9454817406
20	20·9791154374	70	83·5661054909	20	21·4912189738	70	91·6200728536
21	22·0840110146	71	84·9839360184	21	22·6524031161	71	93·3072234000
22	23·1944310696	72	86·4088556985	22	23·8222961394	72	95·0070275755
23	24·3104032250	73	87·8408999770	23	25·0009633605	73	96·7195802824
24	25·4319552411	74	89·2801044769	24	26·1884705857	74	98·4449771345
25	26·5591150173	75	90·7265049993	25	27·3848841151	75	100·1833144630
26	27·6919105924	76	92·1801375243	26	28·5902707459	76	101·9346893215
27	28·8303701454	77	93·6410382119	27	29·8046977765	77	103·6991994914
28	29·9745219061	78	95·1092434030	28	31·0282330098	78	105·4769434876
29	31·1243946061	79	96·5847896200	29	32·2609447574	79	107·2680205637
30	32·2800165791	80	98·0677135681	30	33·5029018431	80	109·0725307180
31	33·4414166620	81	99·5580521359	31	34·7541736069	81	110·8905746984
32	34·6086237453	82	101·0558423966	32	36·0148299090	82	112·7222540086
33	35·7816686640	83	102·5611216086	33	37·2849411333	83	114·5676709137
34	36·9605751984	84	104·0739272166	34	38·5645781918	84	116·4269284455
35	38·1453780744	85	105·5952968527	35	39·8538125282	85	118·3001304089
36	39·3361049647	86	107·1222683370	36	41·1527161216	86	120·1873813869
37	40·5327854896	87	108·6578796787	37	42·4613614925	87	122·0887867473
38	41·7354494170	88	110·2011690771	38	43·7798217037	88	124·0044526480
39	42·9441266641	89	111·7521749224	39	45·1081703665	89	125·9344860428
40	44·1588472974	90	113·3109357971	40	46·4464816442	90	127·8789946881
41	45·3796415339	91	114·8774904760	41	47·7948302566	91	129·8380871483
42	46·6065397416	92	116·4518779284	42	49·1532914835	92	131·8118728019
43	47·8395724403	93	118·0341373181	43	50·5219411696	93	133·8004618479
44	49·0787703025	94	119·6243080047	44	51·9008557284	94	135·8039653118
45	50·3241641540	95	121·2224295447	45	53·2901121464	95	137·8224950517
46	51·5757849748	96	122·8285416924	46	54·6897879875	96	139·8561637645
47	52·8336638996	97	124·4426844009	47	56·0999613974	97	141·9050849928
48	54·0978322191	98	126·0648978229	48	57·5207111079	98	143·9693731302
49	55·3683213802	99	127·6952223120	49	58·9521164412	99	146·0491434287
50	56·6451629871	100	129·3336984235	50	60·3942573145	100	148·1445120044

Amount of £1 per annum in n years at the following rates per cent.

Years	1 per cent.	Years	1 per cent.	Years	1¼ per cent.	Years	1¼ per cent.
1	1·	51	66·1078140061	1	1·	51	70·7428122595
2	2·01	52	67·7688921462	2	2·0125	52	72·6270974128
3	3·0.01	53	69·4465810676	3	3·03765625	53	74·5349361304
4	4·060401	54	71·1410468783	4	4·0756269531	54	76·4666228320
5	5·10100501	55	72·8524573471	5	5·1265722900	55	78·4224556174
6	6·1520150601	56	74·5809819206	6	6·1906544437	56	80·4027363127
7	7·2135352107	57	76·3267917398	7	7·2680376242	57	82·4077705166
8	8·2856705628	58	78·0900596571	8	8·3588880945	58	84·4378676480
9	9·3685272684	59	79·8709602537	9	9·4633741957	59	86·4933409936
10	10·4622125411	60	81·6696698563	10	10·5816663731	60	88·5745077560
11	11·5668346665	61	83·4863665548	11	11·7139372028	61	90·6816891030
12	12·6825030132	62	85·3212302204	12	12·8603614178	62	92·8152102168
13	13·8093280433	63	87·1744425226	13	14·0211159356	63	94·9754003445
14	14·9474213238	64	89·0461869478	14	15·1963798848	64	97·1625928488
15	16·0968955370	65	90·9366488173	15	16·3863346333	65	99·3771252594
16	17·2578644924	66	92·8460153054	16	17·5911638162	66	101·6193393252
17	18·4304431373	67	94·7744754585	17	18·8110533639	67	103·8895810667
18	19·6147475687	68	96·7222202131	18	20·0461915310	68	106·1882008300
19	20·8108950443	69	98·6894424152	19	21·2967689251	69	108·5155533404
20	22·0190039948	70	100·6763368393	20	22·5629785367	70	110·8719977572
21	23·2391949347	71	102·6831002077	21	23·8450157684	71	113·2578977291
22	24·4715859751	72	104·7099312098	22	25·1430784655	72	115·6736214508
23	25·7163018348	73	106·7570305219	23	26·4573669463	73	118·1195417189
24	26·9734648532	74	108·8246008271	24	27·7880840331	74	120·5960359904
25	28·2431995017	75	110·9128468354	25	29·1354350836	75	123·1034864403
26	29·5256314967	76	113·0219753037	26	30·4996280221	76	125·6422800208
27	30·8208878117	77	115·1521950568	27	31·8808733724	77	128·2128085210
28	32·1290966898	78	117·3037170074	28	33·2793842895	78	130·8154686275
29	33·4503876567	79	119·4767541774	29	34·6953765932	79	133·4506619854
30	34·7848915333	80	121·6715217192	30	36·1290688006	80	136·1187952602
31	36·1327404486	81	123·8882369364	31	37·5806821606	81	138·8202802010
32	37·4940678531	82	126·1271193058	32	39·0504406876	82	141·5555337035
33	38·8690085316	83	128·3883904988	33	40·5385711962	83	144·3249778748
34	40·2576986169	84	130·6722744038	34	42·0453033361	84	147·1290400982
35	41·6602756031	85	132·9789971478	35	43·5708696278	85	149·9681530994
36	43·0768783591	86	135·3087871193	36	45·1155054982	86	152·8427550132
37	44·5076471427	87	137·6618749905	37	46·6794493169	87	155·7532894508
38	45·9527236141	88	140·0384847404	38	48·2629424334	88	158·7002055690
39	47·4122508503	89	142·4388786778	39	49·8662292138	89	161·6839581386
40	48·8863733588	90	144·8632674646	40	51·4895570789	90	164·7050076153
41	50·3752370923	91	147·3119001393	41	53·1331765424	91	167·7638202105
42	51·8789894633	92	149·7850191406	42	54·7973412492	92	170·8608679631
43	53·3977793579	93	152·2828693321	43	56·4823080148	93	173·9966288127
44	54·9317571515	94	154·8056980254	44	58·1883368650	94	177·1715866728
45	56·4810747230	95	157·3537550056	45	59·9156910758	95	180·3862315062
46	58·0458854702	96	159·9272925557	46	61·6646372143	96	183·6410594001
47	59·6263443249	97	162·5265654812	47	63·4354451794	97	186·9365726426
48	61·2226077681	98	165·1518311361	48	65·2283882442	98	190·2732798c06
49	62·8348338458	99	167·8033494474	49	67·0437430972	99	193·6516957981
50	64·4631821843	100	170·4813829419	50	68·8817898859	100	197·0723419956

TABLE III.   xxxi

Amount of £1 per annum in n years at the following rates per cent.

Years	1¼ per cent.	Years	1¼ per cent.	Years	1¾ per cent.	Years	1¾ per cent.
1	1·	51	75·7880704611	1	1·	51	81·2830136099
2	2·015	52	77·9248915180	2	2·0175	52	83·7054663481
3	3·045225	53	80·0937648907	3	3·05280625	53	86·1703120092
4	4·090903375	54	82·2951713641	4	4·1062343594	54	88·6782924693
5	5·1522669256	55	84·5295989346	5	5·1780893907	55	91·2301625875
6	6·2295509295	56	86·7975429186	6	6·2687059550	56	93·8266904328
7	7·3229941935	57	89·0995060624	7	7·3784083092	57	96·4686575154
8	8·4328391064	58	91·4359986533	8	8·5075304546	58	99·1568590219
9	9·5593316922	59	93·8075386331	9	9·6564123376	59	101·8921040548
10	10·7027216683	60	96·2146517126	10	10·8253994517	60	104·6752158758
11	11·8632624934	61	98·6578714883	11	12·0418439421	61	107·5070321536
12	13·0412114308	62	101·1377395606	12	13·2251037111	62	110·3884052163
13	14·2368296022	63	103·6548056540	13	14·4565430261	63	113·3202023076
14	15·4503820463	64	106·2096277388	14	15·7095325290	64	116·3033305480
15	16·6821377770	65	108·8027721549	15	16·9844493483	65	119·3386137003
16	17·9323698436	66	111·4348137372	16	18·2816772119	66	122·4270394401
17	19·2013553913	67	114·1063359433	17	19·6016065631	67	125·5695126303
18	20·4893757221	68	116·8179309824	18	20·9446346780	68	128·7669791013
19	21·7967163580	69	119·5701999472	19	22·3111657848	69	132·0204012356
20	23·1236671033	70	122·3637529464	20	23·7016111861	70	135·3307582572
21	24·4705221099	71	125·1992092406	21	25·1163893818	71	138·6990465267
22	25·8375799415	72	128·0771973792	22	26·5559261960	72	142·1262798409
23	27·2251436406	73	130·9983553399	23	28·0206549044	73	145·6134897381
24	28·6335207953	74	133·9633306700	24	29·5110163653	74	149·1617258086
25	30·0630236072	75	136·9727806300	25	31·0274591517	75	152·7720560102
26	31·5139689613	76	140·0273723395	26	32·5704396868	76	156·4455669904
27	32·9866784957	77	143·1277829246	27	34·1404223813	77	160·1833644137
28	34·4814786731	78	146·2746996684	28	35·7378797730	78	163·9865732910
29	35·9987008532	79	149·4688201635	29	37·3632926691	79	167·8563383235
30	37·5386813660	80	152·7108524659	30	39·0171502908	80	171·7938242442
31	39·1017615865	81	156·0015152529	31	40·6999504209	81	175·8002161684
32	40·6882880103	82	159·3415379817	32	42·4121995532	82	179·8767199514
33	42·2986123305	83	162·7316610514	33	44·1544130454	83	184·0245624505
34	43·9330915154	84	166·1726359672	34	45·9271152737	84	188·2449923951
35	45·5920878882	85	169·6652255067	35	47·7308397910	85	192·5392797620
36	47·2759692065	86	173·2102038893	36	49·5661294873	86	196·9087171579
37	48·9851087446	87	176·8083569476	37	51·4335367534	87	201·3546197081
38	50·7198853757	88	180·4604823018	38	53·3336236466	88	205·8783255530
39	52·4806836564	89	184·1673895364	39	55·2669620604	89	210·4811962502
40	54·2678939112	90	187·9299003794	40	57·2341338964	90	215·1646171846
41	56·0819123199	91	191·7488488851	41	59·2357312396	91	219·9299979853
42	57·9231410047	92	195·6250816184	42	61·2723565363	92	224·7787729500
43	59·7919881198	93	199·5594578427	43	63·3446227757	93	229·7124014766
44	61·6888679416	94	203·5528497103	44	65·4531536743	94	234·7323685025
45	63·6142009607	95	207·6061424560	45	67·5985838636	95	239·8401849513
46	65·5684139751	96	211·7202345928	46	69·7815590812	96	245·0373881879
47	67·5519401847	97	215·8960381117	47	72·0027363651	97	250·3255424812
48	69·5652192875	98	220·1344786834	48	74·2627842515	98	255·7062394746
49	71·6086975768	99	224·4364958636	49	76·5623829759	99	261·1810986654
50	73·6828280405	100	228·8030433016	50	78·9022246780	100	266·7517678920

Amount of £1 per annum in n years at the following rates per cent.

Years	2 per cent.	Years	2 per cent.	Years	2¼ per cent.	Years	2¼ per cent.
1	1·	51	87·2709894828	1	1·	51	93·7996641635
2	2·02	52	90·0164092724	2	2·0225	52	96·9101566072
3	3·0604	53	92·8167374579	3	3·06800625	53	100·0906351308
4	4·121608	54	95·6730722070	4	4·1370363906	54	103·3426744213
5	5·20404016	55	98·5865336512	5	5·2301197094	55	106·6678845958
6	6·3081209632	56	101·5582643242	6	6·3477974029	56	110·0679119992
7	7·4342833825	57	104·5894296107	7	7·4906228444	57	113·5444400192
8	8·5829690501	58	107·6812182029	8	8·6591618584	58	117·0991899196
9	9·7546284311	59	110·8348425670	9	9·8539930003	59	120·7339216928
10	10·9497209997	60	114·0515394183	10	11·0757078428	60	124·4504349309
11	12·1687154197	61	117·3325702067	11	12·3249112692	61	128·2505697168
12	13·4120897281	62	120·6792216108	12	13·6022217728	62	132·1362075354
13	14·6803315267	63	124·0928060430	13	14·9082717627	63	136·1092722050
14	15·9739381531	64	127·5746621639	14	16·2437078773	64	140·1717308296
15	17.2934169162	65	131·1261554073	15	17·6091913046	65	144·3255947733
16	18·6392852545	66	134·7486785154	16	19·0053981089	66	148·5729206557
17	20·0120709596	67	138·4436520857	17	20·4330195664	67	152·9158113704
18	21·4123123788	68	142·2125251275	18	21·8927625066	68	157·3564171262
19	22·8405586264	69	146·0567756300	19	23·3853496630	69	161·8969365116
20	24·2973697989	70	149·9779111426	20	24·9115200304	70	166·5396175831
21	25·7833171949	71	153·9774693655	21	26·4720292311	71	171·2867589787
22	27·3989835388	72	158·0570187528	22	28·0676498888	72	176·1407110557
23	28·8449632096	73	162·2181591278	23	29·6991720113	73	181·1038770545
24	30·4218624738	74	166·4625223104	24	31·3674033816	74	186·1787142882
25	32·0302997232	75	170·7917727566	25	33·0731699577	75	191·3677353597
26	33·6709057177	76	175·2076082117	26	34·8173162817	76	196·6735094053
27	35·3443238320	77	179·7117603759	27	36·6007058980	77	202·0986633669
28	37·0512103087	78	184·3059955835	28	38·4242217808	78	207·6458832927
29	38·7922345149	79	188·9921154951	29	40·2887667708	79	213·3179156667
30	40·5680792052	80	193·7719578050	30	42·1952640232	80	219·1175687692
31	42·3794407893	81	198·6473969611	31	44·1446574637	81	225·0477140666
32	44·2270296051	82	203·6203449003	32	46·1379122566	82	231·1112876331
33	46·1115701972	83	208·6927517984	33	48·1760152824	83	237·3112916048
34	48·0338016011	84	213·8666068343	34	50·2599756262	84	243·6507956659
35	49·9944776331	85	219·1439389710	35	52·3908250778	85	250·1329385684
36	51·9943671858	86	224·5268177504	36	54·5696186421	86	256·7609296862
37	54·0342545295	87	230·0173541054	37	56·7974350615	87	263·5380506041
38	56·1149396201	88	235·6177011875	38	59·0753773504	88	270·4676567427
39	58·2372384125	89	241·3300552113	39	61·4045733408	89	277·5531790194
40	60·4019831808	90	247·1566563155	40	63·7861762410	90	284·7981255474
41	62·6100228444	91	253·0997894418	41	66·2213652064	91	292·2060833722
42	64·8622233013	92	259·1617852306	42	68·7113459235	92	299·7807202481
43	67·1594677673	93	265·3450209353	43	71·2573512068	93	307·5257864536
44	69·5026571226	94	271·6519213540	44	73·8606416089	94	315·4451166489
45	71·8927102651	95	278·0849597810	45	76·5225060451	95	323·5426317735
46	74·3305644704	96	284·6466589766	46	79·2442624312	96	331·8223409884
47	76·8171757598	97	291·3395921562	47	82·0272583359	97	340·2883436606
48	79·3535192750	98	298·1663839993	48	84·8728716484	98	348·9448313930
49	81·9405896605	99	305·1297116793	49	87·7825112605	99	357·7960900993
50	84·5794014537	100	312·2323059129	50	90·7576177639	100	366·8465021265

TABLE III. xxxiii

Amount of £1 per annum in n years at the following rates per cent.

Years	2½ per cent.	Years	2½ per cent.	Years	2¾ per cent.	Years	2¾ per cent.
1	1·	51	100·9214575078	1	1·	51	108·6940225574
2	2·025	52	104·4444939455	2	2·0275	52	112·6831081777
3	3·075625	53	108·0556062941	3	3·08325625	53	116·7818936526
4	4·152515625	54	111·7569964515	4	4·1630457969	54	120·9933957281
5	5·2563285156	55	115·5509213628	5	5·2826670563	55	125·3207141106
6	6·3877367285	56	119·4396943968	6	6·4279404003	56	129·7670337486
7	7·5474301467	57	123·4256867568	7	7·6047087613	57	134·3356271767
8	8·7361159004	58	127·5113289257	8	8·8138382523	58	139·0298569241
9	9·9545187979	59	131·6991121488	9	10·0562188042	59	143·8531779855
10	11·2033817679	60	135·9915899525	10	11·3327648213	60	148·8091403842
11	12·4834663121	61	140·3913797014	11	12·6444158539	61	153·9013917448
12	13·7955529699	62	144·9011641939	12	13·9921372899	62	159·1336800177
13	15·1404417941	63	149·5236932987	13	15·3769210654	63	164·5098562182
14	16·5189528390	64	154·2617856312	14	16·7997863947	64	170·0338772642
15	17·9319266599	65	159·1183302720	15	18·2617805205	65	175·7098088890
16	19·3802248264	66	164·0962885288	16	19·7639794849	66	181·5418286334
17	20·8647304471	67	169·1986957420	17	21·3074889207	67	187·5342289209
18	22·3863487083	68	174·4286631356	18	22·8934448660	68	193·6914202162
19	23·9460074260	69	179·7893797139	19	24·5230145998	69	200·0179342721
20	25·5446576116	70	185·2841142068	20	26·1973975013	70	206·5184274646
21	27·1832740519	71	190·9162170620	21	27·9178259326	71	213·1976842199
22	28·8628559032	72	196·6891224885	22	29·6855661458	72	220·0606205359
23	30·5844273008	73	202·6063505507	23	31·5019192148	73	227·1122876008
24	32·3490379833	74	208·6715093145	24	33·3682219932	74	234·3578755098
25	34·1577639329	75	214·8882970474	25	35·2858480980	75	241·8027170863
26	36·0117080312	76	221·2605044735	26	37·2562089207	76	249·4522918062
27	37·9120007320	77	227·7920170854	27	39·2807546660	77	257·3122298308
28	39·8598007503	78	234·4868175125	28	41·3609754193	78	265·3883161511
29	41·8562957690	79	241·3489879503	29	43·4984022433	79	273·6864948452
30	43·9027031633	80	248·3827126491	30	45·6946083050	80	282·2128734535
31	46·0002707423	81	255·5922804653	31	47·9512100334	81	290·9737274734
32	48·1502775109	82	262·9820874770	32	50·2698683093	82	299·9755049789
33	50·3540344487	83	270·5566396639	33	52·6522896878	83	309·2248313659
34	52·6128853099	84	278·3205556555	34	55·1002276543	84	318·7285142284
35	54·9282074426	85	286·2785695469	35	57·6154839148	85	328·4935483697
36	57·3014126287	86	294·4355337855	36	60·1999097224	86	338·5271209499
37	59·7339479444	87	302·7964221302	37	62·8554072398	87	348·8366167760
38	62·2272966430	88	311·3663326834	38	65·5839309389	88	359·4296237373
39	64·7829790591	89	320·1504910005	39	68·3874890397	89	370·3139383901
40	67·4025535356	90	329·1542532755	40	71·2681449883	90	381·4975716958
41	70·0876173740	91	338·3831096074	41	74·2280189755	91	392·9887549174
42	72·8398078083	92	347·8426873476	42	77·2692894973	92	404·7959456777
43	75·6608030035	93	357·5387545313	43	80·3941949585	93	416·9278341838
44	78·5523230786	94	367·4772233946	44	83·6050353198	94	429·3933496238
45	81·5161311556	95	377·6641539794	45	86·9041737911	95	442·2016667385
46	84·5540344344	96	388·1057578289	46	90·2940385704	96	455·3622125738
47	87·6678852953	97	398·8084017747	47	93·7771246311	97	468·8846734196
48	90·8595824277	98	409·7786118190	48	97·3559955584	98	482·7790019386
49	94·1310719884	99	421·0230771145	49	101·0332854363	99	497·0554244919
50	97·4843487881	100	432·5486540424	50	104·8117007858	100	511·7244486654

Amount of £1 per annum in n years at the following rates

Years	3 per cent.	Years	3 per cent.	Years	3½ per cent.	Years
1	1·	51	117·1807733090	1	1·	51
2	2·03	52	121·6961965083	2	2·035	52
3	3·0909	53	126·3470824035	3	3·106225	53
4	4·183627	54	131·1374948756	4	4·214942875	54
5	5·30913581	55	136·0716197219	5	5·3624658746	55
6	6·4684098843	56	141·1537683135	6	6·5501521813	56
7	7·6624621808	57	146·3883813629	7	7·7794075076	57
8	8·8923360463	58	151·7800328038	8	9·0516867704	58
9	10·1591061276	59	157·3334337879	9	10·3684958073	59
10	11·4638793115	60	163·0534368016	10	11·7313931606	60
11	12·8077956908	61	168·9450399056	11	13·1419919212	61
12	14·1920295615	62	175·0133911028	12	14·6019616385	62
13	15·6177904484	63	181·2637928359	13	16·1130302958	63
14	17·0863241618	64	187·7017066209	14	17·6769863562	64
15	18·5989138867	65	194·3327578196	15	19·2956808786	65
16	20·1568813033	66	201·1627405541	16	20·9710297094	66
17	21·7615877424	67	208·1976227708	17	22·7050157492	67
18	23·4144353747	68	215·4435514539	18	24·4996913004	68
19	25·1168684359	69	222·9068579975	19	26·3571804960	69
20	26·8703744890	70	230·5940637374	20	28·2796818133	70
21	28·6764857237	71	238·5118856496	21	30·2694706768	71
22	30·5367802954	72	246·6672422190	22	32·3289021505	72
23	32·4528837042	73	255·0672594856	23	34·4604137267	73
24	34·4264702154	74	263·7192772702	24	36·6665282071	74
25	36·4592643218	75	272·6308555883	25	38·9498566944	75
26	38·5530422515	76	281·8097812559	26	41·3131016786	76
27	40·7096335190	77	291·2640746936	27	43·7590602373	77
28	43·9309225246	78	301·0019969344	28	46·2906273456	78
29	45·2188502003	79	311·0320568424	29	48·9107993027	79
30	47·5754157063	80	321·3630185477	30	51·6226772782	80
31	50·0026781775	81	332·0039091041	31	54·4294709829	81
32	52·5027585229	82	342·9640263773	32	57·3345024673	82
33	55·0778412785	83	354·2529471686	33	60·3412100536	83
34	57·7301765169	84	365·8805355836	34	63·4531524055	84
35	60·4620818124	85	377·8569516512	35	66·6740127396	85
36	63·2759442668	86	390·1926602007	36	70·0076031855	86
37	66·1742225948	87	402·8984400067	37	73·4578692969	87
38	69·1594492726	88	415·9853932069	38	77·0288947223	88
39	72·2342327508	89	429·4649550031	39	80·7249060376	89
40	75·4012597333	90	443·3489036532	40	84·5502777488	90
41	78·6632975253	91	457·6493707628	41	88·5095374700	91
42	82·0231964511	92	472·3788518857	42	92·6073712814	92
43	85·4838923446	93	487·5502174423	43	96·8486292763	93
44	89·0484091150	94	503·1767239655	44	101·2383313009	94
45	92·7198613884	95	519·2720256845	45	105·7816728964	95
46	96·5014572301	96	535·8501864550	46	110·4840314477	96
47	100·3965009470	97	552·9256920487	47	115·3509725484	97
48	104·4083959754	98	570·5134628101	48	120·3882565875	98
49	108·5406478546	99	588·6288666944	49	125·6018455681	99
50	112·7968672903	100	607·2877326953	50	130·9979101629	100

TABLE III.   XXXV

. Amount of £1 per annum in n years at the following rates per cent.

Years	4 per cent.	Years	4 per cent.	Years	4½ per cent.	Years	4½ per cent.
1	1·	51	159·7737670032	1	1·	51	187·5356645512
2	2·04	52	167·1647176833	2	2·045	52	196·9747694560
3	3·1216	53	174·8513063906	3	3·137025	53	206·8386340816
4	4·246464	54	182·8453586462	4	4·278191125	54	217·1463726152
5	5·41632256	55	191·1591729921	5	5·4707097256	55	227·9179593829
6	6·6329754624	56	199·8055399118	6	6·7168916633	56	239·1742675552
7	7·8982944809	57	208·7977615082	7	8·0191517881	57	250·9371095951
8	9·2142262601	58	218·1496719686	8	9·3800136186	58	263·2292795269
9	10·5827953105	59	227·8756588473	9	10·8021142314	59	276·0745971056
10	12·0061071230	60	237·9906852012	10	12·2882093718	60	289·4979539754
11	13·4863514079	61	248·5103126092	11	13·8411787936	61	303·5253619043
12	15·0258054642	62	259·4507251136	12	15·4640318393	62	318·1840031900
13	16·6268376828	63	270·8287541182	13	17·1599132721	63	333·5022833335
14	18·2919111901	64	282·6619042829	14	18·9321093693	64	349·5098860835
15	20·0235876377	65	294·9683804542	15	20·7840542909	65	366·2378309573
16	21·8245311432	66	307·7671156724	16	22·7193367340	66	383·7185333503
17	23·6975123889	67	321·0778002993	17	24·7417068870	67	401·9858673511
18	25·6454128845	68	334·9209123112	18	26·8550836970	68	421·0752313819
19	27·6712293998	69	349·3177488037	19	29·0635624633	69	441·0236167941
20	29·7780785858	70	364·2904587558	20	31·3714227742	70	461·8696795498
21	31·9692017189	71	379·8620771061	21	33·7831367990	71	483·6538151296
22	34·2479697876	72	396·0565601903	22	36·3033779550	72	506·4182368104
23	36·6178885791	73	412·8988225979	23	38·9370299629	73	530·2070574669
24	39·0826041223	74	430·4147755018	24	41·6891963113	74	555·0663750529
25	41·6459082872	75	448·6313665219	25	44·5652101453	75	581·0443619302
26	44·3117446187	76	467·5766211828	26	47·5706446018	76	608·1913582171
27	47·0842144034	77	487·2796860301	27	50·7113236089	77	636·5599693369
28	49·9675829795	78	507·7708734713	28	53·9933331713	78	666·2051679570
29	52·9662862987	79	529·0817084102	29	57·4230331640	79	697·1844005151
30	56·0849377507	80	551·2449767466	30	61·0070696564	80	729·5576985383
31	59·3283352607	81	574·2947758164	31	64·7523877909	81	763·3877949725
32	62·7014686711	82	598·2665668491	32	68·6662452415	82	798·7402457462
33	66·2095274180	83	623·1972295231	33	72·7562262774	83	835·6835568048
34	69·8579085147	84	649·1251187040	34	77·0302564599	84	874·2893168610
35	73·6522248553	85	676·0901234521	35	81·4966180005	85	914·6323361199
36	77·5983138495	86	704·1337283902	36	86·1639658106	86	956·7907912453
37	81·7022464035	87	733·2990775258	37	91·0413442720	87	1000·8463768513
38	85·9703362596	88	763·6310406269	38	96·1382047643	88	1046·8844638096
39	90·4091497100	89	795·1762822519	39	101·4644239787	89	1094·9942646810
40	95·0255156984	90	827·9833335420	40	107·0303230577	90	1145·2690965917
41	99·8265363263	91	862·1026668837	41	112·8466875953	91	1197·8061118883
42	104·8195977794	92	897·5867735591	42	118·9247885371	92	1252·7073869233
43	110·0123816905	93	934·4902445014	43	125·2764040213	93	1310·0792193348
44	115·4128769582	94	972·8698542815	44	131·9138422022	94	1370·0327842049
45	121·0293920365	95	1012·7846484527	45	138·8499651013	95	1432·6842594941
46	126·8705677179	96	1054·2960343908	46	146·0982135309	96	1498·1550511713
47	132·9453904267	97	1097·4678757665	47	153·6726331398	97	1566·5720284740
48	139·2632060437	98	1142·3665907971	48	161·5879016311	98	1638·0677697553
49	145·8337342855	99	1189·0612544290	49	169·8593572045	99	1712·7808193943
50	152·6670836569	100	1237·6237046062	50	178·5030282787	100	1790·8559562671

Amount of £1 per annum in n years at the following rates per cent.

Years	5 per cent.	Years	5 per cent.	Years	5½ per cent.	Years	5½ per cent.
1	1.	51	220·8153955009	1	1·	51	260·759.
2	2·05	52	232·8561652759	2	2·055	52	276·101
3	3·1525	53	245·4989735397	3	3·168025	53	292·286
4	4·310125	54	258·7739222167	4	4·342266375	54	309·362
5	5·52563125	55	272·7126183276	5	5·5810910256	55	327·377
6	6·8019128125	56	287·3482492439	6	6·8880510320	56	346·383
7	8·1420084531	57	302·7156617061	7	8·2668938388	57	366·434
8	9·5491088758	58	318·8514447914	8	9·7215729999	58	387·588
9	11·0265643196	59	335·7940170310	9	11·2562595149	59	409·905
10	12·5778925355	60	353·5837178826	10	12·8753537882	60	433·450
11	14·2067871623	61	372·2629037767	11	14·5834982466	61	458·290
12	15·9171265204	62	391·8760489655	12	16·3855906502	62	484·496
13	17·7129828465	63	412·4698514138	13	18·2867981359	63	512·143
14	19·5986319888	64	434·0933439845	14	20·2925720334	64	541·311
15	21·5785635882	65	456·7980111837	15	22·4086634952	65	572·083
16	23·6574917676	66	480·6379117429	16	24·6411399875	66	604·547
17	25·8403663560	67	505·6698073301	17	26·9964026868	67	638·798
18	28·1323846738	68	531·9532976966	18	29·4812048345	68	674·932
19	30·5390039075	69	559·5509625814	19	32·1026711004	69	713·053
20	33·0659541029	70	588·5285107105	20	34·8683180110	70	753·271
21	35·7192518080	71	618·9549362460	21	37·7860755016	71	795·701
22	38·5052143984	72	650·9026830583	22	40·8643096542	72	840·464
23	41·4304751184	73	684·4478172112	23	44·1118466851	73	887·690
24	44·5019988743	74	719·6702080718	24	47·5379982528	74	937·513
25	47·7270988180	75	756·6537184754	25	51·1525881567	75	
26	51·1134537589	76	795·4864043992	26	54·9659805053	76	1045·530
27	54·6691264468	77	836·2607246191	27	58·9891094331	77	1104·034
28	58·4025827692	78	879·0737608501	28	63·2335104519	78	1165·756
29	62·3227119076	79	924·0274488926	29	67·7113535268	79	1230·873
30	66·4388475030	80	971·2288123372	30	72·4354779708	80	1299·571
31	70·7607898782	81	1020·7902624041	31	77·4194292592	81	1372·047
32	75·2988293721	82	1072·8297755243	32	82·6774978684	82	1448·510
33	80·0637708407	83	1127·4712643005	33	88·2247602512	83	1529·178
34	85·0669593827	84	1184·8448275156	34	94·0771220650	84	1614·283
35	90·3203073518	85	1245·0870688914	35	100·2513637786	85	1704·068
36	95·8363227194	86	1308·3414223359	36	106·7651887864	86	1798·792
37	101·6281388554	87	1374·7584934527	37	113·6372741696	87	1898·726
38	107·7095457982	88	1444·4964181254	38	120·8873242490	88	2004·156
39	114·0950230881	89	1517·7212390316	39	128·5361270827	89	2115·384
40	120·7997742425	90	1594·6073009832	40	136·6056140722	90	2232·731
41	127·8397629546	91	1675·3376660324	41	145·1189228462	91	2356·531
42	135·2317511024	92	1760·1045493340	42	154·1004636027	92	2487·140
43	142·9933386575	93	1849·1097768007	43	163·5759891009	93	2624·933
44	151·1430055904	94	1942·5652656408	44	173·5726685014	94	2770·304
45	159·7001558699	95	2040·6935289228	45	184·1191652690	95	2923·671
46	168·6851636634	96	2143·7282053689	46	195·2457193588	96	3085·473
47	178·1194218465	97	2251·9416156374	47	206·9842339235	97	3256·174
48	188·0253929389	98	2365·5103464193	48	219·3683667893	98	3436·263
49	198·4266625858	99	2484·7858637402	49	232·4336269627	99	3626·258
50	209·3479957151	100	2610·0251569272	50	246·2174764457	100	3826·702

Amount of £1 per annum in n years at the following rates per cent.

Years	6 per cent.	Years	6 per cent.	Years	7 per cent.	Years	7 per cent.
1	1·	51	308·7560588582	1	1·	51	435·9859545351
2	2·06	52	328·2814223897	2	2·07	52	467·5049713526
3	3·1836	53	348·9783077331	3	3·2149	53	501·2303193473
4	4·374616	54	370·9170061970	4	4·439943	54	537·3164417016
5	5·63709296	55	394·1720265689	5	5·75073901	55	575·9285926207
6	6·9753185376	56	418·8223481630	6	7·1532907407	56	617·2435941042
7	8·3938376499	57	444·9516890528	7	8·6540210925	57	661·4506456914
8	9·8974679088	58	472·6487903960	8	10·2598025690	58	708·7521908898
9	11·4913159834	59	502·0077178197	9	11·9779887489	59	759·3648442521
10	13·1807949424	60	533·1281808889	10	13·8164479613	60	813·5203833498
11	14·9716426389	61	566·1158717422	11	15·7835993186	61	871·4668101843
12	16·8699411973	62	601·0828240468	12	17·8884512709	62	933·4694868972
13	18·8821376691	63	638·1477934896	13	20·1406428598	63	999·8123509800
14	21·0150659292	64	677·4366610990	14	22·5504878600	64	1070·7992155486
15	23·2759698850	65	719·0828607649	15	25·1290220102	65	1146·7551606370
16	25·6725280781	66	763·2278324108	16	27·8880535509	66	1228·0280218815
17	28·2128797628	67	810·0215023555	17	30·8402172995	67	1314·9899834132
18	30·9056525485	68	859·6227924968	18	33·9990325104	68	1408·0392822522
19	33·7599170014	69	912·2001600466	19	37·3789647862	69	1507·6020320098
20	36·7855912035	70	967·9321696494	20	40·9954923212	70	1614·1341742505
21	39·9927266757	71	1027·0080998284	21	44·8651767837	71	1728·1235664480
22	43·3922902763	72	1089·6285858181	22	49·0057391585	72	1850·0922160994
23	46·9958276929	73	1156·0063009672	23	53·4361408996	73	1980·5986712264
24	50·8155773544	74	1226·3666790252	24	58·1766707626	74	2120·2405782122
25	54·8645119957	75	1300·9486797667	25	63·2490377160	75	2269·6574186871
26	59·1563827154	76	1380·0056005527	26	68·6764703561	76	2429·5334379952
27	63·7057656784	77	1463·8059365859	27	74·4832832810	77	2600·6007786548
28	68·5281116191	78	1552·6342927810	28	80·6976909107	78	2783·6428331606
29	73·6397983162	79	1646·7923503479	29	87·3465292745	79	2979·4978314819
30	79·0581862152	80	1746·5998913688	30	94·4607863237	80	3189·0626796856
31	84·8016773881	81	1852·3958848509	31	102·0730413663	81	3413·2970672636
32	90·8897780314	82	1964·5396379420	32	110·2181542620	82	3653·2278619721
33	97·3431647133	83	2083·4120162185	33	118·9334250603	83	3909·9538123101
34	104·1837545961	84	2209·4167371916	34	128·2587648145	84	4184·6505791718
35	111·4347798719	85	2342·9817414231	35	138·2368783515	85	4478·5761197139
36	119·1208666642	86	2484·5606459085	36	148·9134598361	86	4793·0764480938
37	127·2681186640	87	2634·6342846630	37	160·3374020247	87	5129·5917994604
38	135·9042057839	88	2793·7123417428	38	172·5610201664	88	5489·6632254226
39	145·0584581309	89	2962·3350822473	39	185·6402915780	89	5874·9396512022
40	154·7619656188	90	3141·0751871822	40	199·6351119885	90	6287·1854267864
41	165·0476835559	91	3330·5396984131	41	214·6095698277	91	6728·2884066614
42	175·9505445692	92	3531·3720803179	42	230·6322397156	92	7200·2685951277
43	187·5075772434	93	3744·2544051369	43	247·7764964957	93	7705·2873967866
44	199·7580318780	94	3969·9096694452	44	266·1208512504	94	8245·6575145617
45	212·7435137907	95	4209·1042496119	45	285·7493108380	95	8823·8535405810
46	226·5081246181	96	4462·6505045886	46	306·7517625966	96	9442·5232884217
47	241·0986120952	97	4731·4095348639	47	329·2243859784	97	10104·4999186112
48	256·5645288209	98	5016·2941069558	48	353·2700929969	98	10812·8149129140
49	272·9584005502	99	5318·2717533731	49	378·9989995066	99	11570·7119568180
50	290·3359045832	100	5638·3680585755	50	406·5289294721	100	12381·6617937952

Amount of £1 per annum in n years at the following rates per cent.

Years	8 per cent.	Years	8 per cent.	Years	9 per cent.	Years	9 per cent.
1	1·	51	620·671769	1	1·	51	889·441076
2	2·08	52	671·325510	2	2·09	52	970·490773
3	3·2464	53	726·031551	3	3·2781	53	1058·834943
4	4·506112	54	785·114075	4	4·573129	54	1155·130088
5	5·866601	55	848·923201	5	5·984711	55	1260·091796
6	7·335929	56	917·837058	6	7·523335	56	1374·500057
7	8·992803	57	992·264022	7	9·200435	57	1499·205063
8	10·636628	58	1072·645144	8	11·028474	58	1635·133518
9	12·487558	59	1159·456755	9	13·021036	59	1783·295535
10	14·486562	60	1253·213296	10	15·192930	60	1944·792133
11	16·645487	61	1354·470360	11	17·560293	61	2120·823425
12	18·977126	62	1463·827988	12	20·140720	62	2312·697533
13	21·495297	63	1581·934227	13	22·953385	63	2521·840331
14	24·214920	64	1709·488966	14	26·019189	64	2749·805939
15	27·152114	65	1847·248083	15	29·360916	65	2998·288474
16	30·324283	66	1996·027929	16	33·003399	66	3269·134436
17	33·750226	67	2156·710164	17	36·973705	67	3564·356535
18	37·450244	68	2330·246977	18	41·301338	68	3886·148624
19	41·446263	69	2517·666735	19	46·018458	69	4236·902000
20	45·761964	70	2720·080074	20	51·160120	70	4619·223180
21	50·422921	71	2938·686480	21	56·764530	71	5035·953266
22	55·456755	72	3174·781398	22	62·873338	72	5490·189060
23	60·893296	73	3429·763910	23	69·531939	73	5985·306075
24	66·764759	74	3705·145023	24	76·789813	74	6524·983622
25	73·105940	75	4002·556624	25	84·700896	75	7113·232148
26	79·954415	76	4323·761154	26	93·323977	76	7754·423041
27	87·350768	77	4670·662047	27	102·723135	77	8453·321115
28	95·338830	78	5045·315011	28	112·968217	78	9215·120015
29	103·965936	79	5449·940211	29	124·135356	79	10045·480817
30	113·283211	80	5886·935428	30	136·307539	80	10950·574090
31	123·345868	81	6358·890263	31	149·575217	81	11937·125758
32	134·213537	82	6868·601484	32	164·036987	82	13012·467077
33	145·950620	83	7419·089602	33	179·800315	83	14184·589114
34	158·626670	84	8013·616770	34	196·982344	84	15462·202134
35	172·316804	85	8655·706112	35	215·710755	85	16854·800326
36	187·102148	86	9349·162601	36	236·124723	86	18372·732355
37	203·070320	87	10098·095609	37	258·375948	87	20027·278267
38	220·315945	88	10906·943258	38	282·629783	88	21830·733311
39	238·941221	89	11780·498719	39	309·066463	89	23796·499309
40	259·056519	90	12723·938616	40	337·882445	90	25939·184247
41	280·781040	91	13742·853705	41	369·291865	91	28274·710829
42	304·243523	92	14843·282002	42	403·528133	92	30820·434804
43	329·583005	93	16031·744562	43	440·845665	93	33595·273936
44	356·949646	94	17315·284127	44	481·521775	94	36619·848591
45	386·505617	95	18701·506857	45	525·858734	95	39916·634964·
46	418·426067	96	20198·627405	46	574·186021	96	43510·132110
47	452·900152	97	21815·517598	47	626·862762	97	47427·044000
48	490·132164	98	23561·759006	48	684·280411	98	51696·477960
49	530·342737	99	25447·699726	49	746·865648	99	56350·160977
50	573·770156	100	27484·515704	50	815·083556	100	61422·675465

Amount of £1 per annum in n years at the following rates per cent.

Years	10 per cent.	Years	10 per cent.	Years	10 per cent.	Years	10 per cent.
1	1·	26	109·181765	51	1281·299382	76	13980·849085
2	2·10	27	121·099942	52	1410·429320	77	15379·933994
3	3·31	28	134·209936	53	1552·472252	78	16918·927393
4	4·641	29	148·630930	54	1708·719477	79	18611·820133
5	6·1051	30	164·494023	55	1880·591425	80	20474·002146
6	7·71561			56	2069·650567		
7	9·487171	31	181·943425	57	2277·615624	81	22522·402360
8	11·435888	32	201·137767	58	2506·377186	82	24775·642596
9	13·579477	33	222·251544	59	2758·014905	83	27254·206856
10	15·937425	34	245·476699	60	3034·816395	84	29980·627542
		35	271·024368			85	32979·690296
11	18·531167	36	299·126805	61	3339·298035	86	36278·659326
12	21·384284	37	330·039486	62	3674·227838	87	39907·525258
13	24·522712	38	364·043434	63	4042·650622	88	43899·277784
14	27·974983	39	401·447778	64	4447·915685	89	48290·205562
15	31·772482	40	442·592556	65	4893·707253	90	53120·226118
16	35·949730			66	5384·077978		
17	40·544703	41	487·851811	67	5923·485776	91	58433·248730
18	45·599173	42	537·636992	68	6516·834354	92	64277·573603
19	51·159090	43	592·400692	69	7169·517789	93	70706·330964
20	57·274999	44	652·640761	70	7887·469568	94	77777·964060
		45	718·904837			95	85556·760466
21	64·002499	46	791·795321	71	8677·216525	96	94113·436513
22	71·402749	47	871·974853	72	9545·938177	97	103525·780164
23	79·543024	48	960·172338	73	10501·531995	98	113879·358180
24	88·497327	49	1057·189572	74	11552·685195	99	125268·293998
25	98·347059	50	1163·908529	75	12708·953714	100	137796·123398

TABLE IV.

Present value of £1, due n years hence, to one hundred years, at the rates of 3, 3½, 4, 4½, 5, 6, 7, 8, 9, *and* 10 *per cent., and at the rates of* 11, 12, 13, 14, 15, 16, 17, 18, 19, 20, 21, 22, 23, 24, *and* 25 *per cent. to* **50** *years.*

Calculated to **8** *decimal places for each percentage.*

f

TABLE IV. xliii

Present value of £1 due n years hence at the following rates per cent.

Years	3 per cent.	Years	3 per cent.	Years	3½ per cent.	Years	3½ per cent.
1	·97087379	51	·22146318	1	·96618357	51	·17299843
2	·94259591	52	.21501280	2	·93351070	52	·16714824
3	·91514166	53	·20875029	3	·90194270	53	·16149589
4	·88848705	54	·20267019	4	·87144223	54	·15603467
5	·86260878	55	·19576717	5	·84197317	55	·15075814
6	·83748426	56	·19103609	6	·81350064	56	·14566004
7	·81309151	57	·18547193	7	·78599096	57	·14073433
8	·78940923	58	·18006984	8	·75941156	58	·13597520
9	·76641673	59	·17482508	9	·73373097	59	·13137701
10	·74409391	60	·16973309	10	·70891881	60	·12693431
11	·72242126	61	·16478941	11	·68494571	61	·12264184
12	·70137988	62	·15998972	12	·66178330	62	·11849453
13	·68095134	63	·15532982	13	·63940415	63	·11448747
14	·66111781	64	·15080565	14	·61778179	64	·11061591
15	·64186195	65	·14641325	15	·59689062	65	·10687528
16	·62316694	66	·14214879	16	·57670591	66	·10326114
17	·60501645	67	·13800853	17	·55720378	67	·09976922
18	·58739461	68	·13398887	18	·53836114	68	·09639538
19	·57028603	69	·13008628	19	·52015569	69	·09313563
20	·55367575	70	·12629736	20	·50256588	70	·08998612
21	·53754928	71	·12261880	21	·48557090	71	·08694311
22	·52189250	72	·11904737	22	·46915063	72	·08400300
23	·50669175	73	·11557998	23	·45328563	73	·08116232
24	·49193374	74	·11221357	24	·43795713	74	·07841770
25	·47760556	75	·10894521	25	·42314699	75	·07576590
26	·46369473	76	·10577205	26	·40883767	76	·07320376
27	·45018906	77	·10269131	27	·39501224	77	·07072827
28	·43707675	78	·09970030	28	·38165434	78	·06833650
29	·42434636	79	·09679641	29	·36874815	79	·06602560
30	·41198676	80	·09397710	30	·35627841	80	·06379285
31	·39998714	81	·09123990	31	·34423035	81	·06163561
32	·38833703	82	·08858243	32	·33258971	82	·05955131
33	·37702625	83	·08600236	33	·32134271	83	·05753750
34	·36604490	84	·08349743	34	·31047605	84	·05559178
35	·35538340	85	·08106547	35	·29997686	85	·05371187
36	·34503243	86	·07870434	36	·28983272	86	·05189553
37	·33498294	87	·07641198	37	·28003161	87	·05014060
38	·32522615	88	·07418639	38	·27056194	88	·04844503
39	·31575355	89	·07202562	39	·26141250	89	·04680679
40	·30655684	90	·06992779	40	·25257247	90	·04522395
41	·29762800	91	·06789105	41	·24403137	91	·04369464
42	·28895922	92	·06591364	42	·23577910	92	·04221704
43	·28054294	93	·06399383	43	·22780590	93	·04078941
44	·27237178	94	·06212993	44	·22010231	94	·03941006
45	·26443862	95	·06032032	45	·21265924	95	·03807735
46	·25673652	96	·05856342	46	·20546787	96	·03678971
47	·24925877	97	·05685769	47	·19851968	97	·03554562
48	·24199880	98	·05520164	48	·19180645	98	·03434358
49	·23495029	99	·05359383	49	·18532024	99	·03318221
50	·22810708	100	·05203284	50	·17905337	100	·03206011

THE ENGINEER'S VALUING ASSISTANT.

Present value of £1 due n years hence at the following rates per cent.

Years	4 per cent.	Years	4 per cent.	Years	4½ per cent.	Years	4½ per cent.
1	·96153846	51	·13530059	1	·95693780	51	·10594225
2	·92455621	52	·13009672	2	·91572995	52	·10138014
3	·88899636	53	·12509300	3	·87629660	53	·09701449
4	·85480419	54	·12028173	4	·83856134	54	·09283683
5	·82192711	55	·11565551	5	·80245105	55	·08883907
6	·79031453	56	·11120722	6	·76789574	56	·08501347
7	·75991781	57	·10693002	7	·73482846	57	·08135260
8	·73069020	58	·10281733	8	·70318513	58	·07784938
9	·70258674	59	·09886282	9	·67290443	59	·07449701
10	·67556417	60	·09506040	10	·64392768	60	·07128901
11	·64958093	61	·09140423	11	·61619874	61	·06821915
12	·62459705	62	·08788868	12	·58966386	62	·06528148
13	·60057409	63	·08450835	13	·56427164	63	·06247032
14	·57747508	64	·08125803	14	·53997286	64	·05978021
15	·55526450	65	·07813272	15	·51672044	65	·05720594
16	·53390818	66	·07512760	16	·49446932	66	·05474253
17	·51337325	67	·07223809	17	·47317639	67	·05238519
18	·49362812	68	·06945970	18	·45280037	68	·05012397
19	·47464242	69	·06678818	19	·43330179	69	·04797069
20	·45638695	70	·06421940	20	·41464286	70	·04590497
21	·43883360	71	·06174942	21	·39678743	71	·04392820
22	·42195539	72	·05937445	22	·37970089	72	·04203655
23	·40572633	73	·05709081	23	·36335013	73	·04022637
24	·39012147	74	·05489501	24	·34770347	74	·03849413
25	·37511680	75	·05278367	25	·33273060	75	·03683649
26	·36068923	76	·05075353	26	·31840248	76	·03525023
27	·34681657	77	·04880147	27	·30469137	77	·03373228
28	·33347747	78	·04692449	28	·29157069	78	·03227969
29	·32065141	79	·04511970	29	·27901502	79	·03088966
30	·30831867	80	·04338433	30	·26700001	80	·02955947
31	·29646026	81	·04171570	31	·25550241	81	·02828658
32	·28505794	82	·04011125	32	·24449991	82	·02706850
33	·27409417	83	·03856851	33	·23397121	83	·02590287
34	·26355209	84	·03708510	34	·22389589	84	·02478744
35	·25341547	85	·03565875	35	·21425444	85	·02372003
36	·24366872	86	·03428726	36	·20502817	86	·02269860
37	·23429685	87	·03296852	37	·19619921	87	·02172115
38	·22528543	88	·03170050	38	·18775044	88	·02078579
39	·21662061	89	·03048125	39	·17966549	89	·01989070
40	·20828904	90	·02930890	40	·17192870	90	·01903417
41	·20027792	91	·02818163	41	·16452507	91	·01821451
42	·19257493	92	·02709772	42	·15744026	92	·01743016
43	·18516820	93	·02605550	43	·15066054	93	·01667958
44	·17804635	94	·02505337	44	·14417276	94	·01596132
45	·17119841	95	·02408978	45	·13796437	95	·01527399
46	·16461386	96	·02316325	46	·13202332	96	·01461626
47	·15828256	97	·02227235	47	·12633810	97	·01398685
48	·15219476	98	·02141572	48	·12089771	98	·01338454
49	·14634112	99	·02059204	49	·11569158	99	·01280817
50	·14071262	100	·01980004	50	·11070965	100	·01225663

Present value of £1 due n years hence at the following rates per cent.

Years	5 per cent.	Years	5 per cent.	Years	6 per cent.	Years	6 per cent.
1	·95238095	51	·08305117	1	·94339623	51	·05121544
2	·90702948	52	·07909635	2	·88999644	52	·04831645
3	·86383760	53	·07532986	3	·83961928	53	·04558156
4	·82270247	54	·07174272	4	·79209466	54	·04300147
5	·78352616	55	·06832640	5	·74725817	55	·04056742
6	·74621546	56	·06507276	6	·70496054	56	·03827115
7	·71068133	57	·06197406	7	·66505711	57	·03610486
8	·67683936	58	·05902291	8	·62741237	58	·03406119
9	·64460892	59	·05621230	9	·59189846	59	·03213320
10	·61391325	60	·05353552	10	·55839478	60	·03031434
11	·58467929	61	·05098621	11	·52678753	61	·02859843
12	·55683742	62	·04855830	12	·49696936	62	·02697965
13	·53032135	63	·04624600	13	·46883902	63	·02545250
14	·50506795	64	·04404381	14	·44230096	64	·02401179
15	·48101710	65	·04194648	15	·41726506	65	·02265264
16	·45811152	66	·03994903	16	·39364628	66	·02137041
17	·43629669	67	·03804670	17	·37136442	67	·02016077
18	·41552065	68	·03623495	18	·35034379	68	·01901959
19	·39573396	69	·03450948	19	·33051301	69	·01794301
20	·37688948	70	·03286617	20	·31180473	70	·01692737
21	·35894236	71	·03130111	21	·29415540	71	·01596921
22	·34184987	72	·02981058	22	·27750510	72	·01506530
23	·32557131	73	·02839103	23	·26179726	73	·01421254
24	·31006791	74	·02703908	24	·24697855	74	·01340806
25	·29530277	75	·02575150	25	·23299863	75	·01264911
26	·28124073	76	·02452524	26	·21981003	76	·01193313
27	·26784832	77	·02335737	27	·20736795	77	·01125767
28	·25509364	78	·02224512	28	·19563014	78	·01062044
29	·24294632	79	·02118582	29	·18455674	79	·01001928
30	·23137745	80	·02017698	30	·17411013	80	·00945215
31	·22035947	81	·01921617	31	·16425484	81	·00891713
32	·20986617	82	·01830111	32	·15495740	82	·00841238
33	·19987254	83	·01742963	33	·14618622	83	·00793621
34	·19035480	84	·01659965	34	·13791153	84	·00748699
35	·18129029	85	·01580919	35	·13010522	85	·00706320
36	·17265741	86	·01505637	36	·12274077	86	·00666340
37	·16443563	87	·01433940	37	·11579318	87	·00628622
38	·15660536	88	·01365657	38	·10923885	88	·00593040
39	·14914797	89	·01300626	39	·10305552	89	·00559472
40	·14204568	90	·01238691	40	·09722219	90	·00527803
41	·13528160	91	·01179706	41	·09171905	91	·00497928
42	·12883962	92	·01123530	42	·08652740	92	·00469743
43	·12270440	93	·01070028	43	·08162962	93	·00443154
44	·11686133	94	·01019074	44	·07700908	94	·00418070
45	·11129651	95	·00970547	45	·07265007	95	·00394405
46	·10599668	96	·00924331	46	·06853781	96	·00372081
47	·10094921	97	·00880315	47	·06465831	97	·00351019
48	·09614211	98	·00838395	48	·06099840	98	·00331150
49	·09156391	99	·00798471	49	·05754566	99	·00312406
50	·08720373	100	·00760449	50	·05428836	100	·00294723

Present value of £1 due n years hence at the following rates per cent.

Years	7 per cent.	Y ars	7 per cent.	Years	8 per cent.	Years	8 per cent.
1	·93457944	51	·03172688	1	·92592593	51	·01974188
2	·87343873	52	·02965129	2	·85733882	52	·01827952
3	·81629788	53	·02771148	3	·79383224	53	·01692548
4	·76289521	54	·02589858	4	·73502985	54	·01567174
5	·71298618	55	·02420428	5	·68058320	55	·01451087
6	·66634222	56	·02262083	6	·63016963	56	·01343599
7	·62274974	57	·02114096	7	·58349040	57	·01244073
8	·58200910	58	·01975791	8	·54026888	58	·01151920
9	·54393374	59	·01846533	9	·50024897	59	·01066592
10	·50834929	60	·01725732	10	·46319349	60	·00987585
11	·47509280	61	·01612834	11	·42888286	61	·00914431
12	·44401196	62	·01507321	12	·39711376	62	·00846696
13	·41496445	63	·01408711	13	·36769792	63	·00783977
14	·38781724	64	·01316553	14	·34046104	64	·00725905
15	·36244602	65	·01230423	15	·31524171	65	·00672134
16	·33873460	66	·01149928	16	·29189047	66	·00622346
17	·31657439	67	·01074699	17	·27026895	67	·00576247
18	·29586392	68	·01004392	18	·25024903	68	·00533562
19	·27650833	69	·00938684	19	·23171206	69	·00494039
20	·25841900	70	·00877275	20	·21454821	70	·00457443
21	·24151309	71	·00819883	21	·19865575	71	·00423558
22	·22571317	72	·00766246	22	·18394051	72	·00392184
23	·21094688	73	·00716117	23	·17031528	73	·00363133
24	·19714662	74	·00669269	24	·15769934	74	·00336234
25	·18424918	75	·00625485	25	·14601790	75	·00311328
26	·17219549	76	·00584565	26	·13520176	76	·00288267
27	·16093037	77	·00546323	27	·12518682	77	·00266914
28	·15040221	78	·00510582	28	·11591372	78	·00247142
29	·14056282	79	·00477179	29	·10732752	79	·00228835
30	·13136712	80	·00445962	30	·09937733	80	·00211885
31	·12277301	81	·00416787	31	·09201605	81	·00196190
32	·11474113	82	·00389520	32	·08520005	82	·00181657
33	·10723470	83	·00364038	33	·07888893	83	·00168201
34	·10021934	84	·00340222	34	·07304531	84	·00155742
35	·09366294	85	·00317965	35	·06763454	85	·00144205
36	·08753546	86	·00297163	36	·06262458	86	·00133523
37	·08180884	87	·00277723	37	·05798572	87	·00123633
38	·07645686	88	·00259554	38	·05369048	88	·00114475
39	·07145501	89	·00242574	39	·04971341	89	·00105995
40	·06678038	90	·00226704	40	·04603093	90	·00098144
41	·06241157	91	·00211873	41	·04262123	91	·00090874
42	·05832857	92	·00198012	42	·03946411	92	·00084142
43	·05451268	93	·00185058	43	·03654084	93	·00077910
44	·05094643	94	·00172952	44	·03383411	94	·00072138
45	·04761349	95	·00161637	45	·03132788	95	·00065795
46	·04449859	96	·00151063	46	·02900730	96	·00061842
47	·04158747	97	·00141180	47	·02685861	97	·00057265
48	·03886679	98	·00131944	48	·02486908	98	·00053024
49	·03632410	99	·00123312	49	·02302693	99	·00049096
50	·03394776	100	·00115245	50	·02132123	100	·00045459

TABLE IV.   xlvii

Present value of £1 due n years hence at the following rates per cent.

Years	9 per cent.	Years	9 per cent.	Years	10 per cent.	Years	10 per cent.
1	·91743119	51	·01233811	1	·90909091	51	·00774414
2	·84167999	52	·01131937	2	·82644628	52	·00704013
3	·77218348	53	·01038474	3	·75131480	53	·00640011
4	·70842521	54	·00952728	4	·68301346	54	·00581829
5	·64993139	55	·00874063	5	·62092132	55	·00528935
6	·59626733	56	·00801892	6	·56447393	56	·00480850
7	·54703424	57	·00735681	7	·51315812	57	·00437136
8	·50186628	58	·00674937	8	·46650738	58	·00397397
9	·46042778	59	·00619208	9	·42409762	59	·00361270
10	·42241081	60	·00568081	10	·38554329	60	·00328427
11	·38753285	61	·00521175	11	·35049390	61	·00298570
12	·35553473	62	·00478142	12	·31863082	62	·00271427
13	·32617865	63	·00438663	13	·28966438	63	·00246752
14	·29924647	64	·00402443	14	·26333125	64	·00224320
15	·27453804	65	·00369214	15	·23939205	65	·00203927
16	·25186976	66	·00338728	16	·21762914	66	·00185388
17	·23107318	67	·00310760	17	·19784467	67	·00168535
18	·21199374	68	·00285101	18	·17985879	68	·00153214
19	·19448967	69	·00261560	19	·16350799	69	·00139285
20	·17843089	70	·00239963	20	·14864363	70	·00126623
21	·16369806	71	·00220150	21	·13513057	71	·00115112
22	·15018171	72	·00201972	22	·12284597	72	·00104647
23	·13778139	73	·00185296	23	·11167816	73	·00095134
24	·12640494	74	·00169996	24	·10152560	74	·00086485
25	·11596784	75	·00155960	25	·09229600	75	·00078623
26	·10639251	76	·00143082	26	·08390545	76	·00071475
27	·09760781	77	·00131268	27	·07627768	77	·00064978
28	·08954854	78	·00120430	28	·06934335	78	·00059070
29	·08215454	79	·00110486	29	·06303941	79	·00053700
30	·07537114	80	·00101363	30	·05730855	80	·00048819
31	·06914783	81	·00092994	31	·05209868	81	·00044381
32	·06343838	82	·00085315	32	·04736244	82	·00040346
33	·05820035	83	·00078271	33	·04305676.	83	·00036678
34	·05339481	84	·00071808	34	·03914251	84	·00033344
35	·04898607	85	·00065879	35	·03558410	85	·00030313
36	·04494135	86	·00060440	36	·03234918	86	·00027557
37	·04123059	87	·00055449	37	·02940835	87	·00025052
38	·03782623	88	·00050871	38	·02673486	88	·00022774
39	·03470296	89	·00046670	39	·02430442	89	·00020704
40	·03183758	90	·00042817	40	·02209493	90	·00018822
41	·02920879	91	·00039282	41	·02008630	91	·00017111
42	·02679706	92	·00036038	42	·01826027	92	·00015555
43	·02458446	93	·00033063	43	·01660025	93	·00014141
44	·02255455	94	·00030333	44	·01509113	94	·00012855
45	·02069224	95	·00027828	45	·01371921	95	·00011687
46	·01898371	96	·00025530	46	·01247201	96	·00010624
47	·01741625	97	·00023422	47	·01133819	97	·00009658
48	·01597821	98	·00021488	48	·01030745	98	·00008780
49	·01465891	99	·00019714	49	·00937041	99	·00007982
50	·01344854	100	·00018086	50	·00851855	100	·00007257

Present value of £1 due n years hence at the following rates per cent.

Years	11 per cent.	12 per cent.	13 per cent.	14 per cent.	15 per cent.	Years
1	·90090090	·89285714	·88495575	·87719211	·86956530	1
2	·81162243	·79719388	·78314668	·76946753	·75614367	2
3	·73119138	·71178025	·69305016	·67497152	·65751623	3
4	·65873097	·63551808	·61331873	·59208028	·57175325	4
5	·59345133	·56742686	·54275994	·51936866	·49717674	5
6	·53464084	·50663112	·48031853	·45558655	·43232760	6
7	·48165841	·45234922	·42506064	·39963732	·37593704	7
8	·43392650	·40388323	·37615986	·35055905	·32690177	8
9	·39092477	·36061003	·33288483	·30750794	·28426241	9
10	·35218448	·32197324	·29458835	·26974381	·24718571	10
11	·31728331	·28747610	·26069765	·23661738	·21494322	11
12	·28584082	·25667509	·23070589	·20755910	·18690715	12
13	·25751426	·22917419	·20416450	·18206939	·16252796	13
14	·23199482	·20461981	·18067655	·15970999	·14132866	14
15	·20900435	·18269626	·15989075	·14009648	·12289449	15
16	·18829220	·16312166	·14149624	·12289165	·10686477	16
17	·16963262	·14564434	·12521791	·10779969	·09292589	17
18	·15282218	·13003959	·11081231	·09456113	·08080512	18
19	·13767764	·11610678	·09806399	·08294836	·07026532	19
20	·12403391	·10366677	·08678229	·07276172	·06110028	20
21	·11174226	·09255961	·07679849	·06382607	·05313068	21
22	·10066870	·08264251	·06796327	·05598778	·04620059	22
23	·09069252	·07378796	·06014448	·04911209	·04017443	23
24	·08170498	·06588210	·05322521	·04308078	·03493428	24
25	·07360809	·05882331	·04710195	·03779016	·03037764	25
26	·06631359	·05252081	·04168314	·03314926	·02641534	26
27	·05974197	·04689358	·03688774	·02907830	·02296986	27
28	·05382160	·04186927	·03264402	·02550728	·01997379	28
29	·04848793	·03738327	·02888851	·02237481	·01736851	29
30	·04368282	·03337792	·02556505	·01962702	·01510305 ·	30
31	·03935389	·02980172	·02262394	·01721669	·01313309	31
32	·03545395	·02660868	·02002119	·01510236	·01142008	32
33	·03194050	·02375775	·01771786	·01324768	·00903344	33
34	·02877522	·02121227	·01567953	·01162077	·00863522	34
35	·02592363	·01893953	·01387569	·01019366	·00750889	35
36	·02335462	·01691029	·01227937	·00894181	·00652947	36
37	·02104020	·01509848	·01086670	·00784369	·00567798	37
38	·01895513	·01348078	·00961655	·00688043	·00493722	38
39	·01707670	·01203641	·00851022	·00603547	·00429323	39
40	·01538441	·01074680	·00753117	·00529427	·00373324	40
41	·01385983	·00959536	·00666475	·00464410	·00324630	41
42	·01248633	·00856728	·00589801	·00407377	·00282287	42
43	·01124895	·00764936	·00521948	·00357348	·00245467	43
44	·01013419	·00682978	·00461901	·00313463	·00213449	44
45	·00912990	·00609802	·00408762	·00274968	·00185608	45
46	·00822513	·00544466	·00361736	·00241200	·00161398	46
47	·00741003	·00486131	·00320120	·00211579	·00140346	47
48	·00667670	·00434045	·00283292	·00185595	·00122040	48
49	·00601415	·00387540	·00250701	·00162803	·00106122	49
50	·00541815	·00346018	·00221859	·00142810	·00092280	50

TABLE IV. xlix

Present value of £1 due n years hence at the following rates per cent.

Years	16 per cent.	17 per cent.	18 per cent.	19 per cent.	20 per cent.	Years
1	·86206897	·85470085	·84745763	·84033613	·83333333	1
2	·74316290	·73051355	·71818443	·70616482	·69444444	2
3	·64065767	·62437056	·60863087	·59341581	·57870370	3
4	·55229110	·53365005	·51578888	·49866875	·48225309	4
5	·47611302	·45611115	·43710922	·41904937	·40187757	5
6	·41044225	·38983859	·37043154	·35214233	·33489798	6
7	·35382952	·33319538	·31392503	·29591792	·27908165	7
8	·30502546	·28478237	·26603816	·24867052	·23256804	8
9	·26295298	·24340374	·22545607	·20896683	·19380670	9
10	·22668360	·20803738	·19106447	·17560238	·16150558	10
11	·19541690	·17780973	·16191904	·14756502	·13458799	11
12	·16846284	·15197413	·13721953	·12400422	·11215665	12
13	·14522659	·12989242	·11628773	·10420523	·09346388	13
14	·12519534	·11101916	·09854893	·08756742	·07788657	14
15	·10792701	·09488817	·08351604	·07358606	·06490547	15
16	·09304053	·08110100	·07077628	·06183703	·05408789	16
17	·08020735	·06931709	·05997992	·05196389	·04507324	17
18	·06914427	·05924538	·05083044	·04366713	·03756104	18
19	·05960713	·05063708	·04307664	·03669507	·03130086	19
20	·05138546	·04327955	·03650563	·03083619	·02608405	20
21	·04429781	·03699107	·03093698	·02591277	·02173671	21
22	·03818776	·03161630	·02621778	·02177544	·01811393	22
23	·03292049	·02702248	·02219846	·01829869	·01509494	23
24	·02837973	·02309614	·01882920	·01537705	·01257912	24
25	·02446528	·01974029	·01595695	·01292189	·01048260	25
26	·02109076	·01687204	·01352284	·01085873	·00873550	26
27	·01818169	·01442055	·01146003	·00912498	·00727958	27
28	·01567387	·01232525	·00971189	·00766805	·00606632	28
29	·01351196	·01053440	·00823042	·00644374	·00505526	29
30	·01164824	·00900376	·00697493	·00541491	·00421272	30
31	·01004159	·00769553	·00591096	·00455034	·00351060	31
32	·00865654	·00657737	·00500929	·00382382	·00292550	32
33	·00746253	·00562169	·00424516	·00321329	·00243792	33
34	·00643322	·00480486	·00359759	·00270025	·00203160	34
35	·00554588	·00410672	·00304880	·00226911	·00169300	35
36	·00478093	·00351002	·00258373	·00190682	·00141083	36
37	·00412149	·00300001	·00218960	·00160237	·00117569	37
38	·00355301	·00256411	·00185560	·00134653	·00097974	38
39	·00306294	·00299155	·00157254	·00113154	·00081645	39
40	·00264047	·00187312	·00133266	·00095087	·00068038	40
41	·00227626	·00160096	·00112937	·00079905	·00056698	41
42	·00196230	·00136834	·00095710	·00067147	·00047248	42
43	·00169163	·00116952	·00081110	·00056426	·00039374	43
44	·00145831	·00099959	·00068737	·00047417	·00032811	44
45	·00125716	·00085435	·00058252	·00039846	·00027343	45
46	·00108376	·00073021	·00049366	·00033484	·00022786	46
47	·00093427	·00062411	·00043662	·00028138	·00018988	47
48	·00080541	·00053343	·00035454	·00023645	·00015823	48
49	·00069432	·00045592	·00030046	·00019870	·00013186	49
50	·00059855	·00038968	·00025462	·00016698	·00010988	50

Present value of £1 due n years hence at the following rates per cent.

Years	21 per cent.	22 per cent.	23 per cent.	24 per cent.	25 per cent.	Years
1	·82644628	·81967213	·81300813	·80645161	·80000000	1
2	·68301346	·67186240	·66098222	·65036420	·64000000	2
3	·56447393	·55070689	·53738392	·52448726	·51199869	3
4	·46650738	·45139909	·43689749	·42297360	·40959937	4
5	·38554329	·36999925	·35520122	·34110774	·32768000	5
6	·31863082	·30327808	·28878148	·27508689	·26214400	6
7	·26333125	·24858859	·23478169	·22184426	·20971520	7
8	·21762914	·20376114	·19087942	·17890666	·16777216	8
9	·17985879	·16701733	·15518652	·14427957	·13421773	9
10	·14864363	·13689945	·12616790	·11635449	·10737418	10
11	·12284597	·11221266	·10257553	·09383427	·08589935	11
12	·10152560	·09197759	·08339474	·07567280	·06871948	12
13	·08390545	·07539147	·06780060	·06102645	·05497558	13
14	·06934335	·06179629	·05512244	·04921488	·04398047	14
15	·05730855	·05065269	·04481499	·03968942	·03518437	15
16	·04736244	·04151860	·03643495	·03200759	·02814750	16
17	·03914251	·03403164	·02962191	·02581258	·02251800	17
18	·03234918	·02789479	·02408286	·02081659	·01801440	18
19	·02673486	·02286458	·01957956	·01678758	·01441152	19
20	·02209493	·01874146	·01591834	·01353837	·01152922	20
21	·01826027	·01536185	·01294174	·01091804	·00922337	21
22	·01509113	·01259168	·01052174	·00880487	·00737870	22
23	·01247201	·01032105	·00855426	·00710070	·00590296	23
24	·01030745	·00845988	·00695468	·00572637	·00472237	24
25	·00851855	·00693433	·00565421	·00461804	·00377789	25
26	·00704013	·00568387	·00459692	·00372423	·00302231	26
27	·00581829	·00465891	·00373733	·00300341	·00241785	27
28	·00480850	·00381878	·00303848	·00242210	·00193428	28
29	·00397397	·00313015	·00247031	·00195331	·00154743	29
30	·00328427	·00256570	·00200838	·00157525	·00123794	30
31	·00271427	·00210303	·00163283	·00127036	·00099035	31
32	·00224260	·00172379	·00132751	·00102449	·00079228	32
33	·00185388	·00141295	·00107927	·00082620	·00063383	33
34	·00153214	·00115815	·00087746	·00066629	·00050706	34
35	·00126228	·00094931	·00071338	·00053733	·00040565	35
36	·00104647	·00077812	·00057998	·00043333	·00032452	36
37	·00086485	·00063780	·00047153	·00034946	·00025961	37
38	·00071475	·00052279	·00038336	·00028182	·00020769	38
39	·00059070	·00042852	·00031167	·00022728	·00016615	39
40	·00048819	·00035124	·00025339	·00018329	·00013292	40
41	·00040346	·00028793	·00020601	·00014781	·00010634	41
42	·00033344	·00023599	·00016749	·00011920	·00008507	42
43	·00027557	·00019345	·00013617	·00009613	·00006806	43
44	·00022774	·00015855	·00011071	·00007753	·00005445	44
45	·00018822	·00012996	·00009001	·00006252	·00004356	45
46	·00015555	·00010652	·00007318	·00005042	·00003484	46
47	·00012855	·00008731	·00005949	·00004066	·00002788	47
48	·00010624	·00007157	·00004837	·00003279	·00002230	48
49	·00008780	·00005866	·00003932	·00002644	·00001784	49
50	·00007257	·00004808	·00003197	·00002133	·00001427	50

TABLE V.

FOR THE

REDEMPTION OF CAPITAL,

OR THE

Fund necessary to be annually invested to produce £1 in n *years, at the several rates of* 1½, 2, 2½, 3, 3¼, 3½, 4, 4¼, 4½, *and* 5 *per cent., calculated to* **10** *decimal places, and to* **100** *years for each percentage. And for rates of* 10, 12, 15, 18, *and* 20 *per cent., to* **10** *decimal places, and for* **50** *years for each percentage. Also, for rates of interest of* 3, 3¼, 3½, 3¾, 4, 4¼, 4½, 4¾, *and* 5 *per cent. per annum, payments being made half-yearly and quarterly ; calculated to* **6** *decimal places and to* **100** *years for each percentage.*

TABLE V.   liii

Redemption Fund necessary to produce £1 in n years at the following rates per cent.

Years	1½ per cent.	Years	1½ per cent.	Years	2 per cent.	Years	2 per cent.
1	1·0000000000	51	·0131946887	1	1·0000000000	51	·0114585615
2	·4962779156	52	·0128328700	2	·4950495049	52	·0111090856
3	·3283829602	53	·0124853664	3	·3267546725	53	·0107739189
4	·2444447860	54	·0121513812	4	·2426237526	54	·0104522618
5	·1940893230	55	·0118301756	5	·1921583941	55	·0101433732
6	·1605252147	56	·0115202954	6	·1585258123	56	·0098465645
7	·1365561645	57	·0112234068	7	·1345119561	57	·0095611957
8	·1185840245	58	·0109366116	8	·1165097991	58	·0092866706
9	·1046098234	59	·0106601241	9	·1025154374	59	·0090224335
10	·0934341779	60	·0103934274	10	·0913265279	60	·0087679658
11	·0842938442	61	·0101360387	11	·0821779428	61	·0081744379
12	·0766799929	62	·0098875059	12	·0745595966	62	·0082864306
13	·0702403574	63	·0096474061	13	·0681183527	63	·0080584849
14	·0647233186	64	·0094153423	14	·0626019702	64	·0078385471
15	·0599443556	65	·0091909423	15	·0578254723	65	·0076262436
16	·0557650778	66	·0089738563	16	·0536501259	66	·0074212231
17	·0520796569	67	·0087637552	17	·0499698408	67	·0072231553
18	·0488057818	68	·0085603297	18	·0467021021	68	·0070317294
19	·0458784701	69	·0083632878	19	·0437817663	69	·0068466526
20	·0432457359	70	·0081723548	20	·0411567181	70	·0066676485
21	·0408654950	71	·0079872709	21	·0387847689	71	·0064944567
22	·0387033152	72	·0078077911	22	·0366314005	72	·0063268307
23	·0367307520	73	·0076336836	23	·0346680976	73	·0061645379
24	·0349241020	74	·0074647293	24	·0328710973	74	·0060073582
25	·0332634539	75	·0073007206	25	·0312204384	75	·0058550830
26	·0317319599	76	·0071414609	26	·0296992308	76	·0057075147
27	·0303152680	77	·0069867637	27	·0282930862	77	·0055644661
28	·0290010765	78	·0068364523	28	·0269896716	78	·0054257595
29	·0277787802	79	·0066903586	29	·0257783552	79	·0052912260
30	·0266391883	80	·0065483231	30	·0246499223	80	·0051607055
31	·0255742954	81	·0064101941	31	·0235963472	81	·0050340453
32	·0245770970	82	·0062758275	32	·0226106073	82	·0049111006
33	·0236414375	83	·0061450857	33	·0216865311	83	·0047917333
34	·0227618855	84	·0060178380	34	·0208186728	84	·0046758118
35	·0219336303	85	·0058939597	35	·0200022092	85	·0045632109
36	·0211523955	86	·0057733319	36	·0192328526	86	·0044538110
37	·0204143673	87	·0056558413	37	·0185067789	87	·0043474981
38	·0197161329	88	·0055413794	38	·0178205663	88	·0042441633
39	·0190546298	89	·0054298429	39	·0171711439	89	·0041437027
40	·0184271017	90	·0053211330	40	·0165557478	90	·0040460169
41	·0178310610	91	·0052151552	41	·0159718836	91	·0039510108
42	·0172642571	92	·0051118190	42	·0154172945	92	·0038585936
43	·0167246488	93	·0050110379	43	·0148899334	93	·0037686782
44	·0162103801	94	·0049127291	44	·0143879391	94	·0036811814
45	·0157197604	95	·0048168132	45	·0139096161	95	·0035960233
46	·0152512458	96	·0047232141	46	·0134534159	96	·0035131275
47	·0148034238	97	·0046318590	47	·0130172220	97	·0034324205
48	·0143749996	98	·0045426778	48	·0126018355	98	·0033538321
49	·0139647841	99	·0044556033	49	·0122039639	99	·0032772947
50	·0135716832	100	·0043705712	50	·0118232097	100	·0032027435

Redemption Fund necessary to produce £1 in n years at the following rates per cent.

Years	2½ per cent.	Years	2½ per cent.	Years	3 per cent.	Years	3 per cent.
1	1·0000000000	51	·0099086955	1	1·0000000000	51	·0085338232
2	·4938271604	52	·0095744635	2	·4926108374	52	·0082171837
3	·3251371672	53	·0092544943	3	·3235303633	53	·0079147059
4	·2408178777	54	·0089479856	4	·2390270452	54	·0076255841
5	·1902468603	55	·0086541932	5	·1883545714	55	·0073490710
6	·1565499709	56	·0083724260	6	·1545975005	56	·0070844726
7	·1324954297	57	·0081020412	7	·1305063538	57	·0068311432
8	·1144673456	58	·0078424404	8	·1124563888	58	·0065884819
9	·1004568897	59	·0075930656	9	·0984338570	59	·0063559281
10	·0892587631	60	·0073533959	10	·0872305066	60	·0061329587
11	·0801059558	61	·0071229444	11	·0780774478	61	·0059190847
12	·0724871271	62	·0069012558	12	·0704620855	62	·0057138575
13	·0660482710	63	·0066879033	13	·0640295440	63	·0055168216
14	·0605365249	64	·0064824869	14	·0585263390	64	·0053276021
15	·0557664561	65	·0062846310	15	·0537665805	65	·0051458128
16	·0515989886	66	·0060939830	16	·0496108493	66	·0049710995
17	·0479277699	67	·0059102110	17	·0459525294	67	·0048031288
18	·0446700805	68	·0057330027	18	·0427086959	68	·0046415871
19	·0417606151	69	·0055620638	19	·0398138806	69	·0044861787
20	·0391471287	70	·0053971168	20	·0372157076	70	·0043336251
21	·0367873272	71	·0052378997	21	·0348717765	71	·0041926632
22	·0346466060	72	·0050841652	22	·0327473948	72	·0040540446
23	·0326963781	73	·0049356794	23	·0308139027	73	·0039205345
24	·0309128203	74	·0047922210	24	·0290474159	74	·0037919109
25	·0292759209	75	·0046535805	25	·0274278710	75	·0036679633
26	·0277687466	76	·0045195594	26	·0259382903	76	·0035484929
27	·0263768721	77	·0043899655	27	·0245642103	77	·0034333105
28	·0250879326	78	·0042646320	28	·0232932334	78	·0033222371
29	·0238912684	79	·0041433776	29	·0221146711	79	·0032151027
30	·0227776407	80	·0040260451	30	·0210192593	80	·0031117457
31	·0217390024	81	·0039124812	31	·0199989288	81	·0030120127
32	·0207683122	82	·0038025403	32	·0190466183	82	·0029157577
33	·0198593818	83	·0036960837	33	·0181561219	83	·0028228417
34	·0190067507	84	·0035929793	34	·0173219634	84	·0027331326
35	·0182055822	85	·0034931011	35	·0165392916	85	·0026465042
36	·0174515767	86	·0033963292	36	·0158037942	86	·0025628365
37	·0167408991	87	·0033025489	37	·0151116244	87	·0024820151
38	·0160701179	88	·0032116510	38	·0144593401	88	·0024039306
39	·0154361533	89	·0031235310	39	·0138438516	89	·0023284787
40	·0148362331	90	·0030380892	40	·0132623779	90	·0022555599
41	·0142678555	91	·0029552302	41	·0127124089	91	·0021850789
42	·0137287567	92	·0028748628	42	·0121916731	92	·0021169449
43	·0132168832	93	·0027968996	43	·0116981103	93	·0020510708
44	·0127303682	94	·0027212571	44	·0112298469	94	·0019873733
45	·0122675105	95	·0026478552	45	·0107851757	95	·0019257729
46	·0118267567	96	·0025766173	46	·0103625378	96	·0018661933
47	·0114066855	97	·0025074697	47	·0099605065	97	·0018085613
48	·0110059938	98	·0024403421	48	·0095777738	98	·0017528070
49	·0106234846	99	·0023751667	49	·0092131383	99	·0016988633
50	·0102580569	100	·0023118786	50	·0088654944	100	·0016466659

TABLE V. lv

Redemption Fund necessary to produce £1 in n years at the following rates per cent.

Years	3¼ per cent.	Years	3¼ per cent.	Years	3½ per cent.	Years	3½ per cent.
1	1·0000000000	51	·0079081725	1	1·0000000000	51	·0073215641
2	·4920045476	52	·0076010287	2	·4914004914	52	·0070242854
3	·3227307792	53	·0073079716	3	·3219341806	53	·0067409979
4	·2381372828	54	·0070281934	4	·2372511395	54	·0064708979
5	·1874155909	55	·0067609454	5	·1864813732	55	·0062132297
6	·1536299447	56	·0065055321	6	·1526682087	56	·0059672981
7	·1295220120	57	·0062613065	7	·1285444938	57	·0057324549
8	·1114626472	58	·0060276663	8	·1104760465	58	·0055080966
9	·0974355561	59	·0058040498	9	·0964460051	59	·0052936605
10	·0862310733	60	·0055899327	10	·0852413679	60	·0050886213
11	·0770793519	61	·0053848252	11	·0760919658	61	·0048924882
12	·0694671846	62	·0051882685	12	·0684839493	62	·0047048020
13	·0630392523	63	·0049998334	13	·0620615726	63	·0045251325
14	·0575417594	64	·0048191173	14	·0565707287	64	·0043530765
15	·0527885769	65	·0046457423	15	·0518250694	65	·0041882558
16	·0486401341	66	·0044793534	16	·0476848306	66	·0040303148
17	·0449896669	67	·0043196168	17	·0440431317	67	·0038789193
18	·0417541470	68	·0041662182	18	·0408168408	68	·0037337550
19	·0388680383	69	·0040188617	19	·0379403252	69	·0035945255
20	·0362788848	70	·0038772683	20	·0353610768	70	·0034609517
21	·0339442356	71	·0037411746	21	·0330365869	71	·0033327702
22	·0318293586	72	·0036103320	22	·0309320742	72	·0032097323
23	·0299055555	73	·0034845054	23	·0290188043	73	·0030916030
24	·0281489054	74	·0033634725	24	·0272728303	74	·0029781601
25	·0265393258	75	·0032470228	25	·0256740354	75	·0028691934
26	·0250598100	76	·0031349574	26	·0242053963	76	·0027645039
27	·0236958807	77	·0030270874	27	·0228524103	77	·0026639029
28	·0224351188	78	·0029230333	28	·0216026452	78	·0025672117
29	·0212668234	79	·0028232256	29	·0204453825	79	·0024832606
30	·0201817174	80	·0027269026	30	·0193713316	80	·0023848887
31	·0191717180	81	·0026341111	31	·0183723998	81	·0022989429
32	·0182297550	82	·0025447051	32	·0174415048	82	·0022162781
33	·0173496132	83	·0024585460	33	·0165724220	83	·0021367560
34	·0165258003	84	·0023755019	34	·0157596583	84	·0020602452
35	·0157534809	85	·0022954470	35	·0149983473	85	·0019866205
36	·0150283131	86	·0022182616	36	·0142841628	86	·0019157629
37	·0143464505	87	·0021438315	37	·0136132454	87	·0018475589
38	·0137044457	88	·0020720479	38	·0129821414	88	·0017819002
39	·0130992039	89	·0020028067	39	·0123877506	89	·0017186838
40	·0125279401	90	·0019360090	40	·0118272823	90	·0016578111
41	·0119881387	91	·0018715599	41	·0112982174	91	·0015991884
42	·0114775251	92	·0018093692	42	·0107982765	92	·0015427259
43	·0109940346	93	·0017493500	43	·0103253914	93	·0014883379
44	·0105357906	94	·0016914200	44	·0098776816	94	·0014359428
45	·0101010826	95	·0016354999	45	·0094534334	95	·0013854621
46	·0096883484	96	·0015815142	46	·0090510817	96	·0013368213
47	·0092961589	97	·0015293902	47	·0086691944	97	·0012899487
48	·0089232032	98	·0014790587	48	·0083064580	98	·0012447758
49	·0085682777	99	·0014304533	49	·0079616665	99	·0012012372
50	·0082302744	100	·0013835101	50	·0076337096	100	·0011592702

Redemption Fund necessary to produce £1 in n years at the following rates per cent.

Years	4 per cent.	Years	4 per cent.	Years	4½ per cent.	Years	4½ per cent.
1	1·0000000000	51	·0062588497	1	1·0000000000	51	·0057793980
2	·4901960784	52	·0059821236	2	·4895858012	52	·0055132157
3	·3203485392	53	·0057191451	3	·3195596844	53	·0052606425
4	·2354900454	54	·0054691025	4	·2346150491	54	·0050208438
5	·1846271135	55	·0052312426	5	·1837070439	55	·0047930729
6	·1507619025	56	·0050048662	6	·1498173286	56	·0045766300
7	·1266096121	57	·0047893234	7	·1256522089	57	·0043708645
8	·1085278320	58	·0045840087	8	·1075649275	58	·0041751705
9	·0944929927	59	·0043883581	9	·0935294356	59	·0039889840
10	·0832909443	60	·0042018451	10	·0823301166	60	·0038117785
11	·0741490393	61	·0040239779	11	·0731933807	61	·0036430616
12	·0665521727	62	·0038542964	12	·0656034888	62	·0034823743
13	·0601437278	63	·0036923701	13	·0592033981	63	·0033292857
14	·0546689731	64	·0035377955	14	·0537380572	64	·0031833933
15	·0499411004	65	·0033901939	15	·0490204277	65	·0030443184
16	·0458199992	66	·0032492100	16	·0449102239	66	·0029117068
17	·0421985221	67	·0031145099	17	·0413001642	67	·0027852249
18	·0389933282	68	·0029857795	18	·0379785883	68	·0026645598
19	·0361386184	69	·0028527231	19	·0352642692	69	·0025494164
20	·0335817503	70	·0027450623	20	·0327198351	70	·0024395176
21	·0312801054	71	·0026325344	21	·0304308333	71	·0023346017
22	·0291988111	72	·0025248919	22	·0283623442	72	·0022344222
23	·0273090568	73	·0024219008	23	·0264855182	73	·0021387467
24	·0255868313	74	·0023233403	24	·0247763107	74	·0020473553
25	·0240119628	75	·0022290015	25	·0232145232	75	·0019600406
26	·0225673805	76	·0021386869	26	·0217830598	76	·0018766066
27	·0212385106	77	·0020522095	27	·0204673559	77	·0017968678
28	·0200129752	78	·0019693922	28	·0192549241	78	·0017206554
29	·0188799342	79	·0018900672	29	·0181349985	79	·0016477824
30	·0178300991	80	·0018140755	30	·0170983084	80	·0015781123
31	·0168553524	81	·0017412661	31	·0161365371	81	·0015114887
32	·0159485897	82	·0016714957	32	·0152427549	82	·0014477702
33	·0151035665	83	·0016046284	33	·0144106446	83	·0013868224
34	·0143147715	84	·0015405351	34	·0136346858	84	·0013285180
35	·0135773224	85	·0014790927	35	·0129099878	85	·0012727359
36	·0128868780	86	·0014201848	36	·0122322015	86	·0012193611
37	·0122395655	87	·0013637001	37	·0115974477	87	·0011682845
38	·0116319191	88	·0013095329	38	·0110022538	88	·0011194021
39	·0110608274	89	·0012575828	39	·0104435029	89	·0010726152
40	·0105234893	90	·0012077538	40	·0099183887	90	·0010278300
41	·0100173765	91	·0011599547	41	·0094243778	91	·0009849569
42	·0095402007	92	·0011140984	42	·0089591781	92	·0009439110
43	·0090898859	93	·0010701020	43	·0085207094	93	·0009046112
44	·0086645444	94	·0010278867	44	·0081070805	94	·0008669802
45	·0082624558	95	·0009873767	45	·0077165675	95	·0008309447
46	·0078820488	96	·0009485002	46	·0073544897	96	·0007964344
47	·0075218855	97	·0009111884	47	·0069987268	97	·0007633827
48	·0071806476	98	·0008753757	48	·0066686377	98	·0007317257
49	·0068571240	99	·0008409996	49	·0063561161	99	·0007014029
50	·0065502004	100	·0008080000	50	·0060600458	100	·0006723562

TABLE V.   lvii

Redemption Fund necessary to produce £1 in n years at the following rates per cent.

Years	4½ per cent.	Years	4½ per cent.	Years	5 per cent.	Years	5 per cent.
1	1·0000000000	51	·0053323191	1	1·0000000000	51	·0045286697
2	·4889975550	52	·0050767923	2	·4878048780	52	·0042944966
3	·3107733582	53	·0048346867	3	·3172085646	53	·0040733368
4	·2337436479	54	·0046051886	4	·2320118326	54	·0038643770
5	·1827916395	55	·0043875437	5	·1809747981	55	·0036668637
6	·1488783875	56	·0041810518	6	·1470174681	56	·0034800978
7	·1247014680	57	·0039850622	7	·1228198184	57	·0033034300
8	·1066096533	58	·0037989695	8	·1047218136	58	·0031362568
9	·0925744700	59	·0036222094	9	·0906900800	59	·0029780161
10	·0813788217	60	·0034542558	10	·0795045750	60	·0028281845
11	·0722481817	61	·0032946176	11	·0703888915	61	·0026862736
12	·0646661886	62	·0031428356	12	·0628254100	62	·0025518273
13	·0582753528	63	·0029984802	13	·0564557652	63	·0024244196
14	·0528203160	64	·0028611494	14	·0510239695	64	·0023036520
15	·0481138081	65	·0027304661	15	·0463422876	65	·0021891514
16	·0440153695	66	·0026060769	16	·0422699080	66	·0020805683
17	·0404175833	67	·0024876496	17	·0386991417	67	·0019775751
18	·0372368975	68	·0023748725	18	·0355462223	68	·0018798643
19	·0344073443	69	·0022674523	19	·0327450104	69	·0017871473
20	·0318761443	70	·0021651129	20	·0302425872	70	·0016991530
21	·0296005669	71	·0020675946	21	·0279961071	71	·0016156265
22	·0275456461	72	·0019746524	22	·0259705086	72	·0015363280
23	·0256824930	73	·0018860556	23	·0241368219	73	·0014610318
24	·0239870299	74	·0018015863	24	·0224709007	74	·0013895254
25	·0224390280	75	·0017210390	25	·0209524573	75	·0013216085
26	·0210213675	76	·0016442194	26	·0195643207	76	·0012570925
27	·0197194616	77	·0015709439	27	·0182918599	77	·0011957993
28	·0185208051	78	·0015010391	28	·0171225304	78	·0011375610
29	·0174146147	79	·0014343408	29	·0160455149	79	·0011061609
30	·0163915429	80	·0013706935	30	·0150514351	80	·0010296235
31	·0154434459	81	·0013099502	31	·0141321204	81	·0009796332
32	·0145631962	82	·0012519715	32	·0132804189	82	·0009321143
33	·0137445281	83	·0011966252	33	·0124900437	83	·0008869406
34	·0129819119	84	·0011437861	34	·0117554454	84	·0008439924
35	·0122704478	85	·0010933355	35	·0110717072	85	·0008031567
36	·0116057796	86	·0010451606	36	·0104344571	86	·0007643265
37	·0109840206	87	·0009915434	37	·0098397945	87	·0007274005
38	·0104016920	88	·0009452152	38	·0092842282	88	·0006922828
39	·0098556712	89	·0009132468	39	·0087646242	89	·0006588825
40	·0093431466	90	·0008731573	40	·0082781611	90	·0006271136
41	·0088615804	91	·0008348597	41	·0078222924	91	·0005968946
42	·0084086759	92	·0007982710	42	·0073947131	92	·0005681481
43	·0079823492	93	·0007633126	43	·0069933328	93	·0005408008
44	·0075807056	94	·0007299095	44	·0066162506	94	·0005147832
45	·0072020184	95	·0006979905	45	·0062617347	95	·0004900295
46	·0068447107	96	·0006674877	46	·0059282036	96	·0004664770
47	·0065073395	97	·0006383364	47	·0056142109	97	·0004440666
48	·0061885821	98	·0006104754	48	·0053184306	98	·0004227418
49	·0058872235	99	·0005838459	49	·0050396453	99	·0004024492
50	·0056021459	100	·0005583922	50	·0047767355	100	·0003831381

Redemption Fund necessary to produce £1 in **n** years at the following rates per cent

Years	10 per cent.	12 per cent.	15 per cent.	18 per cent.	20 per cent.	Year
1	1·0000000000	1·0000000000	1·0000000000	1·0000000000	1·0000000000	1
2	·4761904761	·4716981132	·4651162790	·4587155063	·4545454545	2
3	·3021148036	·2963489805	·2879769618	·2799238607	·2747252747	3
4	·2154708037	·2092344363	·2002653515	·1917387036	·1862891207	4
5	·1637974807	·1574097319	·1483155524	·1397778418	·1343797033	5
6	·1296073803	·1262257184	·1142369065	·1059101292	·1007057459	6
7	·1054054997	·0991177359	·0903603634	·0823619993	·0774239263	7
8	·0874440175	·0813028414	·0728500896	·0652443589	·0606094224	8
9	·0736405391	·0676788887	·0599574015	·0523948239	·0480794617	9
10	·0627453949	·0569841642	·0492520625	·0425146413	·0385227569	10
11	·0539631420	·0484154043	·0410689830	·0347763862	·0311037942	11
12	·0467633151	·0414368076	·0344807761	·0286278089	·0252649649	12
13	·0407785238	·0356771951	·0291104565	·0236862073	·0206200011	13
14	·0357462232	·0308712461	·0246884898	·0196780583	·0168930552	14
15	·0314737769	·0268242396	·0210170526	·0164027825	·0138821198	15
16	·0278166207	·0233990180	·0179476914	·0137100839	·0114361350	16
17	·0246641344	·0204567275	·0153668623	·0114852711	·0094401469	17
18	·0219302222	·0179373114	·0131862873	·0096894570	·0078053857	18
19	·0195468682	·0157630049	·0113863504	·0081028390	·0063624532	19
20	·0174596248	·0138787800	·0097614704	·0068199812	·0053865307	20
21	·0156243898	·0122400915	·0084167914	·0057464327	·0044439388	21
22	·0140050629	·0108105088	·0072657713	·0048462577	·0036896187	22
23	·0125718127	·0095599650	·0062783947	·0040901996	·0030652575	23
24	·0112997764	·0084634417	·0054298296	·0034542973	·0025478730	24
25	·0101680722	·0074999698	·0046994023	·0029188261	·0021187290	25
26	·0091590386	·0066518581	·0040698058	·0024674779	·0017624956	26
27	·0082576423	·0059040937	·0035264815	·0020867195	·0014665923	27
28	·0074510132	·0052438691	·0030571309	·0017652846	·0012206684	28
29	·0067280747	·0046602068	·0026513265	·0014937692	·0010161900	29
30	·0060792483	·0041436576	·0022801982	·0012643056	·0008461085	30
31	·0054962193	·0036860570	·0019961796	·0010702987	·0007045936	31
32	·0049411167	·0032803263	·0017328006	·0009062108	·0005868168	32
33	·0044994063	·0029203096	·0015045161	·0007673859	·0004875834	33
34	·0040737064	·0026006383	·0013065655	·0006499044	·0004071466	34
35	·0036897051	·0023166193	·0011348546	·0005504633	·0003391738	35
36	·0033430638	·0020641406	·0009858572	·0004662768	·0002825649	36
37	·0030299405	·0018395924	·0008565329	·0003989937	·0002354154	37
38	·0027469250	·0016397998	·0007442569	·0003346284	·0001961410	38
39	·0024909840	·0014619665	·0006467613	·0002835030	·0001634241	39
40	·0022594144	·0013036256	·0005620850	·0002401991	·0001361682	40
41	·0020498028	·0011625982	·0004885308	·0002035171	·0001134606	41
42	·0018599911	·0010369577	·0004246290	·0001724424	·0000945416	42
43	·0016880466	·0009249987	·0003691063	·0001461163	·0000787784	43
44	·0015322365	·0008252102	·0003208590	·0001238120	·0000656444	44
45	·0013910047	·0007362523	·0002789300	·0001049144	·0000547007	45
46	·0012629527	·0006569363	·0002424890	·0000889026	·0000455818	46
47	·0011868221	·0005862064	·0002108156	·0000758355	·0000379834	47
48	·0010414797	·0005231248	·0001832843	·0000638396	·0000316518	48
49	·0009459041	·0004668576	·0001593523	·0000540984	·0000263758	49
50	·0008591740	·0004166635	·0001385480	·0000458440	·0000219794	50

N.B. The above Table for rates of interest of 10, 12, 15, 18, and 20 per cent. was employed in calculating the Old Present Value Table of £1 per annum given in Table XII., but it is evident that it could not be applied practically for the *Redemption of Capital.*

TABLE V.   263

Redemption Fund necessary to produce £1 in n years, payments being made Half-yearly and Quarterly, at the following rates per cent.

Years' Duration	3 per cent. Half-yearly	Quarterly	Years' Duration	3 per cent. Half-yearly	Quarterly	Years' Duration	3½ per cent. Half-yearly	Quarterly	Years' Duration	3½ per cent. Half-yearly	Quarterly
1	·992556	·988820	51	·008413	·008352	1	·991941	·987895	51	·007781	·007718
2	·488890	·487022	52	·008099	·008040	2	·487976	·485954	52	·007478	·007416
3	·321050	·319806	53	·007800	·007743	3	·320046	·318699	53	·007188	·007128
4	·237168	·236235	54	·007514	·007458	4	·236125	·235115	54	·006911	·006853
5	·186868	·186123	55	·007240	·007186	5	·185807	·185000	55	·006647	·006590
6	·153360	·152739	56	·006978	·006925	6	·152291	·151619	56	·006395	·006339
7	·129447	·128915	57	·006728	·006676	7	·128375	·127800	57	·006154	·006100
8	·111530	·111065	58	·006488	·006437	8	·110461	·109958	58	·005923	·005870
9	·097612	·097199	59	·006258	·006208	9	·096546	·096100	59	·005702	·005651
10	·086492	·086121	60	·006037	·005989	10	·085432	·085031	60	·005490	·005441
11	·077407	·077070	61	·005826	·005779	11	·076354	·075990	61	·005288	·005240
12	·069848	·069540	62	·005623	·005577	12	·068804	·068471	62	·005094	·005047
13	·063464	·063180	63	·005428	·005383	13	·062428	·062122	63	·004908	·004862
14	·058002	·057739	64	·005241	·005197	14	·056976	·056692	64	·004730	·004685
15	·053278	·053033	65	·005061	·005019	15	·052262	·051997	65	·004558	·004515
16	·049154	·048925	66	·004888	·004847	16	·048147	·047900	66	·004394	·004352
17	·045524	·045309	67	·004722	·004682	17	·044527	·044295	67	·004237	·004195
18	·042305	·042102	68	·004563	·004523	18	·041319	·041100	68	·004085	·004045
19	·039432	·039241	69	·004409	·004371	19	·038457	·038250	69	·003940	·003901
20	·036854	·036673	70	·004262	·004224	20	·035889	·035694	70	·003800	·003762
21	·034529	·034356	71	·004119	·004083	21	·033575	·033389	71	·003666	·003629
22	·032421	·032257	72	·003982	·003947	22	·031478	·031302	72	·003537	·003501
23	·030503	·030346	73	·003851	·003816	23	·029571	·029403	73	·003413	·003378
24	·028750	·028601	74	·003724	·003689	24	·027829	·027669	74	·003294	·003259
25	·027143	·027001	75	·003601	·003568	25	·026234	·026080	75	·003179	·003146
26	·025666	·025529	76	·003483	·003451	26	·024767	·024620	76	·003069	·003036
27	·024303	·024172	77	·003370	·003338	27	·023415	·023275	77	·002963	·002931
28	·023042	·022916	78	·003260	·003229	28	·022166	·022031	78	·002860	·002829
29	·021873	·021752	79	·003154	·003124	29	·021008	·020878	79	·002762	·002731
30	·020787	·020670	80	·003052	·003023	30	·019933	·019808	80	·002667	·002637
31	·019775	·019663	81	·002954	·002925	31	·018932	·018812	81	·002576	·002547
32	·018831	·018722	82	·002859	·002831	32	·017999	·017883	82	·002488	·002460
33	·017948	·017843	83	·002767	·002740	33	·017127	·017015	83	·002403	·002375
34	·017121	·017020	84	·002679	·002652	34	·016311	·016203	84	·002321	·002294
35	·016345	·016247	85	·002594	·002567	35	·015546	·015442	85	·002243	·002216
36	·015616	·015521	86	·002511	·002485	36	·014828	·014727	86	·002167	·002141
37	·014930	·014838	87	·002432	·002406	37	·014153	·014055	87	·002094	·002069
38	·014283	·014195	88	·002355	·002330	38	·013517	·013423	88	·002023	·001999
39	·013673	·013587	89	·002280	·002256	39	·012918	·012827	89	·001955	·001931
40	·013097	·013014	90	·002208	·002185	40	·012352	·012264	90	·001889	·001866
41	·012552	·012471	91	·002139	·002116	41	·011818	·011732	91	·001826	·001803
42	·012036	·011958	92	·002072	·002050	42	·011312	·011229	92	·001765	·001743
43	·011547	·011471	93	·002007	·001985	43	·010834	·010753	93	·001706	·001685
44	·011083	·011009	94	·001944	·001923	44	·010380	·010302	94	·001649	·001628
45	·010642	·010571	95	·001884	·001863	45	·009950	·009875	95	·001594	·001574
46	·010224	·010154	96	·001825	·001805	46	·009542	·009468	96	·001541	·001521
47	·009826	·009758	97	·001768	·001748	47	·009154	·009083	97	·001490	·001471
48	·009446	·009381	98	·001714	·001694	48	·008785	·008716	98	·001441	·001422
49	·009085	·009021	99	·001661	·001641	49	·008434	·008367	99	·001393	·001375
50	·008741	·008679	100	·001609	·001591	50	·008100	·008034	100	·001347	·001329

Redemption Fund necessary to produce £1 in n years, payments being made Half-yearly and Quarterly, at the following rates per cent.

Years' Duration	3½ per cent. Redemption Fund. Payments being made		Years' Duration	3½ per cent. Redemption Fund. Payments being made		Years' Duration	3¾ per cent. Redemption Fund. Payments being made		Years' Duration	3¾ per cent. Redemption Fund. Payments being made	
	Half-yearly	Quarterly		Half-yearly	Quarterly		Half-yearly	Quarterly		Half-yearly	Quarterly
1	·991326	·986970	51	·007189	·007123	1	·990712	·986047	51	·006636	·006568
2	·487065	·484888	52	·006896	·006832	2	·486155	·483823	52	·006353	·006287
3	·319045	·317594	53	·006616	·006554	3	·318047	·316493	53	·006084	·006019
4	·235086	·233999	54	·006350	·006289	4	·234051	·232887	54	·005827	·005765
5	·184751	·183882	55	·006096	·006037	5	·183699	·182769	55	·005583	·005523
6	·151228	·150504	56	·005853	·005796	6	·150170	·149396	56	·005350	·005292
7	·127311	·126692	57	·005621	·005566	7	·126254	·125591	57	·005129	·005072
8	·109399	·108858	58	·005400	·005346	8	·108346	·107767	58	·004917	·004862
9	·095490	·095010	59	·005189	·005136	9	·094443	·093929	59	·004715	·004661
10	·084382	·083951	**60**	·004986	·004935	**10**	·083343	·082882	**60**	·004522	·004470
11	·075313	·074922	61	·004793	·004743	11	·074282	·073864	61	·004338	·004288
12	·067771	·067414	62	·004608	·004560	12	·066751	·066368	62	·004163	·004113
13	·061405	·061076	63	·004431	·004384	13	·060396	·060044	63	·003995	·003947
14	·055963	·055658	64	·004262	·004216	14	·054964	·054639	64	·003834	·003788
15	·051260	·050976	65	·004099	·004055	15	·050272	·049969	65	·003680	·003635
16	·047156	·046891	66	·003944	·003900	16	·046181	·045898	66	·003534	·003490
17	·043547	·043298	67	·003795	·003752	17	·042584	·042319	67	·003393	·003351
18	·040350	·040116	68	·003652	·003611	18	·039399	·039150	68	·003259	·003217
19	·037500	·037279	69	·003515	·003475	19	·036561	·036326	69	·003130	·003090
20	·034944	·034735	**70**	·003383	·003344	**20**	·034018	·033796	**70**	·003007	·002968
21	·032641	·032443	71	·003257	·003219	21	·031728	·031517	71	·002888	·002851
22	·030556	·030368	72	·003136	·003099	22	·029656	·029455	72	·002775	·002739
23	·028661	·028481	73	·003020	·002984	23	·027773	·027582	73	·002667	·002631
24	·026931	·026760	74	·002908	·002874	24	·026056	·025874	74	·002563	·002522
25	·025348	·025184	75	·002801	·002767	25	·024486	·024312	75	·002463	·002430
26	·023893	·023737	76	·002698	·002665	26	·023044	·022878	76	·002368	·002335
27	·022553	·022403	77	·002599	·002567	27	·021717	·021558	77	·002276	·002244
28	·021316	·021172	78	·002505	·002473	28	·020492	·020340	78	·002188	·002157
29	·020170	·020032	79	·002413	·002383	29	·019359	·019213	79	·002104	·002074
30	·019107	·018974	**80**	·002325	·002296	**30**	·018308	·018168	**80**	·002023	·001994
31	·018118	·017990	81	·002241	·002212	31	·017332	·017197	81	·001946	·001917
32	·017196	·017074	82	·002160	·002132	32	·016423	·016293	82	·001871	·001844
33	·016336	·016218	83	·002082	·002055	33	·015575	·015450	83	·001800	·001773
34	·015532	·015418	84	·002007	·001980	34	·014783	·014663	84	·001731	·001705
35	·014779	·014668	85	·001935	·001909	35	·014042	·013926	85	·001665	·001640
36	·014072	·013966	86	·001865	·001840	36	·013348	·013236	86	·001602	·001577
37	·013408	·013306	87	·001798	·001774	37	·012696	·012588	87	·001541	·001517
38	·012784	·012685	88	·001734	·001710	38	·012084	·011980	88	·001482	·001459
39	·012196	·012100	89	·001672	·001649	39	·011508	·011407	89	·001426	·001404
40	·011642	·011549	**90**	·001612	·001590	**40**	·010965	·010868	**90**	·001372	·001351
41	·011119	·011029	91	·001555	·001533	41	·010454	·010360	91	·001321	·001299
42	·010625	·010537	92	·001500	·001478	42	·009971	·009880	92	·001271	·001250
43	·010157	·010073	93	·001446	·001425	43	·009515	·009427	93	·001223	·001203
44	·009715	·009633	94	·001395	·001375	44	·009084	·008999	94	·001177	·001157
45	·009295	·009216	95	·001346	·001326	45	·008676	·008594	95	·001133	·001114
46	·008898	·008821	96	·001298	·001279	46	·008290	·008210	96	·001090	·001072
47	·008520	·008446	97	·001252	·001233	47	·007924	·007846	97	·001049	·001031
48	·008162	·008090	98	·001208	·001190	48	·007576	·007501	98	·001010	·000993
49	·007822	·007751	99	·001165	·001148	49	·007247	·007174	99	·000972	·000955
50	·007498	·007430	**100**	·001124	·001107	**50**	·006933	·006863	**100**	·000936	·000920

TABLE V.   lxi

Redemption Fund necessary to produce £1 in n years, payments being made Half-yearly and Quarterly, at the following rates per cent.

Years' Duration	4 per cent. Half-yearly	4 per cent. Quarterly	Years' Duration	4 per cent. Half-yearly	4 per cent. Quarterly	Years' Duration	4¼ per cent. Half-yearly	4¼ per cent. Quarterly	Years' Duration	4¼ per cent. Half-yearly	4¼ per cent. Quarterly
1	·990099	·985124	51	·006199	·006049	1	·989487	·984203	51	·005636	·005565
2	·485248	·482761	52	·005846	·005779	2	·484342	·481701	52	·005375	·005306
3	·317052	·315395	53	·005588	·005522	3	·316059	·314300	53	·005127	·005060
4	·233020	·231778	54	·005342	·005278	4	·231992	·230674	54	·004891	·004827
5	·182653	·181661	55	·005108	·005046	5	·181612	·180559	55	·004668	·004605
6	·149119	·148294	56	·004885	·004825	6	·148074	·147198	56	·004456	·004395
7	·125204	·124498	57	·004673	·004615	7	·124161	·123411	57	·004254	·004195
8	·107300	·106684	58	·004472	·004416	8	·106263	·105608	58	·004062	·004005
9	·093404	·092857	59	·004279	·004225	9	·092375	·091795	59	·003879	·003824
10	·082313	·081822	60	·004096	·004043	10	·081294	·080773	60	·003706	·003653
11	·073263	·072818	61	·003922	·003870	11	·072254	·071782	61	·003540	·003489
12	·065742	·065335	62	·003755	·003705	12	·064746	·064314	62	·003383	·003333
13	·059399	·059024	63	·003596	·003548	13	·058414	·058018	63	·003233	·003185
14	·053979	·053633	64	·003444	·003398	14	·053008	·052641	64	·003090	·003044
15	·049300	·048978	65	·003300	·003254	15	·048342	·048001	65	·002954	·002909
16	·045221	·044921	66	·003161	·003118	16	·044277	·043959	66	·002824	·002781
17	·041637	·041356	67	·003029	·002987	17	·040707	·040409	67	·002701	·002658
18	·038466	·038201	68	·002903	·002862	18	·037550	·037270	68	·002583	·002542
19	·035641	·035391	69	·002783	·002743	19	·034739	·034475	69	·002470	·002431
20	·033112	·032875	70	·002667	·002629	20	·032224	·031975	70	·002363	·002324
21	·030835	·030641	71	·002557	·002520	21	·029961	·029725	71	·002260	·002223
22	·028776	·028564	72	·002452	·002415	22	·027917	·027693	72	·002162	·002127
23	·026907	·026705	73	·002351	·002316	23	·026062	·025850	73	·002069	·002034
24	·025204	·025011	74	·002255	·002220	24	·024373	·024171	74	·001980	·001946
25	·023646	·023463	75	·002162	·002129	25	·022830	·022637	75	·001895	·001862
26	·022218	·022043	76	·002074	·002042	26	·021416	·021232	76	·001813	·001782
27	·020905	·020737	77	·001989	·001958	27	·020117	·019941	77	·001736	·001705
28	·019693	·019533	78	·001908	·001878	28	·018920	·018751	78	·001661	·001632
29	·018573	·018420	79	·001831	·001802	29	·017814	·017653	79	·001590	·001562
30	·017536	·017388	80	·001757	·001728	30	·016790	·016636	80	·001522	·001495
31	·016573	·016431	81	·001686	·001658	31	·015841	·015693	81	·001458	·001431
32	·015677	·015541	82	·001618	·001591	32	·014959	·014817	82	·001396	·001370
33	·014842	·014712	83	·001552	·001526	33	·014138	·014002	83	·001336	·001311
34	·014064	·013938	84	·001490	·001465	34	·013373	·013241	84	·001280	·001255
35	·013335	·013214	85	·001430	·001405	35	·012658	·012532	85	·001225	·001202
36	·012654	·012537	86	·001372	·001349	36	·011990	·011868	86	·001174	·001151
37	·012015	·011902	87	·001317	·001294	37	·011364	·011247	87	·001124	·001102
38	·011415	·011306	88	·001265	·001242	38	·010777	·010664	88	·001077	·001055
39	·010852	·010747	89	·001214	·001192	39	·010226	·010118	89	·001031	·001011
40	·010321	·010220	90	·001166	·001145	40	·009709	·009604	90	·000988	·000968
41	·009822	·009724	91	·001119	·001099	41	·009223	·009122	91	·000946	·000927
42	·009352	·009257	92	·001074	·001055	42	·008764	·008667	92	·000906	·000888
43	·008908	·008816	93	·001032	·001012	43	·008333	·008239	93	·000868	·000850
44	·008488	·008400	94	·000991	·000972	44	·007926	·007835	94	·000832	·000814
45	·008092	·008007	95	·000951	·000933	45	·007541	·007454	95	·000797	·000780
46	·007717	·007635	96	·000913	·000896	46	·007178	·007093	96	·000763	·000747
47	·007362	·007283	97	·000877	·000860	47	·006835	·006753	97	·000731	·000716
48	·007026	·006949	98	·000842	·000826	48	·006510	·006431	98	·000701	·000686
49	·006708	·006633	99	·000809	·000793	49	·006203	·006127	99	·000672	·000657
50	·006406	·006333	100	·000777	·000762	50	·005912	·005838	100	·000643	·000629

Redemption Fund necessary to produce £1 in n years, payments being made Half-yearly and Quarterly, at the following rates per cent.

Years' Duration	4½ per cent. Half-yearly	Quarterly	Years' Duration	4½ per cent. Half-yearly	Quarterly	Years' Duration	4¾ per cent. Half-yearly	Quarterly	Years' Duration	4¾ per cent. Half-yearly	Quarterly
1	·988875	·983282	51	·005187	·005115	1	·988264	·982363	51	·004770	·004696
2	·483438	·480643	52	·004937	·004867	2	·482536	·479587	52	·004530	·004459
3	·315070	·313208	53	·004699	·004632	3	·314084	·312119	53	·004303	·004235
4	·230969	·229575	54	·004475	·004409	4	·229950	·228479	54	·004089	·004024
5	·180575	·179461	55	·004261	·004198	5	·179544	·178369	55	·003886	·003823
6	·147035	·146108	56	·004059	·003998	6	·146002	·145024	56	·003694	·003633
7	·123125	·122332	57	·003867	·003809	7	·122095	·121260	57	·003512	·003453
8	·105233	·104541	58	·003685	·003628	8	·104212	·103483	58	·003340	·003283
9	·091354	·090741	59	·003512	·003457	9	·090343	·089697	59	·003176	·003122
10	·080284	·079734	60	·003348	·003295	10	·079284	·078705	60	·003021	·002969
11	·071256	·070758	61	·003192	·003141	11	·070269	·069745	61	·002874	·002824
12	·063761	·063305	62	·003044	·002994	12	·062787	·062308	62	·002735	·002686
13	·057443	·057024	63	·002903	·002855	13	·056484	·056044	63	·002603	·002556
14	·052051	·051664	64	·002768	·002722	14	·051107	·050700	64	·002477	·002432
15	·047399	·047039	65	·002641	·002596	15	·046470	·046092	65	·002358	·002314
16	·043348	·043013	66	·002520	·002476	16	·042435	·042083	66	·002245	·002202
17	·039793	·039479	67	·002404	·002362	17	·038895	·038566	67	·002137	·002096
18	·036650	·036356	68	·002294	·002254	18	·035768	·035459	68	·002035	·001996
19	·033855	·033578	69	·002189	·002151	19	·032989	·032698	69	·001938	·001900
20	·031355	·031093	70	·002090	·002052	20	·030504	·030230	70	·001845	·001809
21	·029107	·028860	71	·001995	·001958	21	·028260	·028014	71	·001758	·001722
22	·027078	·026843	72	·001904	·001869	22	·026259	·026014	72	·001674	·001640
23	·025238	·025016	73	·001818	·001784	23	·024436	·024203	73	·001595	·001562
24	·023565	·023353	74	·001736	·021703	24	·022778	·022556	74	·001519	·001488
25	·022037	·021835	75	·001657	·001626	25	·021265	·021055	75	·001448	·001417
26	·020638	·020445	76	·001583	·001552	26	·019882	·019681	76	·001379	·001350
27	·019353	·019169	77	·001512	·001482	27	·018613	·018422	77	·001314	·001286
28	·018171	·017995	78	·001444	·001415	28	·017446	·017263	78	·001252	·001225
29	·017080	·016911	79	·001379	·001351	29	·016370	·016195	79	·001194	·001167
30	·016071	·015910	80	·001317	·001291	30	·015376	·015209	80	·001138	·001112
31	·015136	·014982	81	·001258	·001233	31	·014457	·014297	81	·001084	·001060
32	·014268	·014120	82	·001202	·001177	32	·013604	·013451	82	·001033	·001010
33	·013461	·013320	83	·001148	·001124	33	·012812	·012665	83	·000985	·000962
34	·012710	·012574	84	·001097	·001074	34	·012075	·011934	84	·000939	·000917
35	·012009	·011878	85	·001048	·001026	35	·011388	·011253	85	·000895	·000874
36	·011355	·011229	86	·001002	·000980	36	·010748	·010618	86	·000853	·000833
37	·010742	·010621	87	·000957	·000936	37	·010150	·010025	87	·000813	·000794
38	·010169	·010053	88	·000915	·000894	38	·009590	·009470	88	·000776	·000757
39	·009632	·009520	89	·000874	·000855	39	·009066	·008951	89	·000739	·000721
40	·009128	·009020	90	·000835	·000817	40	·008576	·008465	90	·000705	·000688
41	·008654	·008550	91	·000798	·000780	41	·008115	·008009	91	·000672	·000655
42	·008208	·008108	92	·000763	·000745	42	·007683	·007580	92	·000641	·000625
43	·007789	·007693	93	·000729	·000712	43	·007276	·007178	93	·000611	·000596
44	·007395	·007302	94	·000697	·000681	44	·006894	·006799	94	·000583	·000568
45	·007023	·006933	95	·000666	·000650	45	·006535	·006443	95	·000556	·000541
46	·006672	·006585	96	·000637	·000622	46	·006196	·006108	96	·000530	·000516
47	·006340	·006257	97	·000609	·000594	47	·005877	·005792	97	·000505	·000492
48	·006027	·005947	98	·000582	·000568	48	·005575	·005494	98	·000482	·000469
49	·005732	·005654	99	·000556	·000543	49	·005291	·005213	99	·000460	·000447
50	·005452	·005377	100	·000532	·000519	50	·005023	·004947	100	·000438	·000426

Redemption Fund necessary to produce £1 in n years, payments being made Half-yearly and Quarterly, at the following rates per cent.

Years' Duration	5 per cent. Half-yearly	5 per cent. Quarterly	Years' Duration	5 per cent. Half-yearly	5 per cent. Quarterly	Years' Duration	5 per cent. Half-yearly	5 per cent. Quarterly	Years' Duration	5 per cent. Half-yearly	5 per cent. Quarterly
1	·987654	·981444	26	·019149	·018941	51	·004381	·004308	76	·001200	·001172
2	·481636	·478533	27	·017896	·017697	52	·004153	·004082	77	·001141	·001114
3	·313100	·311033	28	·016745	·016555	53	·003937	·003869	78	·001085	·001059
4	·228935	·227387	29	·015685	·015504	54	·003733	·003668	79	·001032	·001007
5	·178518	·177282				55	·003540	·003478			
6	·144974	·143947	30	·014707	·014534	56	·003358	·003298	80	·000981	·000957
7	·121073	·120195	31	·013803	·013637	57	·003186	·003128	81	·000933	·000910
8	·103198	·102432	32	·012965	·012807	58	·003023	·002967	82	·000887	·000865
9	·089340	·088661	33	·012188	·012037	59	·002869	·002815	83	·000844	·000822
10	·078294	·077686	34	·011466	·011321	60	·002724	·002672	84	·000802	·000782
			35	·010794	·010655				85	·000763	·000743
11	·069293	·068742	36	·010168	·010035	61	·002586	·002536	86	·000726	·000707
12	·061826	·061323	37	·009584	·009457	62	·002455	·002407	87	·000690	·000672
13	·055538	·055076	38	·009039	·008916	63	·002331	·002285	88	·000657	·000639
14	·050176	·049750	39	·008529	·008412	64	·002214	·002169	89	·000625	·000608
15	·045555	·045160	40	·008052	·007939	65	·002103	·002060	90	·000594	·000578
16	·041537	·041168				66	·001997	·001956			
17	·038014	·037969	41	·007605	·007497	67	·001897	·001858	91	·000565	·000549
18	·034903	·034580	42	·007186	·007082	68	·001803	·001764	92	·000538	·000523
19	·032140	·031836	43	·006793	·006692	69	·001713	·001676	93	·000511	·000497
20	·029673	·029386	44	·006423	·006327	70	·001628	·001592	94	·000487	·000473
			45	·006076	·005984				95	·000463	·000450
21	·027458	·027187	46	·005750	·005661	71	·001547	·001513	96	·000440	·000428
22	·025461	·025205	47	·005443	·005357	72	·001470	·001437	97	·000419	·000407
23	·023654	·023411	48	·005153	·005071	73	·001397	·001366	98	·000399	·000387
24	·022012	·021782	49	·004881	·004801	74	·001328	·001298	99	·000379	·000368
25	·020516	·020297	50	·004624	·004548	75	·001263	·001233	100	·000361	·000350

TABLE VI.

FOR

VALUING MINERAL AND OTHER PROPERTIES,

OR

The present value (or years' purchase) of £1 per annum in n years, allowing interest to a present purchaser upon his purchase money, or capital invested, at the rates of 3½, 4, 4½, 5, 5½, 6, 7, 8, 9, 10, 11, 12, 13, 14, 15, 16, 17, 18, 19, 20, 21, 22, 23, 24, and 25 per cent. per annum, and redeeming the capital so invested by an Annual Redemption Fund, at the rate of 2½ per cent. per annum.

Calculated to 8 places of decimals, and to 100 years for each percentage.

TABLE VI.   lxvii

Present Value of £1 per Annum in n years; Redemption of Capital being at 2½ per cent. with interest allowed to a Purchaser at the following rates per cent.

Years	3¼ per cent.	Years	3¼ per cent.	Years	4 per cent.	Years	4 per cent.
1	0·96618357	51	22·26740253	1	0·96153846	51	20·03658861
2	1·89097701	52	22·43436985	2	1·87326550	52	20·17167569
3	2·77671979	53	22·59657501	3	2·73869682	53	20·30271581
4	3·62558079	54	22·75417147	4	3·56102684	54	20·42984993
5	4·43957354	55	22·90730687	5	4·34316454	55	20·55321308
6	5·22056984	56	23·05612326	6	5·08776468	56	20·67293462
7	5·97031216	57	23·20075737	7	5·79725504	57	20·78913857
8	6·69042456	58	23·34134075	8	6·47386019	58	20·90194379
9	7·38242257	59	23·47800014	9	7·11962227	59	20·81146433
10	8·04772215	60	23·61085761	10	7·73641938	60	21·11780963
11	8·68764777	61	23·74003086	11	8·32598178	61	21·22108482
12	9·30343965	62	23·86563316	12	8·88990612	62	21·32139072
13	9·89626037	63	23·98777393	13	9·42966812	63	21·41882435
14	10·46720091	64	24·10655857	14	9·94663383	64	21·51347888
15	11·01728593	65	24·22208884	15	10·44206960	65	21·60544393
16	11·54747897	66	24·33446278	16	10·91715111	66	21·69480559
17	12·05868675	67	24·44377517	17	11·37297126	67	21·78164679
18	12·55176339	68	24·55011744	18	11·81054741	68	21·86604730
19	13·02751416	69	24·65357791	19	12·23082775	69	21·94808392
20	13·48669891	70	24·75424187	20	12·63469713	70	22·02783063
21	13·93003527	71	24·85219178	21	13·02298226	71	22·10535871
22	14·35820146	72	24·94750720	22	13·39645636	72	22·18073675
23	14·77183903	73	25·04026512	23	13·75584350	73	22·25403095
24	15·17155533	74	25·13053996	24	14·10182243	74	22·32530510
25	15·55792567	75	25·21840367	25	14·43503005	75	22·39462074
26	15·93149544	76	25·30392583	26	14·75606456	76	22·46203721
27	16·29278205	77	25·38717376	27	15·06548845	77	22·52761178
28	16·64227669	78	25·46821272	28	15·36383105	78	22·59139983
29	16·98044595	79	25·54710557	29	15·65159098	79	22·65345459
30	17·30773337	80	25·62391340	30	15·92923832	80	22·71382764
31	17·62456084	81	25·69869536	31	16·19721669	81	22·77256882
32	17·93132983	82	25·77150857	32	16·45594495	82	22·82972616
33	18·22842269	83	25·84240844	33	16·70581904	83	22·88534613
34	18·51620375	84	25·91144861	34	16·94721347	84	22·93947365
35	18·79502035	85	25·97868115	35	17·18048273	85	22·99215220
36	19·06520381	86	26·04415633	36	17·40596268	86	23·04342368
37	19·32707041	87	26·10792307	37	17·62397170	87	23·09332881
38	19·58092210	88	26·17002861	38	17·83481179	88	23·14190680
39	19·82704736	89	26·23051889	39	18·03876966	89	23·18919571
40	20·06572202	90	26·28943832	40	18·23611768	90	23·23523253
41	20·29720981	91	26·34683006	41	18·42711474	91	23·28005217
42	20·52176308	92	26·40273591	42	18·61200708	92	23·32368980
43	20·73962342	93	26·45719651	43	18·79102908	93	23·36917861
44	20·95102212	94	26·51025117	44	18·96440389	94	23·40755090
45	21·15618084	95	26·56193812	45	19·13234417	95	23·44783801
46	21·35531202	96	26·61229435	46	19·29505267	96	23·48707021
47	21·54861933	97	26·66135594	47	19·45272274	97	23·52527702
48	21·73629820	98	26·70915766	48	19·60553899	98	23·56248679
49	21·91853623	99	26·75573351	49	19·75367772	99	23·59872722
50	22·09551334	100	26·80111636	50	19·89730725	100	23·63402508

Present Value of £1 per Annum in n years; Redemption of Capital being at 2½ per cent. with interest allowed to a Purchaser at the following rates per cent.

Years	4½ per cent.	Years	4½ per cent.	Years	5 per cent.	Years	5 per cent.
1	0·95693779	51	18·21205168	1	0·95238095	51	16·69206768
2	1·85588269	52	18·32358828	2	1·83881952	52	16·78571558
3	2·70170112	53	18·43165277	3	2·66569161	53	16·87635701
4	3·49873146	54	18·53637330	4	3·43857815	54	16·96410810
5	4·25085376	55	18·62787228	5	4·16238530	55	17·04907945
6	4·96154872	56	18·73626655	6	4·84144343	56	17·13137638
7	5·63394788	57	18·83166781	7	5·47958928	57	17·21109929
8	6·27087631	58	18·92418277	8	6·08023432	58	17·28834387
9	6·87488920	59	19·01391350	9	6·64642212	59	17·36320145
10	7·44830339	60	19·10095769	10	7·18087665	60	17·43575920
11	7·99322457	61	19·18540887	11	7·68604322	61	17.50610040
12	8·51157080	62	19·26735653	12	8·16412323	62	17·57430457
13	9·00509293	63	19·34688653	13	8·61710383	63	17·64044782
14	9·47539253	64	19·42408108	14	9·04678341	64	17·70460288
15	9·92393738	65	19·49901911	15	9·45479348	65	17·76683941
16	10·35207526	66	19·57177619	16	9·84261767	66	17·82722400
17	10·76104593	67	19·64242497	17	10·21160801	67	17·88582053
18	11·15199177	68	19·71103516	18	10·56299936	68	17·94269018
19	11·52596716	69	19·77767371	19	10·89792172	69	17·99789158
20	11·88394679	70	19·84240495	20	11·21741120	70	18·05148098
21	12·22683311	71	19·90529075	21	11·52241960	71	18·10351236
22	12·55546282	72	19·96639050	22	11·81382275	72	18·15403749
23	12·87061282	73	20·02576138	23	12·09242802	73	18·20310609
24	13·17300551	74	20·08345842	24	12·35898089	74	18·25076593
25	13·46331338	75	20·13953456	25	12·61417072	75	18·29706290
26	13·74216331	76	20·19404074	26	12·85863594	76	18·34204111
27	14·01014041	77	20·24702607	27	13·09296875	77	18·38574298
28	14·26779137	78	20·29853790	28	13·31771918	78	18·42820937
29	14·51562764	79	20·34862170	29	13·53339876	79	18·46947945
30	14·75412820	80	20·39732142	30	13·74048390	80	18·50959103
31	14·98374210	81	20·44467947	31	13·93941882	81	18·54858055
32	15·20489072	82	20·49073663	32	14·13061819	82	18·58648299
33	15·41796996	83	20·53553231	33	14·31446964	83	18·62333211
34	15·62335205	84	20·57910452	34	14·49133585	84	18·65916045
35	15·82138737	85	20·62149001	35	14·66155654	85	18·69399940
36	16·01240598	86	20·66272415	36	14·82545033	86	18·72787914
37	16·19671910	87	20·70284121	37	14·98331628	87	18·76082890
38	16·37462043	88	20·74187420	38	15·13543538	88	18·79287677
39	16·54638731	89	20·77985508	39	15·28207191	89	18·82404993
40	16·71228198	90	20·81681467	40	15·42347469	90	18·85437457
41	16·87255244	91	20·85278281	41	15·55987814	91	18·88387599
42	17·02743351	92	20·88778832	42	15·69150336	92	18·91257863
43	17·17714768	93	20·92185912	43	15·81855905	93	18·94050612
44	17·32190580	94	20·95502216	44	15·94124232	94	18·96768125
45	17·46190800	95	20·98730354	45	16·05973953	95	18·99412609
46	17·59734425	96	21·01872846	46	16·17422704	96	19·01986190
47	17·72839498	97	21·04932143	47	16·28487178	97	19·04490934
48	17·85523177	98	21·07910600	48	16·39183198	98	19·06928826
49	17·97801787	99	21·10810515	49	16·49525768	99	19·09301799
50	18·09690851	100	21·13634101	50	16·59529118	100	19·11611716

TABLE VI.   lxix

Present Value of £1 per Annum in n years; Redemption of Capital being at 2½ per cent. with interest allowed to a Purchaser at the following rates per cent.

Years	5½ per cent.	Years	5½ per cent.	Years	6 per cent.	Years	6 per cent.
1	0·94786729	51	15·40625632	1	0·94339622	51	14·30437219
2	1·82206726	52	15·48599780	2	1·80561748	52	14·37308963
3	2·63062938	53	15·56311369	3	2·59647753	53	14·43949609
4	3·38045830	54	15·63770916	4	3·32427051	54	14·50368696
5	4·07752417	55	15·70988414	5	3·99605413	55	14·56575270
6	4·72701554	56	15·77973360	6	4·61787178	56	14·62577911
7	5·33346334	57	15·84734790	7	5·19492853	57	14·68384768
8	5·90084182	58	15·91281296	8	5·73173161	58	14·74003580
9	6·43265153	59	15·97621063	9	6·23220356	59	14·79441702
10	6·93198790	60	16·03761889	10	6·69977413	60	14·84706133
11	7·40159821	61	16·09711210	11	7·13745532	61	14·89803537
12	7·84392921	62	16·15476111	12	7·54790311	62	14·94740253
13	8·26116715	63	16·21063363	13	7·93346860	63	14·99522328
14	8·65527158	64	16·26479426	14	8·29624050	64	15·04155525
15	9·02800393	65	16·31730474	15	8·63808079	65	15·08645345
16	9·38095204	66	16·36822402	16	8·96065468	66	15·12997030
17	9·71555102	67	16·41760854	17	9·26545597	67	15·17215595
18	10·03310116	68	16·46551225	18	9·55382852	68	15·21305826
19	10·33478341	69	16·51198683	19	9·82698462	69	15·25272302
20	10·62167284	70	16·55708174	20	10·08602078	70	15·29119400
21	10·89475019	71	16·60084440	21	10·33193114	71	15·32851310
22	11·15491199	72	16·64332019	22	10·56561923	72	15·36472039
23	11·40297948	73	16·68455267	23	10·78790801	73	15·39985427
24	11·63970635	74	16·72458362	24	10·99954876	74	15·43395155
25	11·86578550	75	16·76345312	25	11·20122862	75	15·46704749
26	12·08185506	76	16·80119964	26	11·39357731	76	15·49917590
27	12·28850378	77	16·83786014	27	11·57717310	77	15·53036921
28	12·48627562	78	16·87347017	28	11·75254786	78	15·56065862
29	12·67567400	79	16·90806377	29	11·92019168	79	15·59007395
30	12·85716552	80	16·94167377	30	12·08055692	80	15·61864392
31	13·03118321	81	16·97433175	31	12·23406172	81	15·64639615
32	13·19812955	82	17·00606802	32	12·38109319	82	15·67335713
33	13·35837908	83	17·03691178	33	12·52201028	83	15·69955234
34	13·51228085	84	17·06689115	34	12·65714627	84	15·72500630
35	13·66016047	85	17·09603323	35	12·78681102	85	15·74974261
36	13·80232212	86	17·12436404	36	12·91129300	86	15·77378395
37	13·93905028	87	17·15190877	37	13·03086114	87	15·79715221
38	14·07061124	88	17·17869160	38	13·14576640	88	15·81986840
39	14·19725458	89	17·20473589	39	13·25624328	89	15·84195282
40	14·31921448	90	17·23006415	40	13·36251116	90	15·86342500
41	14·43671083	91	17·25469809	41	13·46477548	91	15·88430376
42	14·54995038	92	17·27865867	42	13·56322885	92	15·90460727
43	14·65912762	93	17·30196614	43	13·65805203	93	15·92435306
44	14·76442586	94	17·32464001	44	13·74941479	94	15·94355799
45	14·86601767	95	17·34669914	45	13·83747680	95	15·96223840
46	14·96406603	96	17·36816171	46	13·92238834	96	15·98040999
47	15·05872477	97	17·38904537	47	14·00429096	97	15·99808799
48	15·15013929	98	17·40936707	48	14·08331813	98	16·01528701
49	15·23844712	99	17·42914326	49	14·15959586	99	16·03202128
50	15·32377836	100	17·44838984	50	14·23324305	100	16·04830447

Present Value of £1 per Annum in n years; Redemption of Capital being at 2½ per cent. with interest allowed to a Purchaser at the following rates per cent.

Years	7 per cent.	Years	7 per cent.	Years	8 per cent.	Years	8 per cent.
1	0·93457943	51	12·51428263	1	0·92592592	51	11·12239472
2	1·77359317	52	12·56684564	2	1·74268503	52	11·16389606
3	2·53076674	53	12·61758098	3	2·46829983	53	11·20391760
4	3·21731815	54	12·66656764	4	3·11703327	54	11·24252554
5	3·84250553	55	12·71388033	5	3·70032951	55	11·27978231
6	4·41403720	56	12·75958971	6	4·22743658	56	11·31574684
7	4·93838306	57	12·80376268	7	4·70598357	57	11·35047482
8	5·42101366	58	12·84646261	8	5·14225150	58	11·38401888
9	5·86658598	59	12·88774960	9	5·54148972	59	11·41642884
10	6·27908933	60	12·92768066	10	5·90811360	60	11·44775186
11	6·66196084	61	12·96630993	11	6·24586384	61	11·47803265
12	7·01817785	62	13·00368882	12	6·55793062	62	11·50731357
13	7·35033229	63	13·03986622	13	6·84705127	63	11·53563487
14	7·66069114	64	13·07488865	14	7·11558793	64	11·56303474
15	7·95124575	65	13·10880038	15	7·36558962	65	11·58954948
16	8·22375261	66	13·14164354	16	7·59884259	66	11·61521357
17	8·47976690	67	13·17345831	17	7·81691106	67	11·64005988
18	8·72067060	68	13·20428300	18	8·02117073	68	11·66411963
19	8·94769592	69	13·23415415	19	8·21283630	69	11·68742262
20	9·16194509	70	13·26310663	20	8·39298446	70	11·70999722
21	9·36440705	71	13·29117378	21	8·56257287	71	11·73187049
22	9·55597165	72	13·31838740	22	8·72245621	72	11·75306824
23	9·73744175	73	13·34477792	23	8·87339963	73	11·77361513
24	9·90954367	74	13·37037444	24	9·01609027	74	11·79353469
25	10·07293602	75	13·39520480	25	9·15114686	75	11·81284943
26	10·22821745	76	13·41929566	26	9·27912805	76	11·83158084
27	10·37593334	77	13·44267254	27	9·40053961	77	11·84974951
28	10·51658158	78	13·46535993	28	9·51584045	78	11·86737515
29	10·65061765	79	13·48738124	29	9·62544798	79	11·88447657
30	10·77845904	80	13·50875896	30	9·72974271	80	11·90107185
31	10·90048915	81	13·52951469	31	9·82907220	81	11·91717830
32	11·01706064	82	13·54966910	32	9·92375458	82	11·93281249
33	11·12849855	83	13·56924208	33	10·01408162	83	11·94799035
34	11·23510287	84	13·58825270	34	10·10032137	84	11·96272711
35	11·33715095	85	13·60671934	35	10·18272055	85	11·97703747
36	11·43489961	86	13·62465958	36	10·26150663	86	11·99093545
37	11·52858698	87	13·64209042	37	10·33688966	87	12·00443460
38	11·61843418	88	13·65902812	38	10·40906394	88	12·01754788
39	11·70464682	89	13·67548840	39	10·47820941	89	12·03028779
40	11·78741634	90	13·69148633	40	10·54449304	90	12·04266632
41	11·86692119	91	13·70703646	41	10·60806990	91	12·05469502
42	11·94332795	92	13·72215276	42	10·66908423	92	12·06638498
43	12·01679228	93	13·73684876	43	10·72767041	93	12·07774693
44	12·08745980	94	13·75113742	44	10·78395373	94	12·08879114
45	12·15546689	95	13·76503129	45	10·83805117	95	12·09952754
46	12·22094142	96	13·77854242	46	10·89007209	96	12·10996566
47	12·28400338	97	13·79168249	47	10·94011882	97	12·12011474
48	12·34476553	98	13·80446269	48	10·98828924	98	12·12998363
49	12·40333390	99	13·81689391	49	11·03466728	99	12·13958090
50	12·45980826	100	13·82898660	50	11·07934332	100	12·14891480

TABLE VI.   lxxi

Present Value of £1 per Annum in n years; Redemption of Capital being at 2½ per cent. with interest allowed to a Purchaser at the following rates per cent.

Years	9 per cent.	Years	9 per cent.	Years	10 per cent.	Years	10 per cent.
1	0·91743119	51	10·00913879	1	0·90909091	51	9·09846119
2	1·71283569	52	10·04273551	2	1·68399168	52	9·12621397
3	2·40884238	53	10·07511052	3	2·35218202	53	9·15294155
4	3·02281124	54	10·10631994	4	2·93411838	54	9·17869196
5	3·56828262	55	10·13641658	5	3·44534304	55	9·20351043
6	4·05597290	56	10·16545022	6	3·89787610	56	9·22743946
7	4·49447434	57	10·19346782	7	4·30115982	57	9·25051913
8	4·89075650	58	10·22051367	8	4·66271450	58	9·27278719
9	5·25053203	59	10·24662965	9	4·98860379	59	9·29427928
10	5·57852784	60	10·27185534	10	5·28377119	60	9·31502904
11	5·87868893	61	10·29622821	11	5·55228724	61	9·33506828
12	6·15433369	62	10·31978370	12	5.79753409	62	9·35442706
13	6·40827350	63	10·34255544	13	6·02234515	63	9·37313387
14	6·64290610	64	10·36457529	14	6·22911204	64	9·39121567
15	6·86028889	65	10·38587353	15	6·41986744	65	9·40869805
16	7·06219734	66	10·40647883	16	6·59635008	66	9·42560522
17	7·25017160	67	10·42641852	17	6·76005594	67	9·44196023
18	7·42555433	68	10·44571853	18	6·91227928	68	9·45778493
19	7·58952134	69	10·46440355	19	7·05414546	69	9·47310013
20	7·74310672	70	10·48249710	20	7·18663769	70	9·48792557
21	7·88722361	71	10·50002156	21	7·31061876	71	9·50228010
22	8·02268134	72	10·51699826	22	7·42684892	72	9·51618161
23	8·15019983	73	10·53344755	23	7·53600071	73	9·52964717
24	8·27042159	74	10·54938886	24	7·63867128	74	9·54269306
25	8·38392185	75	10·56484070	25	7·73539259	75	9·55533480
26	8·49121714	76	10·57982079	26	7·82664013	76	9·56758721
27	8·59277262	77	10·59434604	27	7·91284025	77	9·57946443
28	8·68900829	78	10·60843265	28	7·99437627	78	9·59098000
29	8·78030436	79	10·62209606	29	8·07159385	79	9·60214680
30	8·86700585	80	10·63535108	30	8·14480547	80	9·61297720
31	8·94942660	81	10·64821190	31	8·21429435	81	9·62348304
32	9·02785264	82	10·66069210	32	8·28031776	82	9·63367560
33	9·10254530	83	10·67280467	33	8·34310994	83	9·64356574
34	9·17374377	84	10·68456211	34	8·40288466	84	9·65316382
35	9·24166739	85	10·69597637	35	8·45983736	85	9·66247981
36	9·30651770	86	10·70705892	36	8·51414709	86	9·67152323
37	9·36848020	87	10·71782081	37	8·56597823	87	9·68030325
38	9·42772592	88	10·72827258	38	8·61548190	88	9·68882864
39	9·48441278	89	10·73842443	39	8·66279732	89	9·69710783
40	9·53868687	90	10·74828609	40	8·70805296	90	9·70514892
41	9·59068349	91	10·75786696	41	8·75136753	91	9·71295968
42	9·64052816	92	10·76717607	42	8·79285089	92	9·72054759
43	9·68833750	93	10·77622210	43	8·83260492	93	9·72791985
44	9·73421995	94	10·78501340	44	8·87072415	94	9·73508335
45	9·77827655	95	10·79355801	45	8·90729647	95	9·74204476
46	9·82060150	96	10·80186368	46	8·94240367	96	9·74881046
47	9·86128277	97	10·80993787	47	8·97612199	97	9·75538664
48	9·90040256	98	10·81778775	48	9·00852257	98	9·76177919
49	9·93803786	99	10·82542025	49	9·03967185	99	9·76799386
50	9·97426073	100	10·83284205	50	9·06963199	100	9·77403615

Present Value of £1 per Annum in n years; Redemption of Capital being at 2½ per cent. with interest allowed to a Purchaser at the following rates per cent.

Years	11 per cent.	Years	11 per cent.	Years	12 per cent.	Years	12 per cent.
1	0·90090090	51	8·33967875	1	0·89285714	51	7·69771412
2	1·65610305	52	8·36298969	2	1·62912309	52	7·71757006
3	2·29812591	53	8·38542820	3	2·24649855	53	7·73667489
4	2·85048187	54	8·44703602	4	2·77148130	54	7·75506492
5	3·33059270	55	8·42785217	5	3·22323971	55	7·77277425
6	3·75164175	56	8·44791337	6	3·61598302	56	7·78983487
7	4·12378906	57	8·46725416	7	3·96046772	57	7·80627686
8	4·45499098	58	8·48590709	8	4·26498623	58	7·82212853
9	4·75156694	59	8·50390280	9	4·53603424	59	7·83741652
10	5·01859986	60	8·52127024	10	4·77877239	60	7·85216596
11	5·26022447	61	8·53803672	11	4·99735251	61	7·86640055
12	5·47983858	62	8·55422804	12	5·19515261	62	7·88014266
13	5·68026027	63	8·56986861	13	5·37494917	63	7·89341345
14	5·86384647	64	8·58498154	14	5·53904536	64	7·90623291
15	6·03258357	65	8·59958871	15	5·68936771	65	7·91861996
16	6·18815754	66	8·61371084	16	5·82754017	66	7·93059253
17	6·33200862	67	8·62736761	17	5·95494123	67	7·94216761
18	6·46537454	68	8·64057768	18	6·07274860	68	7·95336132
19	6·58932490	69	8·65335879	19	6·18197451	69	7·96418894
20	6·70478881	70	8·66572777	20	6·28349382	70	7·97466501
21	6·81257721	71	8·67770067	21	6·37806650	71	7·98480334
22	6·91340107	72	8·68929273	22	6·46635594	72	7·99461705
23	7·00788635	73	8·70051846	23	6·54894381	73	8·00411864
24	7·09658637	74	8·71139169	24	6·62634227	74	8·01332000
25	7·17999202	75	8·72192561	25	6·69900406	75	8·02223246
26	7·25854031	76	8·73213279	26	6·76733087	76	8·03086684
27	7·33262161	77	8·74202524	27	6·83168034	77	8·03923342
28	7·40258571	78	8·75161441	28	6·89237197	78	8·04734206
29	7·46874693	79	8·76091124	29	6·94969202	79	8·05520213
30	7·53138853	80	8·76992620	30	7·00389777	80	8·06282260
31	7·59076645	81	8·77866929	31	7·05522110	81	8·07021206
32	7·64711254	82	8·78715007	32	7·10387149	82	8·07737868
33	7·70063731	83	8·79537771	33	7·15003875	83	8·08433032
34	7·75153234	84	8·80336097	34	7·19389252	84	8·09107447
35	7·79997238	85	8·81110826	35	7·23559775	85	8·09761834
36	7·84611714	86	8·81862761	36	7·27528941	86	8·10396878
37	7·89011288	87	8·82592677	37	7·31310096	87	8·11013243
38	7·93209380	88	8·83301313	38	7·34915215	88	8·11611558
39	7·97218325	89	8·83989380	39	7·38355288	89	8·12192431
40	8·01049483	90	8·84657558	40	7·41640416	90	8·12756445
41	8·04713332	91	8·85306504	41	7·44779900	91	8·13304158
42	8·08219549	92	8·85936847	42	7·47782320	92	8·13836107
43	8·11577094	93	8·86549190	43	7·50655605	93	8·14352808
44	8·14794264	94	8·87144116	44	7·53407087	94	8·14854756
45	8·17878761	95	8·87722184	45	7·56943564	95	8·15342428
46	8·20837743	96	8·88283930	46	7·58571344	96	8·15816280
47	8·23677869	97	8·88829873	47	7·60996289	97	8·16276756
48	8·26405345	98	8·89360510	48	7·63323853	98	8·16722478
49	8·29025959	99	8·89876322	49	7·65559121	99	8·17159255
50	8·31545117	100	8·90377770	50	7·67706830	100	8·17582079

TABLE VI.

lxxiii

Present Value of £1 per Annum in **n** years; Redemption of Capital being at 2½ per cent. with interest allowed to a Purchaser at the following rates per cent.

Years	13 per cent.	Years	13 per cent.	Years	14 per cent.	Years	14 per cent.
1	0·89285714	51	7·14751858	1	0·87719298	51	6·67072712
2	1·60300811	52	7·16463438	2	1·57771718	52	6·68513321
3	2·19713983	53	7·18109678	3	2·14990345	53	6·69996575
4	2·69674161	54	7·19693773	4	2·62592714	54	6·71375310
5	3·12259111	55	7·21218722	5	3·02803787	55	6·72702181
6	3·48979271	56	7·22687337	6	3·37211296	56	6·73979679
7	3·80959014	57	7·24102259	7	3·66978632	57	6·75210140
8	4·09052586	58	7·25465972	8	3·92977731	58	6·76395761
9	4·33920635	59	7·26780812	9	4·15874963	59	6·77538607
10	4·56082113	60	7·28048982	10	4·36188343	60	6·78640620
11	4·75950335	61	7·29272555	11	4·54326643	61	6·79703634
12	4·93858555	62	7·30453489	12	4·70616744	62	6·80729375
13	5·10078459	63	7·31593635	13	4·85323170	63	6·81719472
14	5·24833756	64	7·32694738	14	4·98662276	64	6·82675466
15	5·38310318	65	7·33758453	15	5·10812741	65	6·83598812
16	5·50663860	66	7·34786342	16	5·21923423	66	6·84490887
17	5·62025816	67	7·35779889	17	5·32119335	67	6·85352994
18	5·72507894	68	7·36740498	18	5·41506235	68	6·86186369
19	5·82205647	69	7·37669501	19	5·50174194	69	6·86992183
20	5·91201286	70	7·38568164	20	5·58200406	70	6·87771547
21	5·99565936	71	7·39437689	21	4·65651405	71	6·88525517
22	6·07361442	72	7·40279217	22	5·72584846	72	6·89255095
23	6·14641833	73	7·41093834	23	5·79050940	73	6·89961233
24	6·21454523	74	7·41882575	24	5·85093616	74	6·90644838
25	6·27841292	75	7·42646424	25	5·90751475	75	6·91306773
26	6·33839098	76	7·43386318	26	5·96058575	76	6·91947861
27	6·39480753	77	7·44103153	27	6·01045078	77	6·92568884
28	6·44795493	78	7·44797781	28	6·05737793	78	6·93170589
29	6·49809447	79	7·45471016	29	6·10160633	79	6·93753689
30	6·54546042	80	7·46123635	30	6·14334988	80	6·94318864
31	6·59026344	81	7·46756382	31	6·18280059	81	6·94866763
32	6·63269347	82	7·47369966	32	6·22013123	82	6·95398008
33	6·67292223	83	7·47965065	33	6·25549773	83	6·95913190
34	6·71110534	84	7·48542330	34	6·28904116	84	6·96412878
35	6·74738418	85	7·49102382	35	6·32088948	85	6·96897616
36	6·78188747	86	7·49645816	36	6·35115901	86	6·97367921
37	6·81473268	87	7·50173202	37	6·37995575	87	6·97824294
38	6·84692720	88	7·50685088	38	6·40737646	88	6·98267210
39	6·87586943	89	7·51181998	39	6·43350970	89	6·98697128
40	6·90434968	90	7·51664434	40	6·45843663	90	6·99114485
41	6·93155101	91	7·52132878	41	6·48223181	91	6·99519701
42	6·95754992	92	7·52587775	42	6·50496382	92	6·99913183
43	6·98241700	93	7·53029629	43	6·51669588	93	7·00295316
44	7·00621748	94	7·53458807	44	6·54748634	94	7·00666474
45	7·02901173	95	7·53875740	45	6·56738917	95	7·01027014
46	7·05085573	96	7·54280823	46	6·58645434	96	7·01377280
47	7·07180142	97	7·54674436	47	6·60472816	97	7·01717603
48	7·09189711	98	7·55056844	48	6·62225369	98	7·02048300
49	7·11118774	99	7·55428699	49	6·63907094	99	7·02369678
50	7·12971520	100	7·55790040	50	6·65521717	100	7·02682032

Present Value of £1 per Annum in n years; Redemption of Capital being at 2½ per cent. with interest allowed to a Purchaser at the following rates per cent.

Years	15 per cent.	Years	15 per cent.	Years	16 per cent.	Years	16 per cent.
1	0·86956521	51	6·25356862	1	0·86206896	51	5·88551396
2	1·55321189	52	6·26666685	2	1·52945619	52	5·89711434
3	2·10465539	53	6·27925764	3	2·06127270	53	5·90826261
4	2·55873658	54	6·29136630	4	2·49489869	54	5·91898149
5	2·93904255	55	6·30301652	5	2·85512909	55	5·92929225
6	3·26211089	56	6·31423048	6	3·15905889	56	5·93921477
7	3·53988028	57	6·32502003	7	3·41885684	57	5·94876774
8	3·78118515	58	6·33543170	8	3·64342067	58	5·95796866
9	3·99270310	59	6·34545687	9	3·83940698	59	5·96683399
10	4·17957523	60	6·35512182	10	4·01189506	60	5·97537919
11	4·34582406	61	6·36444285	11	4·16482797	61	5·98361885
12	4·49464206	62	6·37343529	12	4·30131342	62	5·99156666
13	4·62859525	63	6·38211361	13	4·42383388	63	5·99923558
14	4·74976967	64	6·39049148	14	4·53439629	64	6·00663781
15	4·85987862	65	6·39858183	15	4·63464070	65	6·01378488
16	4·96034235	66	6·40639684	16	4·72592051	66	6·02068770
17	5·05234814	67	6·41394809	17	4·80936241	67	6·02735657
18	5·13689622	68	6·42124651	18	4·88591199	68	6·03380126
19	5·21483517	69	6·42830248	19	4·95636871	69	6·04003101
20	5·28688967	70	6·43512583	20	5·02141310	70	6·04605461
21	5·35368226	71	6·44172591	21	5·08162804	71	6·05188036
22	5·41575078	72	6·44811157	22	5·13751573	72	6·05751617
23	5·47356226	73	6·45429125	23	5·18951114	73	6·06296954
24	5·52752424	74	6·46027296	24	5·23799291	74	6·06824760
25	5·57799394	75	6·46606433	25	5·28329222	75	6·07335715
26	5·62528577	76	6·47167261	26	5·32569993	76	6·07830463
27	5·66967759	77	6·47710472	27	5·36547260	77	6·08309621
28	5·71141589	78	6·48236726	28	5·40283738	78	6·08773774
29	5·75072003	79	6·48746651	29	5·43799610	79	6·09223482
30	5·78778594	80	6·49240847	30	5·47112872	80	6·09659277
31	5·82278915	81	6·49719888	31	5·50239622	81	6·10081668
32	5·85588735	82	6·50184319	32	5·53194300	82	6·10491143
33	5·88722265	83	6·50634665	33	5·55989902	83	6·10888164
34	5·91692341	84	6·51071426	34	5·58638150	84	6·11273176
35	5·94510591	85	6·51495079	35	5·61149650	85	6·11646604
36	5·97187569	86	6·51906082	36	5·63534018	86	6·12008853
37	5·99732882	87	6·52304875	37	5·65799996	87	6·12360313
38	6·02155290	88	6·52691877	38	5·67955546	88	6·12701357
39	6·04462797	89	6·53067490	39	5·70007938	89	6·13032341
40	6·06662735	90	6·53432100	40	5·71963821	90	6·13353606
41	6·08761828	91	6·53786078	41	5·73829291	91	6·13665483
42	6·10766258	92	6·54129778	42	5·75609944	92	6·13968284
43	6·12681716	93	6·54463541	43	5·77310930	93	6·14262312
44	6·14513450	94	6·54787696	44	5·78936993	94	6·14547860
45	6·16266310	95	6·55102555	45	5·80492513	95	6·14825199
46	6·17944783	96	6·55408422	46	5·81981537	96	6·15094604
47	6·19553023	97	6·55705587	47	5·83407816	97	6·15356329
48	6·21094890	98	6·55994329	48	5·84774824	98	6·15610622
49	6·22573967	99	6·56274918	49	5·86085791	99	6·15857720
50	6·23993588	100	6·56547611	50	5·87343717	100	6·16097853

TABLE VI.   lxxv

Present Value of £1 per Annum in **n** years; Redemption of Capital being at 2½ per cent. with interest allowed to a Purchaser at the following rates per cent.

Years	17 per cent.	Years	17 per cent.	Years	18 per cent.	Years	18 per cent.
1	0·85470085	51	5·55837503	1	0·84745762	51	5·26568832
2	1·50641622	52	5·56872052	2	1·48406010	52	5·27497207
3	2·01964237	53	5·57866069	3	1·97966031	53	5·28389037
4	2·43416865	54	5·58821602	4	2·37632490	54	5·29246182
5	2·77587430	55	5·59740570	5	2·70090069	55	5·30070381
6	3·06231845	56	5·60624768	6	2·97132695	56	5·30863259
7	3·30583507	57	5·61475878	7	3·20004680	57	5·31626342
8	3·51534197	58	5·62295478	8	3·39596229	58	5·32371056
9	3·69744694	59	5·63085049	9	3·56561039	59	5·33068745
10	3·85715024	60	5·63845984	10	3·71389955	60	5·33750667
11	3·99830543	61	5·64579694	11	3·84458709	61	5·34408008
12	4·12393026	62	5·65287112	12	3·96059796	62	5·35041884
13	4·23642163	63	5·65969702	13	4·06424315	63	5·35653346
14	4·33770744	64	5·66628461	14	4·15737277	64	5·36243385
15	4·42935597	65	5·67264426	15	4·24148548	65	5·36812938
16	4·51265598	66	5·67878574	16	4·31780815	66	5·37362887
17	4·58867633	67	5·68471832	17	4·38735482	67	5·37894070
18	4·65831101	68	5·69045077	18	4·45097094	68	5·38407276
19	4·72231345	69	5·69599137	19	4·50936700	69	5·38903254
20	4·78132311	70	5·70134799	20	4·56314443	70	5·39382714
21	4·83588629	71	3·70652811	21	4·61281576	71	5·39846328
22	4·88647244	72	5·71153878	22	4·65882046	72	5·40294735
23	4·93348726	73	5·71638675	23	4·79153751	73	5·40728540
24	4·97728317	74	5·72107840	24	4·74129547	74	5·41148320
25	5·01816775	75	5·72561981	25	4·77838060	75	5·41554622
26	5·05641067	76	5·73001676	26	4·81304343	76	5·41947967
27	5·09224935	77	5·73427476	27	4·84550420	77	5·42328849
28	5·12589368	78	5·73839906	28	4·87595729	78	5·42697744
29	5·15752983	79	5·74239465	29	4·90457491	79	5·43055098
30	5·18732357.	80	5·74626631	30	4·93151018	80	5·43401343
31	5·21542298	81	5·75001859	31	4·95689970	81	5·43736887
32	5·24196072	82	5·75365583	32	4·98086570	82	5·44062121
33	5·26705602	83	5·75718219	33	5·00351793	83	5·44377419
34	5·29081631	84	5·76060163	34	5·02495517	84	5·44683138
35	5·31333868	85	5·76391795	35	5·04526658	85	5·44979617
36	5·33471006	86	5·76713478	36	5·06453287	86	5·45267184
37	5·35501331	87	5·77025558	37	5·08282723	87	5·45546161
38	5·37431809	88	5·77328369	38	5·10021624	88	5·45816816
39	5·39269167	89	5·77622230	39	5·11676055	89	5·46079466
40	5·41019465	90	5·77907445	40	5·13251557	90	5·46334375
41	5·42688250	91	5·78184307	41	5·14753199	91	5·46581805
42	5·44280611	92	5·78453098	42	5·16185628	92	5·46822010
43	5·45801229	93	5·78714087	43	5·17553116	93	5·47055230
44	5·47254411	94	5·78967532	44	5·18859591	94	5·47281699
45	5·48644132	95	5·79213683	45	5·20108674	95	5·47501639
46	5·49974062	96	5·79452776	46	5·21303710	96	5·47715263
47	5·51247600	97	5·79685043	47	5·22447791	97	5·47922779
48	5·52467893	98	5·79910703	48	5·23453780	98	5·48124383
49	5·53637863	99	5·80129968	49	5·24594334	99	5·48320266
50	5·54760224	100	5·80343043	50	5·25601920	100	5·48510611

Present Value of £1 per Annum in n years; Redemption of Capital being at 2½ per cent. with interest allowed to a Purchaser at the following rates per cent.

Years	19 per cent.	Years	19 per cent.	Years	20 per cent.	Years	20 per cent.
1	0·84033613	51	5·00228366	1	0·83333333	51	4·76397606
2	1·46235783	52	5·01066110	2	1·44128114	52	4·77157371
3	1·94123054	53	5·01870737	3	1·90426438	53	4·77886988
4	2·32116644	54	5·02643943	4	2·26851054	54	4·78588007
5	2·62987050	55	5·03387310	5	2·56248058	55	4·79261876
6	2·88558674	56	5·04102319	6	2·80465596	56	4·79909947
7	3·10081914	57	5·04790356	7	3·00756014	57	4·80533489
8	3·28442447	58	5·05452722	8	3·17998041	58	4·81133689
9	3·44285171	59	5·06090635	9	3·32826450	59	4·81711659
10	3·58090822	60	5·06705241	10	3·45711220	60	4·82268446
11	3·70225083	61	5·07297617	11	3·57007761	61	4·82805033
12	3·80971064	62	5·07868777	12	3·66989814	62	4·83322344
13	3·90551358	63	5·08419676	13	3·75871640	63	4·83821251
14	3·99143399	64	5·08951213	14	3·83823343	64	4·84302575
15	4·06890353	65	5·09464238	15	3·90981685	65	4·84767089
16	4·13909017	66	5·09959553	16	3·97457878	66	4·85215524
17	4·20295622	67	5·10437917	17	4·03343280	67	4·85648572
18	4·26130164	68	5·10900046	18	4·08713643	68	4·86066886
19	4·31479697	69	5·11346618	19	4·13632303	69	4·86471084
20	4·36400842	70	5·11778278	20	4·18152627	70	4·86861751
21	4·40941746	71	5·12195635	21	4·22319899	71	4·87239443
22	4·45143605	72	5·12599267	22	4·26172795	72	4·87604686
23	4·49041879	73	5·12989722	23	4·29744549	73	4·87957979
24	4·52667255	74	5·13367523	24	4·33063872	74	4·88299797
25	4·56046426	75	5·13733165	25	4·36155701	75	4·88630591
26	4·59202716	76	5·14087120	26	4·39041798	76	4·88950789
27	4·62156602	77	5·14429835	27	4·41741239	77	4·89260800
28	4·64926129	78	5·14761740	28	4·44270818	78	4·89561012
29	4·67527266	79	5·15083240	29	4·46645377	79	4·89851795
30	4·69974193	80	5·15394724	30	4·48878082	80	4·90133502
31	4·72279546	81	5·15696563	31	4·50980653	81	4·90406470
32	4·74454623	82	5·15989109	32	4·52963557	82	4·90671018
33	4·76509552	83	5·16272699	33	4·54836174	83	4·90927455
34	4·78453445	84	5·16547658	34	4·56606930	84	4·91176073
35	4·80294519	85	5·16814292	35	4·58283418	85	4·91417151
36	4·82040202	86	5·17072896	36	4·59872499	86	4·91650957
37	4·83697229	87	5·17323753	37	4·61380387	87	4·91877748
38	4·85271717	88	5·17567132	38	4·62812725	88	4·92097769
39	4·86769239	89	5·17803291	39	4·64174645	89	4·92311253
40	4·88194879	90	5·18032480	40	4·65470831	90	4·92518425
41	4·89553287	91	5·18254934	41	4·66705562	91	4·92719502
42	4·90848723	92	5·18470881	42	4·67882757	92	4·92914689
43	4·92085099	93	5·18680540	43	4·69006012	93	4·93104186
44	4·93266011	94	5·18884121	44	4·70078630	94	4·93288180
45	4·94394773	95	5·19081824	45	4·71103655	95	4·93466856
46	4·95474444	96	5·19273842	46	4·72083893	96	4·93640388
47	4·96507848	97	5·19460363	47	4·73021938	97	4·93808945
48	4·97497602	98	5·19641562	48	4·73920187	98	4·93972688
49	4·98446133	99	5·19817613	49	4·74780864	99	4·94131773
50	4·99355689	100	5·19988681	50	4·75606031	100	4·94286350

TABLE VI.   lxxvii

Present Value of £1 per Annum in n years; Redemption of Capital being at 2½ per cent. with interest allowed to a Purchaser at the following rates per cent.

Years	21 per cent.	Years	21 per cent.	Years	22 per cent.	Years	22 per cent.
1	0·82644628	51	4·54734179	1	0·81967213	51	4·34955276
2	1·42080377	52	4·55426366	2	1·40089934	52	4·35588517
3	1·86867977	53	4·56090993	3	1·83440070	53	4·36196465
4	2·21819065	54	4·56729482	4	2·17005470	54	4·36780432
5	2·49845807	55	4·57343162	5	2·43755674	55	4·37341641
6	2·72814099	56	4·57933274	6	2·65569002	56	4·37881235
7	2·91974699	57	4·58500982	7	2·83691621	57	4·38400286
8	3·08197424	58	4·59047373	8	2·98982849	58	4·38899793
9	3·22105913	59	4·59573469	9	3·12054455	59	4·39380698
10	3·34158970	60	4·60080228	10	3·23353812	60	4·39843881
11	3·44701644	61	4·60568552	11	3·33215646	61	4·40290171
12	3·53998432	62	4·61039286	12	3·41895389	62	4·40720346
13	3·62255484	63	4·61493228	13	3·49591346	63	4·41135140
14	3·69635856	64	4·61931131	14	3·56459823	64	4·41535243
15	3·76270210	65	4·62353703	15	3·62625685	65	4·41921308
16	3·82264475	66	4·62761612	16	3·68189884	66	4·42293946
17	3·87705442	67	4·63155492	17	3·73234921	67	4·42653741
18	3·92664893	68	4·63535939	18	3·77828880	68	4·43001240
19	3·97202716	69	4·63903519	19	3·82028442	69	4·43336962
20	4·01369265	70	4·64258768	20	3·85881181	70	4·43661398
21	4·05207193	71	4·64602192	21	3·89427318	71	4·43975015
22	4·08752860	72	4·64934273	22	3·92701091	72	4·44278254
23	4·12037463	73	4·65255467	23	3·95731829	73	4·44571534
24	4·15087914	74	4·65566209	24	3·98544801	74	4·44855251
25	4·17927553	75	4·65866909	25	4·01161892	75	4·45129785
26	4·20576722	76	4·66157959	26	4·03602155	76	4·45395494
27	4·23053233	77	4·66439732	27	4·05882253	77	4·45652719
28	4·25372748	78	4·66712584	28	4·08016825	78	4·45901786
29	4·27549094	79	4·66976850	29	4·10018779	79	4·46143005
30	4·29594525	80	4·67232855	30	4·11899546	80	4·46376670
31	4·31519938	81	4·67480904	31	4·13669284	81	4·46603063
32	4·33335058	82	4·67721290	32	4·15337048	82	4·46822453
33	4·35048590	83	4·67954294	33	4·16910939	83	4·47035095
34	4·36668350	84	4·68180182	34	4·18398224	84	4·47241234
35	4·38201375	85	4·68399210	35	4·19805443	85	4·47441104
36	4·39654020	86	4·68611622	36	4·21138496	86	4·47634929
37	4·41032034	87	4·68817651	37	4·22402721	87	4·47822922
38	4·42340637	88	4·69017521	38	4·23602957	88	4·48005288
39	4·43584574	89	4·69211445	39	4·24743603	89	4·48182222
40	4·44768170	90	4·69399629	40	4·25828667	90	4·48353913
41	4·45895377	91	4·69582268	41	4·26861806	91	4·48520539
42	4·46969811	92	4·69759551	42	4·27846369	92	4·48682273
43	4·47994787	93	4·69931659	43	4·28785423	93	4·48839280
44	4·48973352	94	4·70098764	44	4·29681785	94	4·48991719
45	4·49908310	95	4·70261033	45	4·30538045	95	4·49139741
46	4·50892245	96	4·70418625	46	4·31356593	96	4·49283493
47	4·51657545	97	4·70571694	47	4·32139632	97	4·49423114
48	4·52476416	98	4·70720387	48	4·32889201	98	4·49558740
49	4·53260904	99	4·70864845	49	4·33607185	99	4·49690500
50	4·54012904	100	4·71005205	50	4·34295335	100	4·49818519

Present Value of £1 per Annum in n years; Redemption of Capital being at 2½ per cent. with interest allowed to a Purchaser at the following rates per cent.

Years	23 per cent.	Years	23 per cent.	Years	24 per cent.	Years	24 per cent.
1	0·81300813	51	4·16825242	1	0·80645161	51	4·00146141
2	1·38154528	52	4·17406758	2	1·36271871	52	4·00682019
3	1·80135660	53	4·17964980	3	1·76948192	53	4·01196377
4	2·12396353	54	4·18501122	4	2·07978956	54	4·01690336
5	2·37955377	55	4·19016313	5	2·32424706	55	4·02164945
6	2·58698765	56	4·19511609	6	2·52175028	56	4·02621183
7	2·75865547	57	4·19988000	7	2·68459670	57	4·03059965
8	2·90303279	58	4·20446409	8	2·82113434	58	4·03482147
9	3·02611333	59	4·20887705	9	2·93722944	59	4·03888533
10	3·13225545	60	4·21312700	10	3·03712494	60	4·04279875
11	3·22470427	61	4·21722159	11	3·12396562	61	4·04656881
12	3·30592581	62	4·22116800	12	3·20013182	62	4·05020216
13	3·37782753	63	4·22497300	13	3·26745842	63	4·05370505
14	3·44190804	64	4·22864295	14	3·32738259	64	4·05708338
15	3·49936103	65	4·23218385	15	3·38104602	65	4·06034269
16	3·55114912	66	4·23560138	16	3·42936718	66	4·06348822
17	3·59805715	67	4·23890088	17	3·47309327	67	4·06652492
18	3·64073145	68	4·24208740	18	3·51283843	68	4·06945746
19	3·67970907	69	4·24516573	19	3·54911207	69	4·07229026
20	3·71543997	70	4·24814039	20	3·58234027	70	4·07502750
21	3·74830398	71	4·25101568	21	3·61288217	71	4·07767315
22	3·77862394	72	4·25379565	22	3·64104263	72	4·08023097
23	3·80667601	73	4·25648417	23	3·66708207	73	4·08270450
24	3·83269783	74	4·25908489	24	3·69122435	74	4·08509713
25	3·85689499	75	4·26160128	25	3·71366291	75	4·08741208
26	3·87944626	76	4·26403667	26	3·73456579	76	4·08965239
27	3·90050784	77	4·26639417	27	3·75407967	77	4·09182096
28	3·92021680	78	4·26867680	28	3·77233317	78	4·09392056
29	3·93869394	79	4·27088739	29	3·78943951	79	4·09595382
30	3·95604610	80	4·27302867	30	3·80549881	80	4·09792323
31	3·97236817	81	4·27510321	31	3·82059988	81	4·09983120
32	3·98774467	82	4·27711349	32	3·83482177	82	4·10167999
33	4·00225116	83	4·27906187	33	3·84823512	83	4·10347177
34	4·01595538	84	4·28095058	34	3·86090323	84	4·10520863
35	4·02891825	85	4·28278178	35	3·87288296	85	4·10689254
36	4·04119470	86	4·28455753	36	3·88422558	86	4·10852540
37	4·05283438	87	4·28627979	37	3·89497740	87	4·11010902
38	4·06388231	88	4·28795043	38	3·90518038	88	4·11164513
39	4·07437937	89	4·28957127	39	3·91487261	89	4·11313539
40	4·08436279	90	4·29114401	40	3·92408877	90	4·11458139
41	4·09386654	91	4·29267031	41	3·93286048	91	4·11598466
42	4·10292168	92	4·29415174	42	3·94121605	92	4·11734664
43	4·11155668	93	4·29558985	43	3·94918375	93	4·11866874
44	4·11979765	94	4·29698607	44	3·95678607	94	4·11995229
45	4·12766862	95	4·29834180	45	3·96404594	95	4·12119860
46	4·13519171	96	4·29965837	46	3·97098391	96	4·12240887
47	4·14238735	97	4·30093709	47	3·97761896	97	4·12358432
48	4·14927440	98	4·30217918	48	3·98396861	98	4·12472607
49	4·15587033	99	4·30338583	49	3·99004906	99	4·12583522
50	4·16219132	100	4·30455819	50	3·99587535	100	4·12691283

TABLE VI. lxxix

Present Value of £1 per Annum in n years; Redemption of Capital being at 2½ per cent. with interest allowed to a Purchaser at the following rates per cent.

Years	25 per cent.	Years	25 per cent.	Years	25 per cent.	Years	25 per cent.
1	0·80000000	26	3·60011705	51	3·84750498	76	3·92897113
2	1·34459834	27	3·61824777	52	3·85245908	77	3·93097260
3	1·73871566	28	3·63520126	53	3·85721375	78	3·93291034
4	2·03741560	29	3·65108390	54	3·86177941	79	3·93478677
5	2·27145288	30	3·66598962	55	3·86616582	80	3·93660422
6	2·45972223			56	3·87038205		
7	2·61441032	31	3·68000173	57	3·87443662	81	3·93836489
8	2·74373003	32	3·69319435	58	3·87833748	82	3·94007089
9	2·85341801	33	3·70563363	59	3·88209208	83	3·94172423
10	2·94760256	34	3·71737883	60	3·88570742	84	3·94332683
		35	3·72848317			85	3·94488053
11	3·02933038	36	3·73899460	61	3·88919006	86	3.94638708
12	3·10089897	37	3·74895640	62	3·89254617	87	3·94784815
13	3·16407363	38	3·75840778	63	3·89578156	88	3·94926535
14	3·22023311	39	3·76738431	64	3·89890168	89	3·95064021
15	3·27046993	40	3·77591838	65	3·90196170	90	3·95197420
16	3·31566099			66	3·90481646		
17	3·35651826	41	3·78403949	67	3·90762055	91	3·95326872
18	3·39362584·	42	3·79177460	68	3·91032831	92	3·95452513
19	3·42746741	43	3·79914840	69	3·91294383	93	3·95574472
20	3·45844693	44	3·80618353	70	3·91547098	94	3·95692872
		45	3·81290080			95	3·95807833
21	3·48690442	46	3·81931936	71	3·91791345	96	3·95919468
22	3·51312813	47	3·82545687	72	3·92027470	97	3·96027888
23	3·53736403	48	3·83132964	73	3·92255804	98	3·96133198
24	3·55982329	49	3·83695277	74	3·92476660	99	3·96235499
25	3·58068822	50	3·84234022	75	3.92690336	100	3·96334887

TABLE VII.

VALUING MINERAL AND OTHER PROPERTIES,

OR

The present value (or years' purchase) of £1 per annum in n years, allowing interest to a present purchaser upon his purchase money or capital invested, at the rates of $3\frac{1}{2}$, 4, $4\frac{1}{2}$, 5, $5\frac{1}{2}$, 6, 7, 8, 9, 10, 11, 12, 13, 14, 15, 16, 17, 18, 19, 20, 21, 22, 23, 24, and 25 per cent. per annum, and redeeming the capital so invested by an Annual Redemption Fund, at the rate of 3 per cent. per annum.

Calculated to 8 places of decimals, and to 100 years for each percentage.

TABLE VII.   lxxxiii

Present Value of £1 per Annum in n years; Redemption of Capital being at **3** per cent. with interest allowed to a Purchaser at the following rates per cent.

Years	3½ per cent.	Years	3½ per cent.	Years	4 per cent.	Years	4 per cent.
1	0·96618357	51	22·97064504	1	0·96153846	51	20·60418764
2	1·89533635	52	23·13894415	2	1·87754347	52	20·73949417
3	2·78916405	53	23·30203549	3	2·75080186	53	20·87041924
4	3·64927483	54	23·46008908	4	3·58388198	54	20·99711781
5	4·47718618	55	23·61326887	5	4·37915472	55	21·11973855
6	5·27433115	56	23·76173297	6	5·13881215	56	21·23842415
7	6·04206411	57	23·90563402	7	5·86488408	57	21·35331174
8	6·77766575	58	24·04511909	8	6·55925283	58	21·46453285
9	7·49434980	59	24·18033027	9	7·22366639	59	21·57221398
10	8·18126365	60	24·31140457	10	7·85974980	60	21·67647661
11	8·84349638	61	24·43847430	11	8·46901774	61	21·77743756
12	9·48208065	62	24·56166184	12	9·05287996	62	21·87520491
13	10·09799661	63	24·68110677	13	9·61265388	63	21·96989959
14	10·69217517	64	24·79691199	14	10·14957026	64	22·06161265
15	11·26550098	65	24·90919800	15	10·66478051	65	22·15044847
16	11·82781528	66	25·01813220	16	11·15936304	66	22·23650325
17	12·35291852	67	25·12365309	17	11·63432894	67	22·31986976
18	12·86857266	68	25·22603339	18	12·09062710	68	22·40063727
19	13·36650354	69	25·32531718	19	12·52914897	69	22·47889185
20	13·84740292	70	25·42160131	20	12·95073284	70	22·55471628
21	14·31193037	71	25·51497955	21	13·35616766	71	22·62819046
22	14·76071520	72	25·60554253	22	13·74619672	72	22·69939137
23	15·19435802	73	25·69337787	23	14·12152080	73	22·76839322
24	15·61343242	74	25·77857024	24	14·48280123	74	22·83526751
25	16·01848636	75	25·86120175	25	14·83066253	75	22·90008337
26	16·41004359	76	25·94135060	26	15·16569501	76	22·96290717
27	16·78860502	77	26·01909611	27	15·48845708	77	23·02380335
28	17·15464972	78	26·09451002	28	15·79947723	78	23·08283383
29	17·50863624	79	26·16766486	29	16·09925614	79	23·14005839
30	17·85100361	80	26·23863016	30	16·38826842	80	23·19553485
31	18·18217230	81	26·30747306	31	16·66696423	81	23·24931890
32	18·50254524	82	26·37425864	32	16·93577090	82	23·30146439
33	18·81250859	83	26·43904992	33	17·19509430	83	23·35202337
34	19·11243262	84	26·50190777	34	17·44532010	84	23·40104596
35	19·40267262	85	26·56289133	35	17·68681516	85	23·44858081
36	19·68356922	86	26·62205768	36	17·91992846	86	23·49467475
37	19·95544970	87	26·67946207	37	18·14499229	87	23·53937302
38	20·21862803	88	26·73515815	38	18·36232312	88	23·58271948
39	20·47340591	89	26·78919781	39	18·57222265	89	23·62475645
40	20·72007314	90	26·84163123	40	18·77497850	90	23·66552478
41	20·95890824	91	26·89250714	41	18·97086513	91	23·70506411
42	21·19017899	92	26·94187258	42	19·16014453	92	23·74341259
43	21·41414275	93	26·98977326	43	19·34306678	93	23·78060727
44	21·63104719	94	27·03625348	44	19·51987094	94	23·81668396
45	21·84113056	95	27·08135596	45	19·69078547	95	23·85167716
46	22·04462203	96	27·12512225	46	19·85602878	96	23·88562038
47	22·24174232	97	27·16759267	47	20·01580989	97	23·91854608
48	22·43270390	98	27·20880612	48	20·17032883	98	23·95048553
49	22·61771135	99	27·24880037	49	20·31977709	99	23·98146906
50	22·79696180	100	27·28761200	50	20·46433812	100	24·01152597

Present Value of £1 per Annum in n years; Redemption of Capital being at 3 per cent. with interest allowed to a Purchaser at the following rates per cent.

Years	4½ per cent.	Years	4½ per cent.	Years	5 per cent.	Years	5 per cent.
1	0·95693780	51	18·67977925	1	0·95238095	51	17·08413948
2	1·86008155	52	18·79092298	2	1·84294144	52	17·17705901
3	2·71348062	53	18·89833805	3	2·67715853	53	17·26677162
4	3·52079148	54	19·00216439	4	3·45988383	54	17·35340328
5	4·28532423	55	19·10253575	5	4·19543034	55	17·43707409
6	5·01008279	56	19·19958003	6	4·88764524	56	17·51789855
7	5·69779941	57	19·29341971	7	5·53997119	57	17·59598600
8	6·35096491	58	19·38417188	8	6·15549815	58	17·67144066
9	6·97185463	59	19·47194875	9	6·73700745	59	17·74436219
10	7·56255138	60	19·55685776	10	7·28700946	60	17·81484574
11	8·12496536	61	19·63900188	11	7·80777582	61	17·88298227
12	8·66085170	62	19·71847636	12	8·30136716	62	17·94885590
13	9·17182594	63	19·79538634	13	8·76965710	63	18·01255856
14	9·65937760	64	19·86981216	14	9·21435302	64	18·07416122
15	10·12488227	65	19·94184448	15	9·63701411	65	18·13374306
16	10·56961234	66	20·01160293	16	10·03906710	66	18·19137700
17	10·99474645	67	20·07905977	17	10·42182010	67	18·24713336
18	11·40137805	68	20·14440026	18	10·78647467	68	18·30107896
19	11·79052288	69	20·20766255	19	11·13413643	69	18·35327828
20	12·16312587	70	20·26891783	20	11·46582453	70	18·40379299
21	12·52006709	71	20·32823464	21	11·78247989	71	18·45268235
22	12·86216731	72	20·38567886	22	12·08497261	72	18·50000324
23	13·19019289	73	20·44131386	23	12·37410850	73	18·54581022
24	13·50486020	74	20·49520057	24	12·65063492	74	18·59015572
25	13·80683963	75	20·54739776	25	12·91524598	75	18·63309018
26	14·09675925	76	20·59796175	26	13·16858723	76	18·67466190
27	14·37520811	77	20·64694710	27	13·41125985	77	18·71491754
28	14·64273912	78	20·69440614	28	13·64382432	78	18·75390183
29	14·89987187	79	20·74038930	29	13·86680387	79	18·79165781
30	15·14709511	80	20·78494524	30	14·08068755	80	18·82822692
31	15·38486893	81	20·82812079	31	14·28593290	81	18·86364890
32	15·61362686	82	20·86996112	32	14·48296853	82	18·89796241
33	15·83377779	83	20·91050980	33	14·67219631	83	18·93120415
34	16·04570757	84	20·94980877	34	14·85399340	84	18·96340973
35	16·24978084	85	20·98789863	35	15·02871425	85	18·99461351
36	16·44634209	86	21·02481840	36	15·19669211	86	19·02484848
37	16·63571747	87	21·06060575	37	15·35824070	87	19·05414642
38	16·81821558	88	21·09529711	38	15·51365556	88	19·08253806
39	16·99412892	89	21·12892760	39	15·66321541	89	19·11005297
40	17·16373475	90	21·16153109	40	15·80718325	90	19·13671965
41	17·32729614	91	21·19314036	41	15·94580750	91	19·16256564
42	17·48506287	92	21·22378694	42	16·07932301	92	19·18761747
43	17·63727212	93	21·25350142	43	16·20795183	93	19·21190065
44	17·78414944	94	21·28231331	44	16·33190430	94	19·23544000
45	17·92590930	95	21·31025102	45	16·45137961	95	19·25825932
46	18·06275579	96	21·33734211	46	16·56656656	96	19·28038162
47	18·19488327	97	21·36361324	47	16·67764431	97	19·30182917
48	18·32247691	98	21·38909007	48	16·78478292	98	19·32262341
49	18·44571319	99	21·41379745	49	16·88814389	99	19·34278505
50	18·56476045	100	21·43775939	50	16·93788076	100	19·36233409

Present Value of £1 per Annum in n years; Redemption of Capita
per cent. with interest allowed to a Purchaser at the following rate

Years	5½ per cent.	Years	5½ per cent.	Years	6 per cent.	Years
1	0·94786729	51	15·73964779	1	0·94339622	51
2	1·82611433	52	15·81848386	2	1·80959173	52
3	2·64179600	53	15·89453508	3	2·60735549	53
4	3·40104768	54	15·96791494	4	3·34417912	54
5	4·10923039	55	16·03873135	5	4·02650128	55
6	4·77104927	56	16·10708698	6	4·65988652	56
7	5·39065094	57	16·17307959	7	5·24916876	57
8	5·97170408	58	16·23680223	8	5·79856743	58
9	6·51746635	59	16·29834363	9	6·31178221	59
10	7·03084046	60	16·35778836	10	6·79207063	60
11	7·51442124	61	16·41521709	11	7·24231231	61
12	7·97053545	62	16·47070440	12	7·66506220	62
13	8·40127557	63	16·52433115	13	8·06259515	63
14	8·80852857	64	16·57616025	14	8·43694329	64
15	9·19400054	65	16·62626130	15	8·78992755	65
16	9·55923794	66	16·67472348	16	9·12318449	66
17	9·90564581	67	16·72153314	17	9·43818902	67
18	10·23450360	68	16·76682410	18	9·73627395	68
19	10·54697892	69	16·81062764	19	10·01864665	69
20	10·84413953	70	16·85299759	20	10·28640355	70
21	11·12696376	71	16·89398560	21	10·54054258	71
22	11·39634974	72	16·93364115	22	10·78197401	72
23	11·65312343	73	16·97201168	23	11·01152985	73
24	11·89804576	74	17·00914266	24	11·22997214	74
25	12·13181886	75	17·04507782	25	11·43800013	75
26	12·35509172	76	17·07985894	26	11·63625663	76
27	12·56846510	77	17·11352637	27	11·82533363	77
28	12·77249587	78	17·14611870	28	12·00577717	78
29	12·96770103	79	17·17767304	29	12·17809177	79
30	13·15456122	80	17·20822508	30	12·34274429	80
31	13·33352377	81	17·23780909	31	12·50016738	81
32	13·50500567	82	17·26645804	32	12·65076257	82
33	13·66939600	83	17·29420365	33	12·79490302	83
34	13·82705824	84	17·32107639	34	12·93293595	84
35	13·97833245	85	17·34710567	35	13·06518494	85
36	14·12353690	86	17·37231973	36	13·19195181	86
37	14·26297006	87	17·39674572	37	13·31351849	87
38	14·39691190	88	17·42040988	38	13·43014857	88
39	14·52562541	89	17·44333746	39	13·54208886	89
40	14·64935783	90	17·46555272	40	13·64957061	90
41	14·76834182	91	17·48707913	41	13·75281077	91
42	14·88276854	92	17·50793922	42	13·85201308	92
43	14·99292852	93	17·52815479	43	13·94736899	93
44	15·09893268	94	17·54774684	44	14·03905867	94
45	15·20099307	95	17·56673558	45	14·12725179	95
46	15·29928356	96	17·58514052	46	14·21210819	96
47	15·39396864	97	17·60298056	47	14·29377873	97
48	15·48520398	98	17·62027383	48	14·37240580	98
49	15·57313700	99	17·63703788	49	14·44812393	99
50	15·65790744	100	17·65328964	50	14·52106035	100

Present Value of £1 per Annum in n years; Redemption of Capital being at 3 per cent. with interest allowed to a Purchaser at the following rates per cent.

Years	7 per cent.	Years	7 per cent.	Years	8 per cent.	Years	8 per cent.
1	0·93457943	51	12·73336709	1	0·92592592	51	11·29511823
2	1·77742755	52	12·78491442	2	1·74638679	52	11·33565999
3	2·54109998	53	12·83454758	3	2·47812827	53	11·37466127
4	3·23596273	54	12·88235073	4	3·13453049	54	11·41219109
5	3·87064953	55	12·92840350	5	3·72641314	55	11·44831867
6	4·45240930	56	12·97278124	6	4·26262001	56	11·48310336
7	4·98737312	57	13·01555539	7	4·75045044	57	11·51660525
8	5·48076177	58	13·05679360	8	5·19598235	58	11·54887815
9	5·93704863	59	13·09656008	9	5·60431757	59	11·57998092
10	6·36008890	60	13·13491577	10	5·97977020	60	11·60995762
11	6·75322282	61	13·17191855	11	6·32601306	61	11·63885769
12	7·11935891	62	13·20762187	12	6·64619261	62	11·66672495
13	7·46104157	63	13·24208274	13	6·94301997	63	11·69360579
14	7·78050638	64	13·27534625	14	7·21884378	64	11·71953712
15	8·07972553	65	13·30746136	15	7·47570879	65	11·74455874
16	8·36044561	66	13·33847318	16	7·71540350	66	11·76870731
17	8·62421894	67	13·36842477	17	7·93949915	67	11·79201775
18	8·87242987	68	13·39735714	18	8·14938169	68	11·81452327
19	9·10631693	69	13·42530947	19	8·34627837	69	11·83625553
20	9·32699156	70	13·45231908	20	8·53127981	70	11·85724469
21	9·53545399	71	13·47842168	21	8·70535854	71	11·87751951
22	9·73260687	72	13·50365136	22	8·86938454	72	11·89710745
23	9·91926682	73	13·52804071	23	9·02413845	73	11·91603469
24	10·09617455	74	13·55162087	24	9·17032276	74	11·93432623
25	10·26400341	75	13·57442171	25	9·30857133	75	11·95200601
26	10·42336690	76	13·59647167	26	9·43945760	76	11·96909681
27	10·57482526	77	13·61779815	27	9·56350167	77	11·98562054
28	10·71889100	78	13·63842730	28	9·68117627	78	12·00159807
29	10·85603399	79	13·65838417	29	9·79291212	79	12·01704940
30	10·98668576	80	13·67769284	30	9·89910248	80	12·03199369
31	11·11124336	81	13·69637630	31	10·00010712	81	12·04644927
32	11·23007273	82	13·71445668	32	10·09625586	82	12·06043372
33	11·34351170	83	13·73195520	33	10·18785156	83	12·07396389
34	11·45187260	84	13·74889221	34	10·27517289	84	12·08705592
35	11·55544472	85	13·76528728	35	10·35847667	85	12·09972533
36	11·65449628	86	13·78115918	36	10·43799996	86	12·11198697
37	11·74927640	87	13·79652592	37	10·51396195	87	12·12385511
38	11·84001673	88	13·81140487	38	10·58656559	88	12·13534346
39	11·92693299	89	13·82581271	39	10·65599912	89	12·14646518
40	12·01022629	90	13·83976543	40	10·72243731	90	12·15723291
41	12·09008435	91	13·85327848	41	10·78604269	91	12·16765882
42	12·16668261	92	13·86636665	42	10·84696661	92	12·17775456
43	12·24018519	93	13·87904425	43	10·90535014	93	12·18753138
44	12·31074584	94	13·89132502	44	10·96132498	94	12·19700010
45	12·37850870	95	13·90322216	45	11·01501421	95	12·20617108
46	12·44360902	96	13·91474842	46	11·06653293	96	12·21505434
47	12·50617391	97	13·92591610	47	11·11598899	97	12·22365953
48	12·56632288	98	13·93673700	48	11·16348350	98	12·23199590
49	12·62416843	99	13·94722251	49	11·20911134	99	12·24007238
50	12·67981654	100	13·95738361	50	11·25296164	100	12·24789756

TABLE VII.   lxxxvii

Present Value of £1 per Annum in n years; Redemption of Capital being at 3 per cent. with interest allowed to a Purchaser at the following rates per cent.

Years	9 per cent.	Years	9 per cent.	Years	10 per cent.	Years	10 per cent.
1	0·91743119	51	10·14879934	1	0·90909090	51	9·21371763
2	1·71641160	52	10·18151776	2	1·68744805	52	9·24067663
3	2·41820212	53	10·21297047	3	2·36110581	53	9·26657763
4	3·03926384	54	10·24321656	4	2·94961719	54	9·29147106
5	3·59254024	55	10·27231169	5	3·46795265	55	9·31540432
6	4·08834922	56	10·30030831	6	3·92776833	56	9·33842205
7	4·53501671	57	10·32725595	7	4·33827521	57	9·36056631
8	4·93933536	58	10·35320134	8	4·70684834	58	9·38187675
9	5·30690193	59	10·37818866	9	5·03946260	59	9·40239080
10	5·64236947	60	10·40225968	10	5·34100996	60	9·42214381
11	5·94963818	61	10·42545395	11	5·61553421	61	9·44116920
12	6·23200176	62	10·44780794	12	5·86640717	62	9·45949778
13	6·49226099	63	10·46936009	13	6·09646272	63	9·47716189
14	6·73281256	64	10·49014113	14	6·30810001	64	9·49418747
15	6·95571945	65	10·51018401	15	6·50336371	65	9·51060221
16	7·16276711	66	10·52951904	16	6·68400724	66	9·52643161
17	7·35550861	67	10·54817507	17	6·85154279	67	9·54169987
18	7·53530123	68	10·56617953	18	7·00728147	68	9·55642998
19	7·70333646	69	10·58355848	19	7·15236567	69	9·57064382
20	7·86066453	70	10·60033681	20	7·28779538	70	9·58436215
21	8·00821473	71	10·61653813	21	7·41444968	71	9·59760476
22	8·14681242	72	10·63218498	22	7·53310452	72	9·61039048
23	8·27719308	73	10·64729886	23	7·64444741	73	9·62273727
24	8·40001433	74	10·66190027	24	7·74908969	74	9·63466219
25	8·51586588	75	10·67600879	25	7·84757677	75	9·64618160
26	8·62527813	76	10·68964308	26	7·94039682	76	9·65731100
27	8·72872949	77	10·70282102	27	8·02798812	77	9·66806530
28	8·82665248	78	10·71555967	28	8·11074519	78	9·67845866
29	8·91943927	79	10·72787532	29	8·18902423	79	9·68850463
30	9·00744615	80	10·73978361	30	8·26314758	80	9·69821618
31	9·09099762	81	10·75129944	31	8·33340772	81	9·70760568
32	9·17038984	82	10·76243712	32	8·40007061	82	9·71668501
33	9·24589364	83	10·77321036	33	8·46337865	83	9·72546550
34	9·31775723	84	10·78363226	34	8·52355323	84	9·73395802
35	9·38620846	85	10·79371541	35	8·58079697	85	9·74217298
36	9·45145689	86	10·80347187	36	8·63529565	86	9·75012036
37	9·51369561	87	10·81291318	37	8·68721995	87	9·75780969
38	9·57310279	88	10·82205046	38	8·73672694	88	9·76525016
39	9·62984312	89	10·83089437	39	8·78396322	89	9·77245057
40	9·68406907	90	10·83945511	40	8·82905708	90	9·77941934
41	9·73592199	91	10·84774252	41	8·87213759	91	9·78616458
42	9·78553310	92	10·85576602	42	8·91331747	92	9·79269406
43	9·83302440	93	10·86353468	43	8·95270294	93	9·79901526
44	9·87850946	94	10·87105723	44	8·99039267	94	9·80513536
45	9·92209413	95	10·87834204	45	9·02647844	95	9·81106124
46	9·96387718	96	10·88539716	46	9·06104571	96	9·81679954
47	10·00395091	97	10·89223037	47	9·09417419	97	9·82235666
48	10·04240165	98	10·89884912	48	9·12593827	98	9·82773872
49	10·07931023	99	10·90526059	49	9·15640751	99	9·83295159
50	10·11475243	100	10·91147169	50	9·18564698	100	9·83800099

Present Value of £1 per Annum in n years; Redemption of Capital being at 3 per cent. with interest allowed to a Purchaser at the following rates per cent.

Years	11 per cent.	Years	11 per cent.	Years	12 per cent.	Years	12 per cent.
1	0·90090090	51	8·43641058	1	0·89285714	51	7·78005334
2	1·65944576	52	8·45900713	2	1·63235767	52	7·79926661
3	2·30664351	53	8·48070639	3	2·25463707	53	7·81770941
4	2·86510749	54	8·50155183	4	2·78530549	54	7·83541958
5	3·35171670	55	8·52158429	5	3·24301986	55	7·85243263
6	3·77932519	56	8·54084216	6	3·64169375	56	7·86878192
7	4·15789431	57	8·55936159	7	3·99191472	57	7·88449883
8	4·49526312	58	8·57717661	8	4·30188219	58	7·89961286
9	4·79768505	59	8·59431931	9	4·57804488	59	7·91415183
10	5·07020956	60	8·61081997	10	4·82554435	60	7·92814194
11	5·31695858	61	8·62670718	11	5·04853032	61	7·94160792
12	5·54133018	62	8·64200729	12	5·25038880	62	7·95457255
13	5·74615078	63	8·65674788	13	5·43391011	63	7·96705961
14	5·93379056	64	8·67095111	14	5·60141437	64	7·97908827
15	6·10625194	65	8·68464059	15	5·75484651	65	7·99067885
16	6·26523826	66	8·69783802	16	5·89584926	66	8·00185006
17	6·41220764	67	8·71056399	17	6·02581957	67	8·01261963
18	6·54841556	68	8·72283807	18	6·14595301	68	8·02300439
19	6·67494892	69	8·73467882	19	6·25727876	69	8·03302030
20	6·79275341	70	8·74610388	20	6·36068759	70	8·04268251
21	6·90265574	71	8·75713003	21	6·45695441	71	8·05200544
22	7·00538179	72	8·76777324	22	6·54675650	72	8·06100279
23	7·10157151	73	8·77804870	23	6·63468843	73	8·06968759
24	7·19179133	74	8·78797089	24	6·70927432	74	8·07807225
25	7·27654436	75	8·79755360	25	6·78297796	75	8·08616859
26	7·35627907	76	8·80680998	26	6·85221129	76	8·09398785
27	7·43139649	77	8·81575258	27	6·91734142	77	8·10154079
28	7·50225630	78	8·82439339	28	6·97869659	78	8·10883766
29	7·56918207	79	8·83274383	29	7·03657119	79	8·11588222
30	7·63246568	80	8·84081484	30	7·09122998	80	8·12270181
31	7·69237108	81	8·84861685	31	7·14291180	81	8·12928736
32	7·74913758	82	8·85615985	32	7·19183258	82	8·13565338
33	7·80298268	83	8·86345340	33	7·23818812	83	8·14180804
34	7·85410446	84	8·87050663	34	7·28215629	84	8·14775912
35	7·90268372	85	8·87732830	35	7·32389914	85	8·15351409
36	7·94888585	86	8·88392680	36	7·36356452	86	8·15908010
37	7·99286241	87	8·89031015	37	7·40128767	87	8·16446398
38	8·03475255	88	8·89648605	38	7·43719253	88	8·16967229
39	8·07468427	89	8·90246188	39	7·47139288	89	8·17471132
40	8·11277550	90	8·90824473	40	7·50399337	90	8·17958709
41	8·14913511	91	8·91384139	41	7·53509041	91	8·18430539
42	8·18386372	92	8·91925838	42	7·56477300	92	8·18887175
43	8·21705446	93	8·92450195	43	7·59312338	93	8·19329149
44	8·24879372	94	8·92957813	44	7·62021768	94	8·19756974
45	8·27916170	95	8·93449269	45	7·64612652	95	8·20171139
46	8·30823293	96	8·93925118	46	7·67091541	96	8·20572115
47	8·33607684	97	8·94385894	47	7·69464530	97	8·20960357
48	8·36275813	98	8·94832109	48	7·71737290	98	8·21336300
49	8·38833718	99	8·95264258	49	7·73915109	99	8·21700362
50	8·41287040	100	8·95682815	50	7·76002920	100	8·22052945

TABLE VII.   lxxxix

Present Value of £1 per Annum in n years; Redemption of Capital being at **3** per cent. with interest allowed to a Purchaser at the following rates per cent.

Years	13 per cent.	Years	13 per cent.	Years	14 per cent.	Years	14 per cent.
1	0·88495575	51	7·21845378	1	0·87719298	51	6·73247331
2	1·60613973	52	7·23499042	2	1·58075066	52	6·74685603
3	2·20492404	53	7·25085837	3	2·15735598	53	6·76065300
4	2·70982849	54	7·26609087	4	2·63833416	54	6·77389360
5	3·14115169	55	7·28071907	5	3·04548828	55	6·78660539
6	3·51373430	56	7·29477220	6	3·39446193	56	6·79881419
7	3·83867796	57	7·30827775	7	3·69677084	57	6·81054426
8	4·12445308	58	7·32126154	8	3·96108019	58	6·82181838
9	4·37763479	59	7·33374789	9	4·19403525	59	6·83265798
10	4·60340500	60	7·34575968	10	4·40081754	60	6·84308324
11	4·80590285	61	7·35731853	11	4·58552689	61	6·85311316
12	4·98847449	62	7·36244430	12	4·75144964	62	6·86276527
13	5·15385430	63	7·37915772	13	4·90125097	63	6·87205774
14	5·30429862	64	7·38947550	14	5·03711500	64	6·88100530
15	5·44168585	65	7·39941534	15	5·16084867	65	6·88962348
16	5·56759240	66	7·40899351	16	5·27395982	66	6·89792658
17	5·68335109	67	7·41822544	17	5·37771658	67	6·90592813
18	5·79009641	68	7·42712576	18	5·47319324	68	6·91364095
19	5·88880012	69	7·43570834	19	5·56130593	69	6·92107722
20	5·98029942	70	7·44398632	20	5·64284066	70	6·92824846
21	6·06531949	71	7·45197223	21	5·71847567	71	6·93516562
22	6·14449160	72	7·45967795	22	5·78879931	72	6·94183910
23	6·21836784	73	7·46711476	23	5·85432441	73	6·94827881
24	6·28743318	74	7·47429343	24	5·91550007	74	6·95449413
25	6·35211538	75	7·48122419	25	5·97272123	75	6·96049402
26	6·41279315	76	7·48791677	26	6·02633665	76	6·96628700
27	6·46980306	77	7·49438050	27	6·07665542	77	6·97188119
28	6·52344515	78	7·50062422	28	6·12395247	78	6·97728434
29	6·57398785	79	7·50665638	29	6·16847318	79	6·98250381
30	6·62167199	80	7·51248505	30	6·21043721	80	6·98754666
31	6·66671428	81	7·51811795	31	6·25004184	81	6·99241960
32	6·70931022	82	7·52356242	32	6·28746471	82	6·99712905
33	6·74963672	83	7·52882552	33	6·32286622	83	7·00168116
34	6·78785415	84	7·53391396	34	6·35639156	84	7·00608178
35	6·82410833	85	7·53883418	35	6·38817251	85	7·01033654
36	6·85853208	86	7·54359236	36	6·41832893	86	7·01445078
37	6·89124668	87	7·54819437	37	6·44697007	87	7·01842965
38	6·92236306	88	7·55264587	38	6·47419573	88	7·02227808
39	6·95198292	89	7·55695229	39	6·50009727	89	7·02600076
40	6·98019965	90	7·56111880	40	6·52475848	90	7·02960222
41	7·00709916	91	7·56515038	41	6·54825634	91	7·03308679
42	7·03276063	92	7·56905180	42	6·57066172	92	7·03645861
43	7·05725714	93	7·57282765	43	6·59203993	93	7·03972166
44	7·08065626	94	7·57648233	44	6·61245131	94	7·04287978
45	7·10302058	95	7·58002002	45	6·63195168	95	7·04593662
46	7·12440809	96	7·58344482	46	6·65059272	96	7·04889570
47	7·14487269	97	7·58676060	47	6·66844240	97	7·05176042
48	7·16446446	98	7·58997112	48	6·68548525	98	7·05453402
49	7·18323006	99	7·59307996	49	6·70182272	99	7·05721963
50	7·20121297	100	7·59609059	50	6·71747341	100	7·05982025

Present Value of £1 per Annum in n years; Redemption of Capital being at 3 per cent. with interest allowed to a Purchaser at the following rates per cent.

Years	15 per cent.	Years	15 per cent.	Years	16 per cent.	Years	16 per cent.
1	0·86956522	51	6·30780221	1	0·86206897	51	5·93352706
2	1·55615178	52	6·32042599	2	1·53230676	52	5·94469589
3	2·11179700	53	6·33253245	3	2·06812245	53	5·95540453
4	2·57051537	54	6·34414778	4	2·50609579	54	5·96567645
5	2·95547950	55	6·35529650	5	2·87063837	55	5·97553362
6	3·28302103	56	6·36600158	6	3·17866480	56	5·98499660
7	3·56498164	57	6·37628458	7	3·44226550	57	5·99408468
8	3·81015682	58	6·38616575	8	3·67031217	58	6·00281597
9	4·02521626	59	6·39596412	9	3·86946204	59	6·01120748
10	4·21530947	60	6·40479761	10	4·04480828	60	6·01927520
11	4·38447558	61	6·41358306	11	4·20031384	61	6·02703421
12	4·53592734	62	6·42203601	12	4·33910853	62	6·03449835
13	4·67225216	63	6·43017257	13	4·46369698	63	6·04168199
14	4·79555727	64	6·43800578	14	4·57610741	64	6·04859677
15	4·99757610	65	6·44554940	15	4·67799970	65	6·05525495
16	5·00974773	66	6·45281606	16	4·77074542	66	6·06166779
17	5·10327681	67	6·45981776	17	4·85548783	67	6·06784596
18	5·18917943	68	6·46656581	18	4·93318748	68	6·07379956
19	5·26831861	69	6·47307098	19	5·00465732	69	6·07953816
20	5·34143215	70	6·47934344	20	5·07059003	70	6·08507081
21	5·40915449	71	6·48539288	21	5·13157943	71	6·09040612
22	5·47203423	72	6·49122847	22	5·18813757	72	6·09555225
23	5·53054818	73	6·49685894	23	5·24070828	73	6·10051696
24	5·58511272	74	6·50229257	24	5·28967823	74	6·10530761
25	5·63609310	75	6·50753728	25	5·33538579	75	6·10993123
26	5·68381106	76	6·51260055	26	5·37812840	76	6·11439447
27	5·72855111	77	6·51748956	27	5·41816855	77	6·11870369
28	5·77056577	78	6·52221112	28	5·45573877	78	6·12286494
29	5·81007995	79	6·52677172	29	5·49104580	79	6·12688399
30	5·84729465	80	6·53117758	30	5·52427407	80	6·13876634
31	5·88239001	81	6·53543459	31	5·55558862	81	6·13451723
32	5·91552798	82	6·53954841	32	5·58513760	82	6·13814166
33	5·94685456	83	6·54352444	33	5·61305438	83	6·14164444
34	5·97650171	84	6·54736784	34	5·63945932	84	6·14503011
35	6·00458901	85	6·55108353	35	5·66446138	85	6·14830306
36	6·03122507	86	6·55467624	36	5·68815937	86	6·15146747
37	6·05650876	87	6·55815048	37	5·71064316	87	6·15442731
38	6·08053030	88	6·56151056	38	5·73199463	88	6·15758644
39	6·10337215	89	6·56476663	39	5·75228857	89	6·16034850
40	6·12510986	90	6·56790465	40	5·77159342	90	6·16311700
41	6·14581277	91	6·57094642	41	5·78997193	91	6·16579532
42	6·16554464	92	6·57388959	42	5·80748176	92	6·16838666
43	6·18436417	93	6·57673764	43	5·82417592	93	6·17089412
44	6·20232556	94	6·57949393	44	5·84010334	94	6·17332067
45	6·21947885	95	6·58216168	45	5·85530914	95	6·17566915
46	6·23587038	96	6·58474397	46	5·86983508	96	6·17794228
47	6·25154310	97	6·58724377	47	5·88371982	97	6·18014271
48	6·26653685	98	6·58966394	48	5·89699922	98	6·18227293
49	6·28088869	99	6·59200721	49	5·90970660	99	6·18433537
50	6·29463310	100	6·59427620	50	5·92187293	100	6·18633236

TABLE VII. xci

Present Value of £1 per Annum in n years; Redemption of Capital being at **3** per cent. with interest allowed to a Purchaser at the following rates per cent.

Years	17 per cent.	Years	17 per cent.	Years	18 per cent.	Years	18 per cent.
1	0·85470085	51	5·60117955	1	0·84745762	51	5·30408806
2	1·50918147	52	5·61113120	2	1·48674381	52	5·31301117
3	2·02621779	53	5·62067084	3	1·98597756	53	5·32156329
4	2·44482611	54	5·62981963	4	2·38648080	54	5·32976355
5	2·79053228	55	5·63859734	5	2·71477559	55	5·33762988
6	3·08073845	56	5·64702249	6	2·98866548	56	5·34517903
7	3·32771666	57	5·65511245	7	3·22054601	57	5·35242670
8	3·54036956	58	5·66288350	8	3·41931323	58	5·35938762
9	3·72531249	59	5·67035092	9	3·59151725	59	5·36607561
10	3·88756378	60	5·67752911	10	3·74208773	60	5·37250365
11	4·03099923	61	5·68443158	11	3·87480583	61	5·37868397
12	4·15865977	62	5·69107078	12	3·99262027	62	5·38462780
13	4·27296478	63	5·69745960	13	4·09786448	63	5·39034677
14	4·37586321	64	5·70360849	14	4·19240912	64	5·39585031
15	4·46894258	65	5·70952844	15	4·27777143	65	5·40114834
16	4·55350910	66	5·71522956	16	4·35519490	66	5·40624996
17	4·63064731	67	5·72072140	17	4·42570837	67	5·41116380
18	4·70126525	68	5·72601301	18	4·49017043	68	5·41589799
19	4·76612890	69	5·73111296	19	4·54930324	69	5·42046026
20	4·82588898	70	5·73602936	20	4·60371863	70	5·42485792
21	4·88110181	71	5·74076991	21	4·65393835	71	5·42909789
22	4·93224587	72	5·74534193	22	4·70041008	72	5·43318677
23	4·97973490	73	5·74975234	23	4·74352017	73	5·43713078
24	5·02392857	74	5·75400774	24	4·78360374	74	5·44093587
25	5·06514098	75	5·75811440	25	4·82095292	75	5·44460766
26	5·10364768	76	5·76207827	26	4·85582355	76	5·44815152
27	5·13969141	77	5·76590505	27	4·88844064	77	5·45157255
28	5·17348684	78	5·76960012	28	4·91900288	78	5·45486561
29	5·20522454	79	5·77316865	29	4·94768635	79	5·45806533
30	5·23507422	80	5·77661554	30	4·97464772	80	5·46114612
31	5·26318757	81	5·77994548	31	5·00002678	81	5·46412219
32	5·28970055	82	5·78316293	32	5·02394870	82	5·46699755
33	5·31473539	83	5·78627217	33	5·04652589	83	5·46977604
34	5·33840230	84	5·78927728	34	5·06785957	84	5·47246132
35	5·36080089	85	5·79218215	35	5·08804113	85	5·47505688
36	5·38202142	86	5·79499051	36	5·10715333	86	5·47756608
37	5·40214588	87	5·79770592	37	5·12527125	87	5·47999209
38	5·42124893	88	5·80033179	38	5·14246320	88	5·48233800
39	5·43939866	89	5·80287140	39	5·15879143	89	5·48460672
40	5·45665734	90	5·80532785	40	5·17431282	90	5·48680106
41	5·47308202	91	5·80770417	41	5·18907945	91	5·48892371
42	5·48872505	92	5·81000320	42	5·20313905	92	5·49097724
43	5·50363456	93	5·81222770	43	5·21653551	93	5·49296412
44	5·51785491	94	5·81438033	44	5·22930921	94	5·49488672
45	5·53142699	95	5·81646360	45	5·24149739	95	5·49674729
46	5·54438861	96	5·81847995	46	5·25313442	96	5·49854804
47	5·55677476	97	5·82043172	47	5·26425212	97	5·50029104
48	5·56861787	98	5·82232115	48	5·27487996	98	5·50197830
49	5·57994804	99	5·82415038	49	5·28504526	99	5·50361176
50	5·59079326	100	5·82592149	50	5·29477342	100	5·50519326

Present Value of £1 per Annum in n years; Redemption of Capital being at 3 per cent. with interest allowed to a Purchaser at the following rates per cent.

Years	19 per cent.	Years	19 per cent.	Years	20 per cent.	Years	20 per cent.
1	0·84033613	51	5·03692511	1	0·83333333	51	4·79538515
2	1·46496356	52	5·04497129	2	1·44381223	52	4·80267758
3	1·94730452	53	5·05268164	3	1·91010889	53	4·80966460
4	2·33085539	54	5·06007360	4	2·27776400	54	4·81636213
5	2·64302344	55	5·06716345	5	2·57496647	55	4·82278505
6	2·90193631	56	5·07396644	6	2·82009884	56	4·82894728
7	3·12006295	57	5·08049683	7	3·02566044	57	4·83486183
8	3·30626179	58	5·08676801	8	3·20044664	58	4·84054092
9	3·46699937	59	5·09279251	9	3·35082624	59	4·84599599
10	3·60710664	60	5·09858214	10	3·48152434	60	4·85123779
11	3·73026530	61	5·10414798	11	3·59612046	61	4·85627644
12	3·83933039	62	5·10950023	12	3·69737591	62	4·86112123
13	3·93654999	63	5·11464943	13	3·78745494	63	4·86578175
14	4·02371839	64	5·11960414	14	3·86807783	64	4·87026581
15	4·10228505	65	5·12437334	15	3·94062921	65	4·87458158
16	4·17343373	66	5·12896528	16	4·00623612	66	4·87873658
17	4·23814062	67	5·13338778	17	4·06582523	67	4·88273799
18	4·29721801	68	5·13764820	18	4·12016552	68	4·88649228
19	4·35134726	69	5·14175355	19	4·16990041	69	4·89030607
20	4·40110418	70	5·14571044	20	4·21557244	70	4·89388527
21	4·44697870	71	5·14952513	21	4·25764225	71	4·89733561
22	4·48939033	72	5·15320359	22	4·29650352	72	4·90066248
23	4·52870036	73	5·15675146	23	4·33249466	73	4·90387102
24	4·56522163	74	5·16017410	24	4·36590824	74	4·90696611
25	4·59922638	75	5·16347662	25	4·39699847	75	4·90995237
26	4·63095266	76	5·16666384	26	4·42598728	76	4·91283421
27	4·66060951	77	5·16974040	27	4·45306934	77	4·91561582
28	4·68838127	78	5·17271068	28	4·47841605	78	4·91830119
29	4·71443109	79	5·17557885	29	4·50217896	79	4·92089410
30	4·73890394	80	5·17834892	30	4·52449259	80	4·92339818
31	4·76192905	81	5·18102467	31	4·54547668	81	4·92581688
32	4·78362199	82	5·18360974	32	4·56523825	82	4·92815349
33	4·80408643	83	5·18610758	33	4·58387320	83	4·93041115
34	4·82341564	84	5·18852149	34	4·60146772	84	4·93259285
35	4·84169376	85	5·19085464	35	4·61809953	85	4·93470146
36	4·85899691	86	5·19311004	36	4·63383882	86	4·93673972
37	4·87539408	87	5·19529058	37	4·64874924	87	4·93871023
38	4·89094800	88	5·19739902	38	4·66288854	88	4·94061552
39	4·90571578	89	5·19943800	39	4·67630934	89	4·94245796
40	4·91974959	90	5·20141004	40	4·68905960	90	4·94423985
41	4·93309712	91	5·20331758	41	4·70118319	91	4·94596340
42	4·94580209	92	5·20516293	42	4·71272027	92	4·94763069
43	4·95790466	93	5·20694832	43	4·72370773	93	4·94924375
44	4·96944174	94	5·20867588	44	4·73417945	94	4·95080452
45	4·98044737	95	5·21034765	45	4·74416665	95	4·95231483
46	4·99095295	96	5·21196560	46	4·75369812	96	4·95377648
47	5·00098753	97	5·21353162	47	4·76280047	97	4·95519117
48	5·01057799	98	5·21504752	48	4·77149834	98	4·95656053
49	5·01974924	90	5·21651502	49	4·77981454	99	4·95788615
50	5·02852445	100	5·21793581	50	4·78777025	100	4·95916952

TABLE VII. xciii

Present Value of £1 per Annum in n years; Redemption of Capital being at 3 per cent. with interest allowed to a Purchaser at the following rates per cent.

Years	21 per cent.	Years	21 per cent.	Years	22 per cent.	Years	22 per cent.
1	0·82644628	51	4·57595069	1	0·81967213	51	4·37571991
2	1·42326299	52	4·58259053	2	2·40329047	52	4·38179099
3	1·87430757	53	4·58895142	3	1·83982362	53	4·38760630
4	2·22703735	54	4·59504798	4	2·17852087	54	4·39317928
5	2·51032641	55	4·60089383	5	2·44885222	55	4·39852248
6	2·74275056	56	4·60650174	6	2·66953196	56	4·40364763
7	2·93680276	57	4·61188363	7	2·85301533	57	4·40856571
8	3·10119456	58	4·61705069	8	3·00791332	58	4·41328699
9	3·24218622	59	4·62201341	9	3·14036959	59	4·41782112
10	3·36439221	60	4·62678162	10	3·25488511	60	4·42217714
11	3·47128874	61	4·63136458	11	3·35483280	61	4·42636354
12	3·56554433	62	4·63577079	12	3·44279013	62	4·43038815
13	3·64924156	63	4·64000904	13	3·52076050	63	4·43425902
14	3·72402947	64	4·64408645	14	3·59032472	64	4·43798270
15	3·79123086	65	4·64801051	15	3·65274680	65	4·44156606
16	3·85191914	66	4·65178809	16	3·70904955	66	4·44501539
17	3·90697448	67	4·65542567	17	3·76007851	67	4·44833666
18	3·95712540	68	4·65892940	18	3·80649752	68	4·45153550
19	4·00298013	69	4·66230508	19	3·84890906	69	4·45461723
20	4·04505041	70	4·66555821	20	3·88778745	70	4·45758689
21	4·08376994	71	4·66869399	21	3·92355153	71	4·46044927
22	4·11950868	72	4·67171738	22	3·95651952	72	4·46320887
23	4·15258417	73	4·67463305	23	3·98701980	73	4·46587001
24	4·18327049	74	4·67744545	24	4·01529964	74	4·46843675
25	4·21180545	75	4·68015881	25	4·04158188	75	4·47091298
26	4·23839640	76	4·68277714	26	4·06606063	76	4·47330236
27	4·26322498	77	4·68530427	27	4·08890573	77	4·47560839
28	4·28645094	78	4·68774383	28	4·11026639	78	4·47783442
29	4·30821540	79	4·69009928	29	4·13027428	79	4·47998360
30	4·32864343	80	4·69237393	30	4·14904603	80	4·48205896
31	4·34784634	81	4·69457092	31	4·16668526	81	4·48406338
32	4·53592344	82	4·69669324	32	4·18328445	82	4·48599960
33	4·38296370	83	4·69874376	33	4·19892628	83	4·48787024
34	4·39904700	84	4·70072521	34	4·21368501	84	4·48967780
35	4·41424529	85	4·70264020	35	4·22762744	85	4·49142466
36	4·42862355	86	4·70449123	36	4·24081387	86	4·49311312
37	4·44224061	87	4·70628067	37	4·25329884	87	4·49474534
38	4·45514987	88	4·70801080	38	4·26513185	88	4·49632341
39	4·46739990	89	4·70968382	39	4·27635789	89	4·49784933
40	4·47903498	90	4·71130179	40	4·28701795	90	4·59932501
41	4·49009557	91	4·71286673	41	4·29714945	91	4·50075228
42	4·50061870	92	4·71438055	42	4·30678666	92	4·50213288
43	4·51063836	93	4·71584508	43	4·31596097	93	4·50346849
44	4·52018574	94	4·71726209	44	4·32470122	94	4·50476072
45	4·52928960	95	4·71863326	45	4·33303394	95	4·50601112
46	4·53797642	96	4·71996020	46	4·34098361	96	4·50722115
47	4·54627067	97	4·72124448	47	4·34857278	97	7·50839226
48	4·55419500	98	4·72248757	48	4·35582236	98	4·50952578
49	4·56177037	99	4·72369093	49	4·36275166	99	4·51062304
50	4·56901625	100	4·72485591	50	4·36937863	100	4·51168528

Present Value of £1 per Annum in n years; Redemption of Capital being at 3 per cent. with interest allowed to a Purchaser at the following rates per cent.

Years	23 per cent.	Years	23 per cent.	Years	24 per cent.	Years	24 per cent.
1	0·81300813	51	4·19227758	1	0·80645161	51	4·02359722
2	1·38387075	52	4·19784998	2	1·36498117	52	4·02872994
3	1·80658563	53	4·20318700	3	1·77452728	53	4·03364535
4	2·13207321	54	4·20830107	4	2·08756480	54	4·03835494
5	2·39031689	55	4·21320377	5	2·33451460	55	4·04286944
6	2·60012090	56	4·21790592	6	2·53422791	56	4·04719888
7	2·77387621	57	4·22241765	7	2·69900904	57	4·05135263
8	2·92007985	58	4·22674845	8	2·83723045	58	4·05533946
9	3·04475309	59	4·23090721	9	2·95478712	59	4·05916759
10	3·15228195	60	4·23490226	10	3·05594980	60	4·06284475
11	3·24593704	61	4·23874144	11	3·14388840	61	4·06637818
12	3·32820695	62	4·24243195	12	3·22100523	62	4·06977453
13	3·40101878	63	4·24598121	13	3·28915403	63	4·07304067
14	3·46588808	64	4·24939527	14	3·34978817	64	4·07618218
15	3·52402315	65	4·25268045	15	3·40406318	65	4·07920490
16	3·57639914	66	4·25584254	16	3·45290932	66	4·08211418
17	3·62381168	67	4·25888703	17	3·49708395	67	4·08491511
18	3·66691644	68	4·26181911	18	3·53720991	68	4·08761246
19	3·70625854	69	4·26464368	19	3·57380412	69	4·09021076
20	3·74229498	70	4·26736537	20	3·69729920	70	4·09271430
21	3·77541168	71	4·26998859	21	3·63805994	71	4·09512713
22	3·80593688	72	4·27251749	22	3·66639616	72	4·09745309
23	3·83415144	73	4·27495603	23	3·69257261	73	4·09969584
24	3·86029714	74	4·27730795	24	3·71781697	74	4·10185882
25	3·88458327	75	4·27957682	25	3·73932603	75	4·10394533
26	3·90719184	76	4·28176602	26	3·76027085	76	4·10595848
27	3·92828198	77	4·28387876	27	3·77980075	77	4·10790125
28	3·94799335	78	4·28591810	28	3·79804671	78	4·10977645
29	3·96644906	79	4·28788697	29	3·81512411	79	4·11158678
30	3·98375807	80	4·28978813	30	3·83113492	80	4·11333478
31	4·00001714	81	4·29162423	31	3·84616969	81	4·11502291
32	4·01531250	82	4·29339779	32	3·86030903	82	4·11665348
33	4·02972126	83	4·29511122	33	3·87362497	83	4·11822872
34	4·04331256	84	4·29676681	34	3·88618207	84	4·11975073
35	4·05614859	85	4·29836676	35	3·89803836	85	4·12122154
36	4·06828545	86	4·29991315	36	3·90924616	86	4·12264308
37	4·07977387	87	4·30140800	37	3·91985274	87	4·12401720
38	4·09065982	88	4·30285322	38	3·92990094	88	4·12534565
39	4·10098509	89	4·30425063	39	3·93742967	89	4·12663012
40	4·11078774	90	4·30560199	40	3·94847434	90	4·12787224
41	4·12010249	91	4·30690897	41	3·95706726	91	4·12907354
42	4·12896111	92	4·30817319	42	3·96523798	92	4·13023550
43	4·13739271	93	4·30939619	43	3·97301354	93	4·13135954
44	4·14542401	94	4·31057943	44	3·98041878	94	4·13244702
45	4·15307960	95	4·31172434	45	2·98747652	95	4·13349925
46	4·16038210	96	4·31283227	46	3·99420779	96	4·13451746
47	4·16735243	97	4·31390452	47	4·00063200	97	4·13550287
48	4·17400990	98	4·31494234	48	4·00676705	98	4·13645662
49	4·18037240	99	4·31594694	49	4·01262954	99	4·13737932
50	4·18645649	100	4·31691946	50	4·01823484	100	4·13827353

TABLE VII. XCV

Present Value of £1 per Annum in n years; Redemption of Capital being at 3 per cent. with interest allowed to a Purchaser at the following rates per cent.

Years	25 per cent.	Years	25 per cent.	Years	25 per cent.	Years	25 per cent.
1	0·80000000	26	3·62399868	51	3·86796585	76	3·94401871
2	1·34660033	27	3·64213529	52	3·87270896	77	3·94581122
3	1·74358685	28	3·65907340	53	3·87725080	78	3·94754133
4	2·04487668	29	3·67492130	54	3·88160207	79	3·94921152
5	2·28125829	30	3·68977468	55	3·88577272	80	3·95082416
6	2·47159213			56	3·88977207		
7	2·62807701	31	3·70171840	57	3·89360880	81	3·95238151
8	2·75895261	32	3·71682798	58	3·89729107	82	3·95388571
9	2·86998516	33	3·72917088	59	3·90082651	83	3·95533882
10	2·96533078	34	3·74080748	60	3·90422226	84	3·95674279
		35	3·75179207			85	3·95808949
11	3·04806078	39	3·76217353	61	3·90748506	86	3·95941071
12	3·12049395	37	3·77199605	62	3·91062107	87	3·96067815
13	3·18441376	38	3·78129961	63	3·91363666	·88	3·96190344
14	3·24121436	39	3·79012054	64	3·91653700	89	3·96308813
15	3·29200137	40	3·79849186	65	3·91932750	90	3·96422373
16	3·33766285			66	3·92201313		
17	3·37892027	41	3·80644372	67	3·92459859	91	3·96534166
18	3·41636588	42	3·81400366	68	3·92708831	92	3·96651329
19	3·45049036	43	3·82119687	69	3·92948649	93	3·96744992
20	3·48170373	44	3·82804650	70	3·93177908	94	3·96845281
		45	3·83457379			95	3·96942317
21	3·51035126	46	3·84079833	71	3·93402385	96	3·97036215
22	3·53672578	47	3·84673816	72	3·93617036	97	3·97127085
23	3·56107725	48	3·85240996	73	3·92823998	98	3·97215035
24	3·58362036	49	3·85782915	74	3·94023591	99	3·97300165
25	3·60454051	50	3·86301003	75	3·94216119	100	3·97382575

TABLE VIII.

VALUING MINERAL AND OTHER PROPERTIES,

OR

The present value (or years' purchase) of £1 per annum in n years, allowing interest to a present purchaser upon his purchase-money or capital invested, at the rates of 4, 5, 6, 8, 10, 12, 15, 18, 20, and 25 per cent. per annum, and redeeming the capital so invested by an Annual Redemption Fund, at the rate of 3½ per cent. per annum.

Calculated to 8 places of decimals, and to 100 years for each percentage.

TABLE VIII. xcix

Present Value of £1 per Annum in n years; Redemption of Capital being at 3½ per cent. with interest allowed to a Purchaser at the following rates per cent.

Years	4 per cent.	Years	4 per cent.	Years	5 per cent.	Years	5 per cent.
1	0·96153846	51	21·13201494	1	0·95238095	51	17·44544161
2	1·88181986	52	21·26560757	2	1·84706149	52	17·53638810
3	2·76293330	53	21·39449319	3	2·68864776	53	17·62394045
4	3·60683820	54	21·51884394	4	3·48127427	54	17·70823622
5	4·41537415	55	21·63882521	5	4·22866286	55	17·78940661
6	5·19026988	56	21·75459601	6	4·93417298	56	17·86757685
7	5·93315140	57	21·86630921	7	5·60084480	57	17·94286655
8	6·64554948	58	21·97411174	8	6·22520486	58	18·01538984
9	7·32890640	59	22·07814490	9	6·82845530	59	18·08525590
10	7·98458223	60	22·17854463	10	7·39418726	60	18·15256901
11	8·61386051	61	22·27544161	11	7·93071941	61	18·21742888
12	9·21795350	62	22·36896161	12	8·43996175	62	18·27993089
13	9·79800697	63	22·45922569	13	8·92366560	63	18·34016634
14	10·35510463	64	22·54635030	14	9·38343964	64	18·39822259
15	10·89027219	65	22·63044743	15	9·82076424	65	18·45418320
16	11·40448117	66	22·71162503	16	10·23700398	66	18·50812832
17	11·89865227	67	22·78998699	17	10·63341875	67	18·56013470
18	12·37365863	68	22·86563319	18	11·01117360	68	18·61027579
19	12·83032881	69	22·93866003	19	11·37134753	69	18·65862213
20	13·26944946	70	23·00916019	20	11·71494125	70	18·70524126
21	13·69176795	71	23·07722297	21	12·04288419	71	18·75019797
22	14·09799461	72	23·14293439	22	12·35604066	72	18·79355442
23	14·48880504	73	23·20637735	23	12·65521553	73	18·83537025
24	14·86484209	74	23·26763169	24	12·94115921	74	18·87570270
25	15·22671774	75	23·32677432	25	13·21457214	75	18·91460670
26	15·57501483	76	23·38387935	26	13·47610887	76	18·95213497
27	15·91028880	77	23·43901828	27	13·72638181	77	18·98833821
28	16·23306916	78	23·49225988	28	13·96596448	78	19·02326502
29	16·54386089	79	23·53868290	29	14·19539456	79	19·05369423
30	16·84314589	80	23·59331428	30	14·41517665	80	19·08947455
31	17·13138406	81	23·64125274	31	14·62578471	81	19·12084536
32	17·40901467	82	23·68754530	32	14·82766440	82	19·15111601
33	17·67645726	83	23·73224934	33	15·02123506	83	19·18032644
34	17·93411277	84	23·77542012	34	15·20689166	84	19·20851498
35	18·18236454	85	23·81711097	35	15·38500656	85	19·23571854
36	18·42157912	86	23·85737323	36	15·55593098	86	19·26197255
37	18·65210718	87	23·89625646	37	15·71999657	87	19·28731113
38	18·87428430	88	23·93380854	38	15·87751667	88	19·31176716
39	19·08843171	89	23·97007548	39	16·02878754	89	19·33537218
40	19·29485698	90	24·00510189	40	16·17408954	90	19·35815666
41	19·49385477	91	24·03893052	41	16·31368810	91	19·38014978
42	19·68570725	92	24·07160286	42	16·44783466	92	19·40137977
43	19·87068499	93	24·10315887	43	16·57676770	93	19·42187378
44	20·04904734	94	24·13363694	44	16·70071340	94	19·44165782
45	20·22104293	95	24·16307440	45	16·81988646	95	19·46075717
46	20·38691024	96	24·19150695	46	16·93449080	96	19·47919591
47	20·54687800	97	24·21896930	47	17·04472015	97	19·49699746
48	20·70116589	98	24·24549486	48	17·15075884	98	19·51418431
49	20·84998443	99	24·27111581	49	17·25278206	99	19·53077805
50	20·99353605	100	24·29586324	50	17·35095670	100	19·54679955

Present Value of £1 per Annum for n years ; Redemption of Capital being at 3½ per cent. with interest allowed to a Purchaser at the following rates per cent.

Years	6 per cent.	Years	6 per cent.	Years	8 per cent.	Years	8 per cent.
1	0·94339623	51	14·85408150	1	0·92592593	51	11·45192496
2	1·81356385	52	14·91996511	2	1·75008600	52	11·49104523
3	2·61825218	53	14·98329369	3	2·48796954	53	11·52857360
4	3·36415868	54	15·04417770	4	3·15207694	54	11·56458443
5	4·05710171	55	15·10272198	5	3·75260750	55	11·59914787
6	4·70216026	56	15·15902619	6	4·29796578	56	11·63233022
7	5·30378787	57	15·21318504	7	4·79513979	57	11·66419416
8	5·86590610	58	15·26528859	8	5·24998743	58	11·69479896
9	6·39198169	59	15·31542254	9	5·66745617	59	11·72420076
10	6·88509076	60	15·36366848	10	6·05175334	60	11·75245273
11	7·34797233	61	15·41010412	11	6·40647963	61	11·77960525
12	7·78307334	62	15·45480349	12	6·73473466	62	11·80570612
13	8·19258657	63	15·49783721	13	7·03920125	63	11·83080073
14	8·57848288	64	15·53927262	14	7·32221326	64	11·85493216
15	8·94253860	65	15·57917390	15	7·58581053	65	11·87814132
16	9·28635904	66	15·61604066	16	7·83178390	66	11·90046713
17	9·61139850	67	15·65461674	17	8·06171197	67	11·92194663
18	9·91897774	68	15·69027276	18	8·27699180	68	11·94261501
19	10·21029895	69	15·72462396	19	8·47886419	69	11·96250584
20	10·48645876	70	15·75772145	20	8·66843504	70	11·98165105
21	10·74845964	71	15·78961407	21	8·84669316	71	12·00008109
22	10·99721972	72	15·82034860	22	9·01452540	72	12·01782498
23	11·23358157	73	15·84996976	23	9·17272948	73	12·03491043
24	11·45831980	74	15·87852040	24	9·32202495	74	12·05136386
25	11·67214775	75	15·90604151	25	9·46306248	75	12·06721049
26	11·87572345	76	15·93257234	26	9·59643200	76	12·08247440
27	12·06965490	77	15·95815060	27	9·72266957	77	12·09717863
28	12·25450471	78	15·98281228	28	9·84226344	78	12·11134516
29	12·43079427	79	16·00428643	29	9·95565924	79	12·12367203
30	12·59900747	80	16·02952286	30	10·06326456	80	12·13814833
31	12·75959397	81	16·05163672	31	10·16545293	81	12·15082435
32	12·91297221	82	16·07296403	32	10·26256729	82	12·16304146
33	13·05953205	83	16·09353407	33	10·35492306	83	12·17481732
34	13·19963714	84	16·11337494	34	10·44281086	84	12·18616880
35	13·33362715	85	16·13251362	35	10·52649891	85	12·19711208
36	13·46181961	86	16·15097598	36	10·60623513	86	12·20766265
37	13·58451178	87	16·16878689	37	10·68224903	87	12·21783536
38	13·70198216	88	16·18597027	38	10·75475338	88	12·22764447
39	13·81449197	89	16·20254902	39	10·82394574	89	12·23710360
40	13·92228646	90	16·21854526	40	10·89000975	90	12·24622588
41	14·02559610	91	16·23398012	41	10·95311637	91	12·25502385
42	14·12463762	92	16·24887402	42	11·01342490	92	12·26350957
43	14·21961513	93	16·26324656	43	11·07108404	93	12·27169464
44	14·31072092	94	16·27711652	44	11·12623269	94	12·27959013
45	14·39813629	95	16·29050211	45	11·17900076	95	12·28720675
46	14·48203236	96	16·30342066	46	11·22950986	96	12·29455471
47	14·56257072	97	16·31588900	47	11·27787397	97	12·30164388
48	14·63990418	98	16·32792327	48	11·32420009	98	12·30848371
49	14·71417714	99	16·33953896	49	11·36858861	99	12·31508329
50	14·78552641	100	16·35075102	50	11·41113396	100	12·32145135

TABLE VIII.   ci

Present Value of £1 per Annum for n years; Redemption of Capital being at 3½ per cent. with interest allowed to a Purchaser at the following rates per cent.

Years	10 per cent.	Years	10 per cent.	Years	12 per cent.	Years	12 per cent.
1	0·90909091	51	9·31779189	1	0·89285714	51	7·85412909
2	1·69090154	52	9·34367369	2	1·63558913	52	7·87251034
3	2·37003790	53	9·36847137	3	2·26278040	53	7·89010661
4	2·96514936	54	9·39223788	4	2·79915132	54	7·90695738
5	3·49062834	55	9·41502300	5	3·26284103	55	7·92349968
6	3·95775949	56	9·43687362	6	3·66746092	56	7·93856830
7	4·37551561	57	9·45783394	7	4·02342448	57	7·95339597
8	4·75112093	58	9·47794560	8	4·33883439	58	7·96761346
9	5·09045730	59	9·49724794	9	4·62008989	59	7·98124978
10	5·39836221	60	9·51577809	10	4·87231210	60	7·99433226
11	5·67885080	61	9·53357116	11	5·09964799	61	8·00688668
12	5·93528347	62	9·55066034	12	5·30549155	62	8·01893739
13	6·17049424	63	9·56707709	13	5·49264727	63	8·03050742
14	6·38688986	64	9·58285116	14	5·66345287	64	8·04161850
15	6·58652753	65	9·59801076	15	5·81987252	65	8·05229121
16	6·77117613	66	9·61258266	16	5·96356866	66	8·06254504
17	6·94236503	67	9·62659226	17	6·09595775	67	8·07239848
18	7·10142334	68	9·64006364	18	6·21825423	68	8·08186901
19	7·24951169	69	9·65301974	19	6·33150526	69	8·09097325
20	7·38764809	70	9·66548233	20	6·43661862	70	8·09972697
21	7·51672922	71	9·67747209	21	6·53438514	71	8·10814513
22	7·63754799	72	9·68900876	22	6·62549697	72	8·11624197
23	7·75080815	73	9·70011108	23	6·71056250	73	8·12403101
24	7·85713650	74	9·71079692	24	6·79011871	74	8·13152514
25	7·95709310	75	9·72108332	25	6·86464130	75	8·13873659
26	8·05117998	76	9·73098650	26	6·93455325	76	8·14567703
27	8·13984844	77	9·74052195	27	7·00023190	77	8·15235759
28	8·22350532	78	9·74970445	28	7·06201497	78	8·15878885
29	8·30251836	79	9·75769110	29	7·12020561	79	8·16438095
30	8·37722078	80	9·76706634	30	7·17507674	80	8·17094341
31	8·44791524	81	9·77527208	31	7·22687473	81	8·17668556
32	8·51487727	82	9·78317758	32	7·27582255	82	8·18221611
33	8·57835827	83	9·79079461	33	7·32212247	83	8·18754348
34	8·63858804	84	9·79813441	34	7·36595843	84	8·19267566
35	8·69577714	85	9·80520773	35	7·40749809	85	8·19762033
36	8·75011979	86	9·81202487	36	7·44689455	86	8·20238480
37	8·80179064	87	9·81859566	37	7·48428793	87	8·20697607
38	8·85095633	88	9·82492956	38	7·51980672	88	8·21140086
39	8·89776684	89	9·83103558	39	7·55356893	89	8·21566557
40	8·94236164	90	9·83692241	40	7·58568320	90	8·21977636
41	8·98486987	91	9·84259831	41	7·61624963	91	8·22373910
42	9·02541115	92	9·84807125	42	7·64536068	92	8·22755942
43	9·06409655	93	9·85334887	43	7·67310184	93	8·23124275
44	9·10102930	94	9·85843846	44	7·69955228	94	8·23479422
45	9·13630545	95	9·86334706	45	7·72478546	95	8·23821883
46	9·17001450	96	9·86808139	46	7·74886957	96	8·24152132
47	9·20223993	97	9·87264791	47	7·77186804	97	8·24470627
48	9·23305977	98	9·87705284	48	7·79383996	98	8·24777805
49	9·26254691	99	9·88130212	49	7·81484039	99	8·25074086
50	9·29076963	100	9·88540149	50	7·83492075	100	8·25359874

Present Value of £1 per Annum in n years ; Redemption of Capital being at 3½ p cent. with interest allowed to a Purchaser at the following rates per cent.

Years	15 per cent.	Years	15 per cent.	Years	18 per cent.	Years	18 per cent.
1	0·86956522	51	6·35640769	1	0·84745763	51	5·33841368
2	1·55908830	52	6·36844166	2	1·48942399	52	5·34689919
3	2·11893955	53	6·37995165	3	1·99229309	53	5·35501042
4	2·58230357	54	6·39096480	4	2·39663815	54	5·36276712
5	2·97193271	55	6·40150647	5	2·72865164	55	5·37018772
6	3·30394790	56	6·41160046	6	3·00599809	56	5·37728950
7	3·59009071	57	6·42126910	7	3·24102365	57	5·38408864
8	3·83911577	58	6·43053334	8	3·44261755	58	5·39060029
9	4·05768395	59	6·43941290	9	3·61734292	59	5·39683871
10	4·25095301	60	6·44792630	10	3·77015097	60	5·40281727
11	4·42297893	61	6·45609101	11	3·90484722	61	5·40854856
12	4·57699526	62	6·46392347	12	4·02440481	62	5·41404441
13	4·71561154	63	6·47143920	13	4·13118030	63	5·41931598
14	4·84095693	64	6·47865286	14	4·22706565	64	5·42437381
15	4·95478586	65	6·48557826	15	4·31359733	65	5·42922781
16	5·05855708	66	6·49222850	16	4·39203612	66	5·43388735
17	5·15349341	67	6·49861595	17	4·46342627	67	5·43836131
18	5·24062759	68	6·50475232	18	4·52864010	68	5·44265805
19	5·32083787	69	6·51064871	19	4·58841198	69	5·44678550
20	5·39487587	70	6·51631564	20	4·64336460	70	5·45075119
21	5·46338859	71	6·52176308	21	4·69402939	71	5·45456221
22	5·52693603	72	6·52700050	22	4·74086269	72	5·45822532
23	5·58600536	73	6·53203690	23	4·78425854	73	5·46174693
24	5·64102236	74	6·53688082	24	4·82455091	74	5·46513310
25	5·69236084	75	6·54154037	25	4·86206243	75	5·46838963
26	5·74035031	76	6·54602329	26	4·89703024	76	5·47152198
27	5·78528236	77	6·55033692	27	4·92969247	77	5·47453539
28	5·82741600	78	6·55448827	28	4·96025238	78	5·47743481
29	5·86698205	79	6·55809691	29	4·98889018	79	5·47995469
30	5·90418692	80	6·56233048	30	5·01576627	80	5·48291038
31	5·93921570	81	6·56603376	31	5·04102386	81	5·48549533
32	5·97223491	82	6·56959960	32	5·06479122	82	5·48798390
33	6·00339473	83	6·57303354	33	5·08718359	83	5·49037999
34	6·03283097	84	6·57634084	34	5·10830479	84	5·49268732
35	6·06066677	85	6·57952652	35	5·12824859	85	5·49490945
36	6·08701401	86	6·58259539	36	5·14709993	86	5·49704976
37	6·11197460	87	6·58555203	37	5·16493589	87	5·49911149
38	6·13564156	88	6·58840085	38	5·18182663	88	5·50109773
39	6·15809996	89	6·59114603	39	5·19783612	89	5·50301146
40	6·17942776	90	6·59379159	40	5·21302282	90	5·50485550
41	6·19969654	91	6·59634138	41	5·22744024	91	5·50663254
42	6·21897213	92	6·59879908	42	5·25363428	92	5·50834518
43	6·23731520	93	6·60116821	43	5·25415969	93	5·50999591
44	6·25478172	94	6·60345214	44	5·26654840	94	5·51158709
45	6·27142344	95	6·60565411	45	5·27834192	95	5·51312100
46	6·28728827	96	6·60777722	46	5·28957566	96	5·51459981
47	6·30242060	97	6·60982444	47	5·30028234	97	5·51602561
48	6·31686169	98	6·61179862	48	5·31049233	98	5·51740041
49	6·33064985	99	6·61370250	49	5·32023374	99	5·51872612
50	6·34382076	100	6·61553869	50	5·32953275	100	5·52000457

TABLE VIII.　　　　ciii

Present Value of £1 per Annum in n years; Redemption of Capital being at 3½ per cent, with interest allowed to a Purchaser at the following rates per cent.

Years	20 per cent.	Years	20 per cent.	Years	25 per cent.	Years	25 per cent.
1	0·83333333	51	4·82342493	1	0·80000000	51	3·88618810
2	1·44633973	52	4·83035117	2	1·34879867	52	3·89068293
3	1·91595040	53	4·83696994	3	1·74845294	53	3·89497588
4	2·28701519	54	4·84329758	4	2·05232973	54	3·89907786
5	2·58744682	55	4·84934939	5	2·29104851	55	3·90299908
6	2·83552635	56	4·85513967	6	2·48343420	56	3·90674906
7	3·04372777	57	4·86086180	7	2·64169739	57	3·91033669
8	3·22085416	58	4·86598833	8	2·77410481	58	3·91377030
9	3·37329558	59	4·87107102	9	2·88645268	59	3·91705771
10	3·50580285	60	4·87594092	10	2·98292543	60	3·92020622
11	3·62198153	61	4·88060840	11	3·06661956	61	3·92322272
12	3·72461744	62	4·88508325	12	3·13987566	62	3·92611365
13	3·81589712	63	4·88937466	13	3·20449580	63	3·92888510
14	3·89756074	64	4·89349129	14	3·26189002	64	3·93154277
15	3·97101052	65	4·89744132	15	3·31317740	65	3·93409206
16	4·03738896	66	4·90123245	16	3·35925750	66	3·93653805
17	4·09763632	67	4·90487199	17	3·40086162	67	3·93888552
18	4·15253351	68	4·90836680	18	3·43859041	68	3·94113901
19	4·20273444	69	4·91172342	19	3·47294183	69	3·94330279
20	4·24879089	70	4·91494801	20	3·50433216	70	3·94538091
21	4·29117167	71	4·91804641	21	3·53311214	71	3·94737720
22	4·33027765	72	4·92102415	22	3·55957931	72	3·94929528
23	4·36645368	73	4·92388649	23	3·58398783	73	3·95113859
24	4·39999801	74	4·92663841	24	3·60655604	74	3·95291040
25	4·43116993	75	4·92928465	25	3·62747256	75	3·95461379
26	4·46019595	76	4·93182969	26	3·64690124	76	3·95625171
27	4·48727478	77	4·93427782	27	3·66498503	77	3·95782693
28	4·51258151	78	4·93663309	28	3·68184927	78	3·95934212
29	4·53627102	79	4·93867985	29	3·69760427	79	3·96065861
30	4·55848079	80	4·94108037	30	3·71234754	80	3·96220235
31	4·57933329	81	4·94317956	31	3·72616558	81	3·96355208
32	4·59893800	82	4·94520030	32	3·73913541	82	3·96485115
33	4·61739307	83	4·94714578	33	3·75132578	83	3·96610163
34	4·63478672	84	4·94901904	34	3·76279834	84	3·96730551
35	4·65119854	85	4·95082297	35	3·77360844	85	3·96846467
36	4·66670045	86	4·95256034	36	3·78380600	86	3·96958090
37	4·68135765	87	4·95423381	37	3·79343609	87	3·97065592
38	4·69522934	88	4·95584589	38	3·80253957	88	3·97169137
39	4·70836947	89	4·95739899	39	3·81115352	89	3·97268882
40	4·72082722	90	4·95889544	40	3·81931169	90	3·97364976
41	4·73264759	91	4·96033743	41	3·82704486	91	3·97457562
42	4·74387180	92	4·96172708	42	3·83438117	92	3·97546777
43	4·75453769	93	4·96306640	43	3·84134638	93	3·97632752
44	4·76468004	94	4·96435733	44	3·84796414	94	3·97715613
45	4·77433090	95	4·96560173	45	3·85425618	95	3·97795478
46	4·78351985	96	4·96680137	46	3·86024252	96	3·97872462
47	4·79227422	97	4·96795795	47	3·86594160	97	3·97946677
48	4·80061929	98	4·96907309	48	3·87137049	98	3·98018226
49	4·80857851	99	4·97014836	49	3·87654497	99	3·98087211
50	4·81617365	100	4·97118527	50	3·88147965	100	3·98153729

TABLE IX.

FOR

VALUING MINERAL AND OTHER PROPERTIES,

OR

*The present value (or years' purchase) of £1 per annum
in n years, allowing interest to a present purchaser upon
his purchase-money or capital invested, at the rates of 5,
6, 8, 10, 12, 15, 16, 18, 20, and 25 per cent. per annum,
and redeeming the capital so invested by an Annual
Redemption Fund, at the rate of 4 per cent. per annum.*

*Calculated to 8 places of decimals, and to 100 years
for each percentage.*

TABLE IX. C\

Present Value of £1 per Annum in n years; Redemption of Capital being at 4 p cent. with interest allowed to a Purchaser at the following rates per cent.

Years	5 per cent.	Years	5 per cent.	Years	6 per cent.	Years	6 per cent.
1	0·95238095	51	17·77498127	1	0·94339623	51	15·09232358
2	1·85117967	52	17·86284506	2	1·17753386	52	15·15562012
3	2·70015916	53	17·94715260	3	2·62916745	53	15·21626610
4	3·50274910	54	18·02805444	4	3·38420876	54	15·27438076
5	4·26208201	55	18·10569440	5	4·08785431	55	15·33007743
6	4·98102472	56	18·18020966	6	4·74469052	56	15·38346370
7	5·66220597	57	18·25173113	7	5·35878077	57	15·43464181
8	6·30804060	58	18·32038400	8	5·93373800	58	15·48370905
9	6·92075084	59	18·38628770	9	6·47278548	59	15·53075788
10	7·50238514	60	18·44955639	10	6·97880808	60	15·57587633
11	8·05483478	61	18·51029920	11	7·45439554	61	15·61914821
12	8·57984864	62	18·56862064	12	7·90187935	62	15·66065333
13	9·07904626	63	18·62462018	13	8·32336418	63	15·70046770
14	9·55392960	64	18·67839328	14	8·72075482	64	15·73866376
15	10·00589343	65	18·73003124	15	9·09577943	65	15·77531063
16	10·43623469	66	18·77962133	16	9·45000952	66	15·81047415
17	10·84616084	67	18·82724705	17	9·78487731	67	15·84421714
18	11·23679741	68	18·87298835	18	10·10169087	68	15·87659957
19	11·60919479	69	18·92050099	19	10·40164729	69	15·91020962
20	11·96433427	70	18·95912065	20	10·68584416	70	15·93750907
21	12·30313365	71	18·99965508	21	10·95528971	71	15·96614298
22	12·62645216	72	19·03859225	22	11·21091175	72	15·99363021
23	12·93509507	73	19·07599657	23	11·45356549	73	16·02001840
24	13·22981771	74	19·11192967	24	11·68404046	74	16·04535307
25	13·51132927	75	19·14645065	25	11·90306673	75	16·06967773
26	13·78029623	76	19·17961612	26	12·11132040	76	16·09303398
27	14·03735130	77	19·21148035	27	12·30943295	77	16·11546160
28	14·28306677	78	19·24209535	28	12·49797295	78	16·13699867
29	14·51801619	79	19·27151098	29	12·67749536	79	16·15768159
30	14·74271766	80	19·29977501	30	12·84849963	80	16·17754519
31	14·95766552	81	19·32693332	31	13·01145553	81	16·19662282
32	15·16332653	82	19·35302987	32	13·16680143	82	16·21494644
33	15·36014159	83	19·37810679	33	13·31494690	83	16·23254657
34	15·54852760	84	19·40220446	34	13·45627497	84	16·24945246
35	15·72887882	85	19·42536178	35	13·59114421	85	16·26569222
36	15·90156852	86	19·44761583	36	13·71989070	86	16·28129259
37	16·06695021	87	19·46900238	37	13·84282966	87	16·29627936
38	16·22535878	88	19·48955571	38	13·96025699	88	16·31067719
39	16·37711185	89	19·50930858	39	14·07245083	89	16·32450962
40	16·52251070	90	19·52829260	40	14·17967275	90	16·33779934
41	16·66184125	91	19·54653802	41	14·28216894	91	16·35056802
42	16·79537502	92	19·56407393	42	14·38017133	92	16·36283650
43	16·92336996	93	19·58092819	43	14·47389855	93	16·37462468
44	17·04607118	94	19·59712748	44	14·56355690	94	16·38595163
45	17·16371179	95	19·61269758	45	14·64934110	95	16·39683577
46	17·27651354	96	19·62766315	46	14·73143515	96	16·40729462
47	17·38468743	97	19·64204787	47	14·81001296	97	16·41734509
48	17·48843432	98	19·65587450	48	14·88523906	98	16·42700334
49	17·58794553	99	19·66916480	49	14·95726917	99	16·43628485
50	17·68340320	100	19·68193985	50	15·02625076	100	16·44520457

Present Value of £1 per Annum in n years; Redemption of Capital being at 4 per cent. with interest allowed to a Purchaser at the following rates per cent.

Years	8 per cent.	Years	8 per cent.	Years	10 per cent.	Years	10 per cent.
1	0·92592593	51	11·59301339	1	0·90909090	51	9·41098085
2	1·75378267	52	11·63032451	2	1·69435216	52	9·43555353
3	2·49782353	53	11·66600528	3	2·37897817	53	9·45902465
4	3·16967212	54	11·70013456	4	2·98071437	54	9·48144979
5	3·77890227	55	11·73278682	5	3·51336873	55	9·50288123
6	4·33347094	56	11·74603239	6	3·98784660	56	9·52336817
7	4·84004587	57	11·79393772	7	4·41287548	57	9·54295693
8	5·30425662	58	11·82256569	8	4·79552293	58	9·56169124
9	5·73088916	59	11·84997578	9	5·14157341	59	9·57961231
10	6·12403832	60	11·87562431	10	5·45580691	60	9·59675905
11	6·48722823	61	11·90136464	11	5·74220796	61	9·61316823
12	6·82350853	62	11·92544738	12	6·00412462	62	9·62887463
13	7·13553161	63	11·94852050	13	6·24439067	63	9·64391111
14	7·42561540	64	11·97062950	14	6·46542083	64	9·65830879
15	7·69579446	65	11·99181766	15	6·66928545	65	9·67209715
16	7·94786208	66	12·01212600	16	6·85776989	66	9·68530413
17	8·18340503	67	12·03159353	17	7·03242189	67	9·69795619
18	8·40383251	68	12·05025735	18	7·19458993	68	9·71007847
19	8·61040035	69	12·06960933	19	7·34545430	69	9·72264000
20	8·80423129	70	12·08531327	20	7·48605253	70	9·73282781
21	8·98633225	71	12·10177089	21	7·61730040	71	9·74349904
22	9·15760886	72	12·11755601	22	7·74000930	72	9·75372889
23	9·31887792	73	12·13269762	23	7·85490070	73	9·76353682
24	9·47087802	74	12·14722333	24	7·96261829	74	9·77294132
25	9·61427862	75	12·16115946	25	8·06373819	75	9·78195997
26	9·74968840	76	12·17453112	26	8·15877761	76	9·79060952
27	9·87766408	77	12·18736224	27	8·24820426	77	9·79890592
28	9·99870265	78	12·19967567	28	8·33243238	78	9·80686438
29	10·11327533	79	12·21149321	29	8·41184853	79	9·81449937
30	10·22180299	80	12·22283566	30	8·48679588	80	9·82182468
31	10·32467463	81	12·23372291	31	8·55758833	81	9·82885351
32	10·42224803	82	12·24417394	32	8·62451197	82	9·83559840
33	10·51485277	83	12·25420689	33	8·68782810	83	9·84207133
34	10·60279301	84	12·26383906	34	8·74777587	84	9·84828253
35	10·68634979	85	12·27308708	35	8·80457453	85	9·85424656
36	10·76578330	86	12·28196671	36	8·85842551	86	9·85997020
37	10·84133467	87	12·29049316	37	8·90951418	87	9·86546465
38	10·91322772	88	12·29868090	38	8·95801137	88	9·87073942
39	10·98167047	89	12·30654377	39	9·00407482	89	9·87580359
40	11·04685654	90	12·31409506	40	9·04785043	90	9·88066588
41	11·10896627	91	12·32134743	41	9·08947324	91	9·88533460
42	11·16816795	92	12·32831307	42	9·12906854	92	9·88981770
43	11·22461870	93	12·33500359	43	9·16675264	93	9·89412279
44	11·27846544	94	12·34143010	44	9·20263372	94	9·89825713
45	11·32984563	95	12·34760330	45	9·23681245	95	9·90222771
46	11·37888810	96	12·35353338	46	9·26938273	96	9·90604118
47	11·42571362	97	12·35923016	47	9·30043214	97	9·90970393
48	11·47043555	98	12·36470299	48	9·33004253	98	9·91322206
49	11·51316039	99	12·36996084	49	9·35829042	99	9·91660142
50	11·55398826	100	12·37501237	50	9·38524748	100	9·91984763

TABLE IX. cix

Present Value of £1 per Annum in n years; Redemption of Capital being at 4 per cent. with interest allowed to a Purchaser at the following rates per cent.

Years	12 per cent.	Years	12 per cent.	Years	15 per cent.	Years	15 per cent.
1	0·89285714	51	7·92023690	1	0·86956522	51	6·39963754
2	1·63881748	52	7·93763410	2	1·56202144	52	6·41099106
3	2·27092839	53	7·95423799	3	2·12608293	53	6·42181794
4	2·81301829	54	7·97008969	4	2·59410071	54	6·43214622
5	3·28270189	55	7·98522780	5	2·98840091	55	6·44200216
6	3·69328178	56	7·99968858	6	3·32488920	56	6·45141036
7	4·05499198	57	8·01350607	7	3·61520336	57	6·46039390
8	4·37583462	58	8·02671234	8	3·86805549	58	6·46897443
9	4·66215697	59	8·03933757	9	4·09009677	59	6·47717232
10	4·91905827	60	8·05141018	10	4·28649300	60	6·48500671
11	5·15068220	61	8·06295699	11	4·46131736	61	6·49249561
12	5·36043073	62	8·07400332	12	4·61782475	62	6·49965600
13	5·55112305	63	8·08457304	13	4·75864786	63	6·50650386
14	5·72511524	64	8·09468872	14	4·88593842	64	6·51305431
15	5·88439170	65	8·10437174	15	5·00147292	65	6·51932157
16	6·03063566	66	8·11364227	16	5·10673069	66	6·52531912
17	6·16528429	67	8·12251944	17	5·20295364	67	6·53105967
18	6·28957209	68	8·13102136	18	5·29119207	68	6·53655525
19	6·40456544	69	8·13982771	19	5·37234029	69	6·54224524
20	6·51119028	70	8·14696723	20	5·44716454	70	6·54685647
21	6·61025452	71	8·15444290	21	5·51632512	71	6·55168313
22	6·70246628	72	8·16160687	22	5·58039417	72	6·55630689
23	6·78844887	73	8·16847307	23	5·63986983	73	6·56073697
24	6·86875311	74	8·17505472	24	5·69518792	74	6·56498208
25	6·94381937	75	8·18136439	25	5·74673134	75	6·56905051
26	7·01422721	76	8·18741404	26	5·79483792	76	6·57295012
27	7·08022193	77	8·19321505	27	5·83980786	77	6·57668838
28	7·14219520	78	8·19877825	28	5·88190401	78	6·58027242
29	7·20046424	79	8·20411394	29	5·92136659	79	6·58370898
30	7·25530930	80	8·20923195	30	5·95846777	80	6·58700451
31	7·30698495	81	8·21414161	31	5·99324994	81	6·59016513
32	7·35572176	82	8·21885187	32	6·02596263	82	6·59319667
33	7·40172910	83	8·22337104	33	6·05680435	83	6·59610469
34	7·44519749	84	8·23777104	34	6·08588011	84	6·59889448
35	7·48630068	85	8·23186919	35	6·11331684	85	6·60157110
36	7·52519748	86	8·23586294	36	6·13922995	86	6·60413934
37	7·56203332	87	8·23969605	37	6·16372459	87	6·60660382
38	7·59694158	88	8·24337524	38	6·18689678	88	6·60896892
39	7·63004492	89	8·24690693	39	6·20883436	89	6·61123880
40	7·66145623	90	8·25029727	40	6·22961789	90	6·61341747
41	7·69127963	91	8·25355211	41	6·24926130	91	6·61550873
42	7·71961132	92	8·25667708	42	6·26801267	92	6·61751624
43	7·74654027	93	8·25967752	43	6·28575471	93	6·61944347
44	7·77214892	94	8·26255855	44	6·30260531	94	6·62129373
45	7·79651375	95	8·26532509	45	6·31861799	95	6·62307023
46	7·81970581	96	8·26798181	46	6·33384231	96	6·62477599
47	7·84179120	97	8·27053322	47	6·34832421	97	6·62641392
48	7·86283148	98	8·27298359	48	6·36210637	98	6·62798681
49	7·88288405	99	8·27533704	49	6·37522845	99	6·62949730
50	7·90200250	100	8·27759751	50	6·38772737	100	6·63094796

Present Value of £1 per Annum in n years; Redemption of Capital being at 4 per cent. with interest allowed to a Purchaser at the following rates per cent.

Years	16 per cent.	Years	16 per cent.	Years	18 per cent.	Years	18 per cent.
1	0·86206897	51	6·01471742	1	0·84512714	51	5·36887241
2	1·53799759	52	6·02474519	2	1·49210064	52	5·37686085
3	2·08182167	53	6·03430581	3	1·99860681	53	5·38447449
4	2·52850865	54	6·04342433	4	2·40679653	54	5·39173364
5	2·90168425	55	6·05212419	5	2·74252781	55	5·39865730
6	3·21789766	56	6·06042733	6	3·02332280	56	5·40526323
7	3·48906651	57	6·06835430	7	3·26147635	57	5·41156806
8	3·72400876	58	6·07592443	8	3·46587015	58	5·41758740
9	3·92938127	59	6·08315584	9	3·64308025	59	5·42333589
10	4·11030506	60	6·09006555	10	3·79807974	60	5·42882727
11	4·27078413	61	6·09666960	11	3·93469911	61	5·43407447
12	4·41399430	62	6·10298309	12	4·05593668	62	5·43908964
13	4·54248690	63	6·10902023	13	4·16417289	63	5·44388424
14	4·65833504	64	6·11479442	14	4·26132175	64	5·44846906
15	4·76324073	65	6·12031834	15	4·34893979	65	5·45285426
16	4·85861434	66	6·12560392	16	4·42830575	66	5·45704945
17	4·94563457	67	6·13066244	17	4·50047998	67	5·46106368
18	5·02529411	68	6·13550460	18	4·56634916	68	5·46490554
19	5·09843502	69	6·14051752	19	4·62666046	69	5·46888219
20	5·16577621	70	6·14457966	20	4·68204797	70	5·47210407
21	5·22793522	71	6·14883119	21	4·73305330	71	5·47547568
22	5·28544547	72	6·15290365	22	4·78014189	72	5·47870479
23	5·33877014	73	6·15680518	23	4·82371593	73	5·48179794
24	5·38831335	74	6·16054351	24	4·86412478	74	5·48476129
25	5·43442929	75	6·16412596	25	4·90167334	75	5·48760072
26	5·47742974	76	6·16755951	26	4·93662898	76	5·49032178
27	5·51759114	77	6·17085076	27	4·96922779	77	5·49292976
28	5·55515512	78	6·17400600	28	4·99967564	78	5·49542969
29	5·59034195	79	6·17703122	29	5·02815935	79	5·49782633
30	5·62334501	80	6·17993210	30	5·05484254	80	5·50012422
31	5·65433834	81	6·18271406	31	5·07987204	81	5·50202192
32	5·68347835	82	6·18538225	32	5·10337942	82	5·50444084
33	5·71090595	83	6·18794158	33	5·12548293	83	5·50646759
34	5·73674848	84	6·19039673	34	5·14628915	84	5·50841166
35	5·76112125	85	6·19275216	35	5·16589437	85	5·51027661
36	5·78412897	86	6·19501211	36	5·18438584	86	5·51206582
37	5·80586695	87	6·19718065	37	5·20184280	87	5·51378252
38	5·82642206	88	6·19926164	38	5·21833735	88	5·51542980
39	5·84587375	89	6·20125877	39	5·23393525	89	5·51701057
40	5·86429473	90	6·20317557	40	5·24869665	90	5·51852765
41	5·88175174	91	6·20501540	41	5·26267660	91	5·51998372
42	5·89830610	92	6·20678147	42	5·27592562	92	5·52138132
43	5·91401428	93	6·20847685	43	5·28849016	93	5·52272291
44	5·92892836	94	6·21010448	44	5·30041298	94	5·52401079
45	5·94309643	95	6·21166715	45	5·31173354	95	5·52524722
46	5·95656300	96	6·21316756	46	5·32248827	96	5·52643431
47	5·96936930	97	6·21460826	47	5·33271089	97	5·52757410
48	5·98155357	98	6·21599170	48	5·34243263	98	5·52866654
49	5·99315136	99	6·21732023	49	5·35168250	99	5·52971949
50	6·00419572	100	6·21859609	50	5·36048741	100	5·53072873

TABLE IX.   cxi

Present Value of £1 per Annum in n years; Redemption of Capital being at 4 per cent. with interest allowed to a Purchaser at the following rates per cent.

Years	20 per cent.	Years	20 per cent.	Years	25 per cent.	Years	25 per cent.
1	0·83333333	51	4·84827682	1	0·80000000	51	3·90230426
2	1·44886364	52	4·85479022	2	1·35099338	52	3·90652279
3	1·92178881	53	4·86099628	3	1·75331386	53	3·91054021
4	2·29626374	54	4·86691180	4	2·05977447	54	3·91436769
5	2·59992072	55	4·87255248	5	2·30082286	55	3·91801564
6	2·85093675	56	4·87793299	6	2·49524716	56	3·92149379
7	3·06175925	57	4·88306706	7	2·65526946	57	3·92481124
8	3·24119867	58	4·88796757	8	2·78918374	58	3·92797649
9	3·39566653	59	4·89264657	9	2·90281666	59	3·93099750
10	3·52993987	60	4·89711540	10	3·00038155	60	3·93388175
11	3·64765094	61	4·90138468	11	3·08500066	61	3·93663625
12	3·75161076	62	4·90546443	12	3·15903692	62	3·93926758
13	3·84402887	63	4·90936405	13	3·22431154	63	3·94178193
14	3·92666601	64	4·91309242	14	3·28225086	64	3·94418512
15	4·00094262	65	4·91665788	15	3·33398790	65	3·94648263
16	4·06801726	66	4·92006832	16	3·38043406	66	3·94867964
17	4·12884435	67	4·92333118	17	3·42233079	67	3·95078101
18	4·18421722	68	4·92645348	18	3·46028750	68	3·95279135
19	4·23480076	69	4·92968485	19	3·49480963	69	3·95487139
20	4·28115638	70	4·93230261	20	3·52632001	70	3·95655603
21	4·32376143	71	4·93504167	21	3·55517500	71	3·95831836
22	4·36302438	72	4·93766465	22	3·58167714	72	3·96000566
23	4·39929677	73	4·94017691	23	3·60608489	73	3·96162138
24	4·43288287	74	4·94258348	24	3·62862041	74	3·96316884
25	4·46404731	75	4·94488917	25	3·64947570	75	3·96465113
26	4·49302138	76	4·94709852	26	3·66881759	76	3·96607126
27	4·52000873	77	4·94921586	27	3·68679211	77	3·96743199
28	4·54518648	78	4·95124528	28	3·70352573	78	3·96873601
29	4·56871482	79	4·95319068	29	3·71913212	79	3·96998584
30	4·59073381	80	4·95505000	30	3·73371930	80	3·97118389
31	4·61136877	81	4·95684408	31	3·74734848	81	3·97233245
32	4·63073179	82	4·95855890	32	3·76012522	82	3·97343369
33	4·64892338	83	4·96020358	33	3·77211070	83	3·97448968
34	4·66603395	84	4·96178101	34	3·78336782	84	3·97550240
35	4·68214504	85	4·96329414	35	3·79395310	85	3·97647371
36	4·69733038	86	4·96474572	36	3·80391752	86	3·97740540
37	4·71165684	87	4·96613838	37	3·81330711	87	3·97829917
38	4·72518514	88	4·96747464	38	3·82216361	88	3·97915665
39	4·73797062	89	4·96875688	39	3·83052490	89	3·97997939
40	4·75006377	90	4·96998739	40	3·83842548	90	3·98076884
41	4·76151077	91	4·97116835	41	3·84589681	91	3·98152644
42	4·77235393	92	4·97230183	42	3·85296766	92	3·98225351
43	4·78263210	93	4·97338983	43	3·85966437	93	3·98295134
44	4·79238101	94	4·97443423	44	3·86601110	94	3·98362115
45	4·80163357	95	4·97543685	45	3·87203009	95	3·98426412
46	4·81042017	96	4·97639942	46	3·87774180	96	3·98488136
47	4·81876886	97	4·97732360	47	3·88316511	97	3·98547393
48	4·82670564	98	4·97821098	48	3·88831738	98	3·98604286
49	4·83425458	99	4·97906305	49	3·89321497	99	3·98658912
50	4·84143805	100	4·97988128	50	3·89787261	100	3·98711365

TABLE X.

FOR

VALUING MINERAL AND OTHER PROPERTIES,

OR

The present value (or years' purchase) of £1 per annum in n years, deferred 1, 2, 3, 4, 5, 6, 7, 8, 9, and 10 years, allowing interest to a present purchaser upon his purchase-money or capital invested, at the rate of 4, 5, 6, 8, 10, 12, 15, 18, and 20 per cent. per annum, and redeeming the capital so invested, by an Annual Redemption Fund at the rate of 3 per cent. per annum.

Calculated to 6 places of decimals, and to 100 years for each percentage.

p

TABLE X.   CXV

Present Value (or Years' Purchase) of £1 per Annum in **n** years, after **t** years' Deferrence. Redemption of Capital being at **3** per cent., with Interest allowed to a Purchaser at **4** per cent.

n Years	Deferred 1 Year	n Years	Deferred 1 Year	n Years	Deferred 2 Years	n Years	Deferred 2 Years
1	·924555	51	19·811710	1	·888996	51	19·049726
2	1·805329	52	19·941812	2	1·735894	52	19·174824
3	2·645001	53	20·067701	3	2·543270	53	19·295871
4	3·446039	54	20·189527	4	3·313500	54	19·413011
5	4·210724	55	20·307432	5	4·048774	55	19·526381
6	4·941163	56	20·421552	6	4·751119	56	19·636112
7	5·639309	57	20·532021	7	5·422414	57	19·742333
8	6·306971	58	20·638964	8	6·064397	58	19·845163
9	6·945829	59	20·742504	9	6·678684	59	19·944720
10	7·557448	60	20·842756	10	7·266779	60	20·041117
11	8·143283	61	20·939834	11	7·830081	61	20·134461
12	8·704688	62	21·033841	12	8·369895	62	20·224852
13	9·242932	63	21·124894	13	8·887437	63	20·312403
14	9·759197	64	21·213079	14	9·383846	64	20·397197
15	10·254592	65	21·298497	15	9·860187	65	20·479330
16	10·730152	66	21·381243	16	10·317456	66	20·558892
17	11·186849	67	21·461403	17	10·756589	67	20·635970
18	11·625597	68	21·539064	18	11·178462	68	20·710643
19	12·047253	69	21·614309	19	11·583900	69	20·782994
20	12·452622	70	21·687217	20	11·973678	70	20·853098
21	12·842463	71	21·757865	21	12·348525	71	20·921029
22	13·217491	72	21·826327	22	12·709129	72	20·986858
23	13·578379	73	21·892675	23	13·056137	73	21·050654
24	13·925764	74	21·956978	24	13·390161	74	21·112484
25	14·260246	75	22·019300	25	13·711778	75	21·172409
26	14·582392	76	22·079708	26	14·021534	76	21·230493
27	14·892740	77	22·138261	27	14·319946	77	21·286795
28	15·191798	78	22·195022	28	14·607501	78	21·341373
29	15·480046	79	22·250045	29	14·884664	79	21·394279
30	15·757942	80	22·303388	30	15·151872	80	21·445571
31	16·025919	81	22·355104	31	15·409542	81	21·495297
32	16·284387	82	22·405243	32	15·658069	82	21·543508
33	16·533736	83	22·453857	33	15·897827	83	21·590253
34	16·774338	84	22·500995	34	16·129175	84	21·635577
35	17·006545	85	22·546702	35	16·352451	85	21·679535
36	17·230692	86	22·591023	36	16·567977	86	21·722143
37	17·447099	87	22·634002	37	16·776061	87	21·763469
38	17·656071	88	22·675680	38	16·976996	88	21·803544
39	17·857898	89	22·716101	39	17·171060	89	21·842410
40	18·052856	90	22·755302	40	17·358519	90	21·880103
41	18·241208	91	22·793320	41	17·539627	91	21·916659
42	18·423208	92	22·830194	42	17·714627	92	21·952115
43	18·599094	93	22·865957	43	17·883749	93	21·986503
44	18·769098	94	22·900647	44	18·047214	94	22·019858
45	18·933438	95	22·934294	45	18·205233	95	22·052211
46	19·042326	96	22·966931	46	18·358011	96	22·083593
47	19·245962	97	22·998591	47	18·505737	97	22·114035
48	19·394538	98	23·029302	48	18·648599	98	22·143566
49	19·538238	99	23·059094	49	18·786772	99	22·172211
50	19·677239	100	23·087995	50	18·920426	100	22·200000

Present Value (or Years' Purchase) of £1 per Annum in n years, after t years' Deferrence. Redemption of Capital being at 3 per cent., with Interest allowed to a Purchaser at 4 per cent.

n Years	Deferred 3 Years	n Years	Deferred 3 Years	n Years	Deferred 4 Years	n Years	Deferred 4 Years
1	·854803	51	18·317041	1	·821927	51	17·612542
2	1·669128	52	18·437327	2	1·604931	52	17·728202
3	2·445452	53	18·553719	3	2·351397	53	17·840118
4	3·186057	54	18·666354	4	3·063517	54	17·948420
5	3·893051	55	18·775363	5	3·743319	55	18·053237
6	4·568383	56	18·880874	6	4·392677	56	18·154690
7	5·213858	57	18·983009	7	5·013326	57	18·252897
8	5·831150	58	19·081884	8	5·606876	58	18·347969
9	6·421810	59	19·177612	9	6·174819	59	18·440015
10	6·987286	60	19·270301	10	6·718546	60	18·529139
11	7·528923	61	19·360055	11	7·239350	61	18·615441
12	8·047974	62	19·446970	12	7·738438	62	18·699013
13	8·545611	63	19·531153	13	8·216935	63	18·779958
14	9·022927	64	19·612686	14	8·675893	64	18·858355
15	9·480948	65	19·691660	15	9·116297	65	18·934292
16	9·920629	66	19·768162	16	9·539068	66	19·007852
17	10·342872	67	19·842275	17	9·945071	67	19·079114
18	10·748519	68	19·914077	18	10·335116	68	19·148154
19	11·138363	69	19·983645	19	10·709967	69	19·215047
20	11·513150	70	20·051052	20	11·070338	70	19·279861
21	11·873580	71	20·116370	21	11·416906	71	19·342667
22	12·220314	72	20·179668	22	11·750304	72	19·403530
23	12·553976	73	20·241010	23	12·071133	73	19·462513
24	12·875152	74	20·300462	24	12·379956	74	19·519678
25	13·184400	75	20·358082	25	12·677310	75	19·575083
26	13·482242	76	20·413932	26	12·963697	76	19·628785
27	13·769176	77	20·468069	27	13·239595	77	19·680839
28	14·045672	78	20·520547	28	13·505456	78	19·731299
29	14·312174	79	20·571419	29	13·761708	79	19·780214
30	14·569105	80	20·620738	30	14·008757	80	19·827636
31	14·816864	81	20·668552	31	14·246987	81	19·873611
32	15·055833	82	20·714908	32	14·476765	82	19·918185
33	15·286370	83	20·759855	33	14·698435	83	19·961403
34	15·508820	84	20·803436	34	14·912329	84	20·003308
35	15·723508	85	20·845695	35	15·118760	85	20·043941
36	15·930744	86	20·886672	36	15·318026	86	20·083342
37	16·130825	87	20·926408	37	15·510412	87	20·121550
38	16·324032	88	20·964943	38	15·696187	88	20·158603
39	16·510632	89	21·002314	39	15·875611	89	20·194536
40	16·690881	90	21·038557	40	16·048927	90	20·229385
41	16·865023	91	21·073707	41	16·216371	91	20·263184
42	17·033292	92	21·107799	42	16·378169	92	20·295964
43	17·195909	93	21·140865	43	16·534531	93	20·327758
44	17·353177	94	21·172937	44	16·685664	94	20·358597
45	17·505029	95	21·204045	45	16·831762	95	20·388509
46	17·651930	96	21·234221	46	16·973013	96	20·417524
47	17·793975	97	21·263492	47	17·109594	97	20·445669
48	17·931342	98	21·291886	48	17·241678	98	20·472971
49	18·064200	99	21·319430	49	17·369427	99	20·499456
50	18·192715	100	21·346151	50	17·492998	100	20·525148

TABLE X. cxvii

Present Value (or Years' Purchase) of £1 per Annum in **n** years, after **t** years' Deferrence. Redemption of Capital being at 3 per cent., with Interest allowed to a Purchaser at 4 per cent.

n Years	Deferred 5 Years	n Years	Deferred 5 Years	n Years	Deferred 6 Years	n Years	Deferred 6 Years
1	·790314	51	16·935138	1	·759918	51	16·283799
2	1·543203	52	17·046350	2	1·483850	52	16·390733
3	2·260958	53	17·153961	3	2·174000	53	16·494205
4	2·945689	54	17·258098	4	2·832396	54	16·594337
5	3·599346	55	17·358884	5	3·460912	55	16·691247
6	4·223728	56	17·456434	6	4·061280	56	16·785045
7	4·820507	57	17·550864	7	4·635106	57	16·875843
8	5·391227	58	17·642279	8	5·183876	58	16·963742
9	5·937326	59	17·730785	9	5·708972	59	17·048844
10	6·460141	60	17·816482	10	6·211678	60	17·131245
11	6·960915	61	17·899464	11	6·693192	61	17·211036
12	7·440806	62	17·979822	12	7·154627	62	17·288303
13	7·900900	63	18·057654	13	7·597025	63	17·363142
14	8·342206	64	18·133035	14	8·021357	64	17·435867
15	8·765671	65	18·206051	15	8·428536	65	17·505831
16	9·172182	66	18·276782	16	8·819412	66	17·573842
17	9·562569	67	18·345304	17	9·194785	67	17·639728
18	9·937613	68	18·411688	18	9·555404	68	17·703559
19	10·298046	69	18·476008	19	9·901974	69	17·765406
20	10·644557	70	18·538330	20	10·235159	70	17·825330
21	10·977795	71	18·598720	21	10·555580	71	17·883398
22	11·298370	72	18·657242	22	10·863826	72	17·939669
23	11·606859	73	18·713957	23	11·160450	73	17·994203
24	11·903805	74	18·768923	24	11·445975	74	18·047055
25	12·189722	75	18·822197	25	11·720895	75	18·098279
26	12·465094	76	18·873833	26	11·985676	76	18·147930
27	12·730380	77	18·923885	27	12·240760	77	18·196057
28	12·986017	78	18·972405	28	12·486564	78	18·242710
29	13·232413	79	19·019438	29	12·723484	79	18·287935
30	13·469960	80	19·065036	30	12·951894	80	18·331779
31	13·699028	81	19·109243	31	13·172152	81	18·374286
32	13·919967	82	19·152102	32	13·384594	82	18·415497
33	14·133112	83	19·193658	33	13·589541	83	18·455454
34	14·338780	84	19·233952	34	13·787298	84	18·494198
35	14·537271	85	19·273022	35	13·978155	85	18·531765
36	14·728873	86	19·310908	36	14·162388	86	18·568194
37	14·913859	87	19·347646	37	14·340259	87	18·603520
38	15·092489	88	19·383273	38	14·512019	88	18·637777
39	15·265012	89	19·417825	39	14·677906	89	18·670999
40	15·431662	90	19·451334	40	14·838148	90	18·703219
41	15·592666	91	19·483832	41	14·992959	91	18·734468
42	15·748240	92	19·515352	42	15·142550	92	18·764775
43	15·898589	93	19·545923	43	15·287116	93	18·794170
44	16·043909	94	19·575576	44	15·426847	94	18 822683
45	16·184388	95	19·604337	45	15·561923	95	18·850338
46	16·320206	96	19·632236	46	15·692518	96	18·877164
47	16·451535	97	19·659299	47	15·818795	97	18·903186
48	16·578538	98	19·685551	48	15·940914	98	18·928428
49	16·701373	99	19·711017	49	16·059025	99	18·952915
50	16·820192	100	19·735722	50	16·173273	100	18·976669

Present Value (or Years' Purchase) of £1 per Annum in n years, after t years' Deferrence. Redemption of Capital being at 3 per cent., with Interest allowed to a Purchaser at 4 per cent.

n Years	Deferred 7 Years	n Years	Deferred 7 Years	n Years	Deferred 8 Years	n Years	Deferred 8 Years
1	·730690	51	15·657493	1	·702586	51	15·055274
2	1·426779	52	15·760315	2	1·371902	52	15·154141
3	2·090384	53	15·859807	3	2·009984	53	15·249806
4	2·723456	54	15·956088	4	2·618707	54	15·342384
5	3·327799	55	16·049270	5	3·199805	55	15·431982
6	3·905076	56	16·139461	6	3·754879	56	15·518704
7	4·456831	57	16·226766	7	4·285412	57	15·602652
8	4·984494	58	16·311285	8	4·792781	58	15·683920
9	5·489394	59	16·393114	9	5·278252	59	15·762601
10	5·972766	60	16·472345	10	5·743041	60	15·838785
11	6·435759	61	16·549067	11	6·188227	61	15·912556
12	6·879446	62	16·623362	12	6·614849	62	15·983994
13	7·304829	63	16·695322	13	7·023870	63	16·053186
14	7·712841	64	16·765017	14	7·416189	64	16·120200
15	8·104359	65	16·832524	15	7·792649	65	16·185111
16	8·480201	66	16·897919	16	8·154035	66	16·247990
17	8·841136	67	16·961271	17	8·501088	67	16·308906
18	9·187885	68	17·022647	18	8·834500	68	16·367921
19	9·521126	69	17·082115	19	9·154924	69	16·425102
20	9·841495	70	17·139735	20	9·462971	70	16·480505
21	10·149592	71	17·195569	21	9·759218	71	16·534192
22	10·445983	72	17·249676	22	10·044209	72	16·586218
23	10·731198	73	17·302112	23	10·318454	73	16·636637
24	10·005741	74	17·352931	24	10·582438	74	16·685502
25	11·270088	75	17·402185	25	10·836617	75	16·732862
26	11·524685	76	17·449926	26	11·081422	76	16·778767
27	11·769957	77	17·496202	27	11·317261	77	16·823263
28	12·006307	78	17·541061	28	11·544520	78	16·866396
29	12·234114	79	17·584547	29	11·763565	79	16·908209
30	12·453740	80	17·626705	30	11·974744	80	16·948745
31	12·665526	81	17·667576	31	12·178384	81	16·988045
32	12·869797	82	17·707202	32	12·374799	82	17·026147
33	13·066861	83	17·745623	33	12·564283	83	17·063090
34	13·257013	84	17·782876	34	12·747121	84	17·098910
35	13·440529	85	17·818999	35	12·923579	85	17·133644
36	13·617676	86	17·854026	36	13·093912	86	17·167324
37	13·788706	87	17·887993	37	13·258364	87	17·199984
38	13·953860	88	17·920933	38	13·417166	88	17·231657
39	14·113367	89	17·952877	39	13·570538	89	17·262373
40	14·267444	90	17·983858	40	13·718689	90	17·292162
41	14·416302	91	18·013905	41	13·861821	91	17·321053
42	14·560139	92	18·043047	42	14·000126	92	17·349074
43	14·699145	93	18·071311	43	14·133786	93	17·376252
44	14·833501	94	18·098727	44	14·262975	94	17·402613
45	14·963382	95	18·125319	45	14·387860	95	17·428182
46	15·088954	96	18·151113	46	14·508602	96	17·452984
47	15·210374	97	18·176134	47	14·625352	97	17·477042
48	15·327796	98	18·200405	48	14·738258	98	17·500381
49	15·441364	99	18·223950	49	14·847458	99	17·523020
50	15·551219	100	18·246791	50	14·953087	100	17·544982

TABLE X.   cxix

Present Value (or Years' Purchase) of £1 per Annum in **n** years, after
t years' Deferrence.   Redemption of Capital being at **3** per cent., with Interest
allowed to a Purchaser at **4** per cent.

n Years	Deferred 9 Years	n Years	Deferred 9 Years	n Years	Deferred 10 Years	n Years	Deferred 10 Years
1	·675564	51	14·476235	1	·649580	51	13·919448
2	1·319137	52	14·571299	2	1·268400	52	14·010856
3	1·932678	53	14·663285	3	1·858343	53	14·099304
4	2·517989	54	14·752302	4	2·421142	54	14·184897
5	3·076737	55	14·838454	5	2·958399	55	14·267735
6	3·610463	56	14·921841	6	3·471596	56	14·347195
7	4·120591	57	15·002559	7	3·862104	57	14·425529
8	4·608446	58	15·080702	8	4·431195	58	14·500666
9	5·075254	59	15·156357	9	4·880049	59	14·573411
10	5·522158	60	15·229611	10	5·309764	60	14·643848
11	5·950222	61	15·300545	11	5·721364	61	14·712053
12	6·360436	62	15·369235	12	6·115800	62	14·778101
13	6·753726	63	15·435766	13	6·493963	63	14·842074
14	7·130956	64	15·500202	14	6·856684	64	14·904032
15	7·492936	65	15·562617	15	7·204742	65	14·964045
16	7·840423	66	15·623078	16	7·538864	66	15·022181
17	8·174128	67	15·681651	17	7·859834	67	15·078501
18	8·494717	68	15·738396	18	8·167972	68	15·133064
19	8·802817	69	15·793377	19	8·464242	69	15·185930
20	9·099017	70	15·846650	20	8·749049	70	15·237154
21	9·383870	71	15·898272	21	9·022946	71	15·286791
22	9·657899	72	15·948297	22	9·286436	72	15·334891
23	9·921597	73	15·996777	23	9·539991	73	15·381507
24	10·175428	74	16·043762	24	9·784059	74	15·426685
25	10·419831	75	16·089301	25	10·019062	75	15·470472
26	10·655220	76	16·133440	26	10·245398	76	15·512913
27	10·881989	77	16·176225	27	10·463444	77	15·554052
28	11·100507	78	16·217699	28	10·673558	78	15·593932
29	11·311128	79	16·257904	29	10·876078	79	15·632590
30	11·514184	80	16·296881	30	11·071324	80	15·670068
31	11·709992	81	16·334669	31	11·259601	81	15·706403
32	11·898853	82	16·371306	32	11·441197	82	15·741630
33	12·081050	83	16·406828	33	11·616386	83	15·775786
34	12·256855	84	16·441271	34	11·785430	84	15·808904
35	12·426526	85	16·474668	35	11·848575	85	15·841017
36	12·590308	86	16·507053	36	12·106058	86	15·872157
37	12·748435	87	16·538457	37	12·258103	87	15·902353
38	12·901129	88	16·568912	38	12·404924	88	15·931636
39	13·048602	89	16·598446	39	12·546725	89	15·960035
40	13·191056	90	16·627090	40	12·683700	90	15·987577
41	13·328683	91	16·654870	41	12·816033	91	16·014288
42	13·461669	92	16·681813	42	12·943904	92	16·040195
43	13·590187	93	16·707945	43	13·067480	93	16·065322
44	13·714408	94	16·733293	44	13·186922	94	16·089694
45	13·834490	95	16·757878	45	13·302385	95	16·113334
46	13·950588	96	16·781726	46	13·414018	96	16·136265
47	14·062848	97	16·804859	47	13·521961	97	16·158509
48	14·171411	98	16·827300	48	13·626348	98	16·180086
49	14·276411	99	16·849068	49	13·727310	99	16·201017
50	14·377978	100	16·870186	50	13·824970	100	16·221323

Present Value (or Years' Purchase) of £1 per Annum in n years, after
t years' Deferrence.   Redemption of Capital being at **3** per cent., with Intere
allowed to a Purchaser at **5** per cent.

n Years	Deferred 1 Year	n Years	Deferred 1 Year	n Years	Deferred 2 Years	n Years	Deferred 2 Yea
1	·907030	51	16·270609	1	·863837	51	15·495810
2	1·755182	52	16·359105	2	1·671601	52	15·580091
3	2·549675	53	16·444546	3	2·428261	53	15·661463
4	3·295128	54	16·527051	4	3·138215	54	15·740040
5	3·995648	55	16·606738	5	3·805377	55	15·815932
6	4·654900	56	16·683714	6	4·433236	56	15·889242
7	5·276163	57	16·758083	7	5·024914	57	15·960070
8	5·862379	58	16·829945	8	5·583215	58	16·028509
9	6·416197	59	16·899393	9	6·110661	59	16·094651
10	6·940009	60	16·966521	10	6·609528	60	16·158582
11	7·435978	61	17·031412	11	7·081879	61	16·220383
12	7·906064	62	17·094149	12	7·529581	62	16·280133
13	8·352055	63	17·154819	13	7·954333	63	16·337913
14	8·775575	64	17·213488	14	8·357685	64	16·393788
15	9·178109	65	17·270232	15	8·741051	65	16·447831
16	9·561017	66	17·325122	16	9·105725	66	16·500106
17	9·925543	67	17·378223	17	9·452893	67	16·550679
18	10·272834	68	17·429600	18	9·783646	68	16·599609
19	10·603940	69	17·479313	19	10·098984	69	16·646955
20	10·919834	70	17·527423	20	10·399836	70	16·692774
21	11·221410	71	17·573984	21	10·687051	71	16·737118
22	11·509499	72	17·619051	22	10·961421	72	16·780039
23	11·784865	73	17·662677	23	11·223675	73	16·821587
24	12·048224	74	17·704911	24	11·474493	74	16·861811
25	12·300235	75	17·745801	25	11·714503	75	16·900753
26	12·541512	76	17·785393	26	11·944290	76	16·938460
27	12·772629	77	17·823732	27	12·164402	77	16·974973
28	12·994119	78	17·860860	28	12·375344	78	17·010333
29	13·206481	79	17·896818	29	12·577593	79	17·044579
30	13·410180	80	17·931646	30	12·771592	80	17·077748
31	13·605651	81	17·965381	31	12·957756	81	17·109877
32	13·793304	82	17·998060	32	13·136473	82	17·141000
33	13·973521	83	18·029719	33	13·308107	83	17·171151
34	14·146661	84	18·060391	34	13·473002	84	17·200363
35	14·313062	85	18·090109	35	13·631479	85	17·228666
36	14·473041	86	18·118904	36	13·783840	86	17·256089
37	14·626897	87	18·146807	37	13·930370	87	17·282663
38	14·774911	88	18·173847	38	14·071336	88	17·308415
39	14·917348	89	18·200051	39	14·206990	89	17·333372
40	15·054461	90	18·225449	40	14·337573	90	17·357560
41	15·186484	91	18·250064	41	14·463309	91	17·381003
42	15·313642	92	18·273922	42	14·584412	92	17·403725
43	15·436146	93	18·297049	43	14·701082	93	17·425751
44	15·554195	94	18·319468	44	14·813511	94	17·447102
45	15·667982	95	18·341200	45	14·921879	95	17·467799
46	15·777684	96	18·362269	46	15·026357	96	17·487866
47	15·883471	97	18·382695	47	15·127107	97	17·507319
48	15·985508	98	18·402499	48	15·224285	98	17·526179
49	16·083947	99	18·421701	49	15·318036	99	17·544467
50	16·178935	100	18·440319	50	15·408501	100	17·562198

TABLE X.   cxxi

Present Value (or Years' Purchase) of £1 per Annum in n years, after t years' Deferrence. Redemption of Capital being at 3 per cent., with Interest allowed to a Purchaser at 5 per cent.

n Years	Deferred 3 Years	n Years	Deferred 3 Years	n Years	Deferred 4 Years	n Years	Deferred 4 Years
1	·822703	51	14·757928	1	·783526	51	14·055155
2	1·592002	52	14·838196	2	1·516191	52	14·131601
3	2·312632	53	14·915694	3	2·202504	53	14·205408
4	2·988779	54	14·990529	4	2·846453	54	14·276679
5	3·624172	55	15·062807	5	3·451589	55	14·345516
6	4·222133	56	15·132627	6	4·021075	56	14·412011
7	4·785637	57	15·200081	7	4·557745	57	14·476253
8	5·317353	58	15·265262	8	5·064141	58	14·538330
9	5·819683	59	15·328254	9	5·542549	59	14·598322
10	6·294795	60	15·389141	10	5·995037	60	14·656309
11	6·744654	61	15·447999	11	6·423473	61	14·712365
12	7·171036	62	15·504904	12	6·829551	62	14·766560
13	7·575563	63	15·559933	13	7·214814	63	14·818968
14	7·959708	64	15·613147	14	7·580667	64	14·869648
15	8·324819	65	15·664616	15	7·928391	65	14·918667
16	8·672128	66	15·714403	16	8·259160	66	14·966082
17	9·002764	67	15·762567	17	8·574052	67	15·011953
18	9·317767	68	15·809167	18	8·874055	68	15·056334
19	9·618090	69	15·854259	19	9·160076	69	15·099279
20	9·904615	70	15·897896	20	9·432957	70	15·140837
21	10·178154	71	15·940128	21	9·693470	71	15·181058
22	10·439459	72	15·981006	22	9·942331	72	15·219989
23	10·689225	73	16·020575	23	10·180203	73	15·257675
24	10·928099	74	16·058883	24	10·407703	74	15·294159
25	11·156680	75	16·095971	25	10·625399	75	15·329480
26	11·375526	76	16·131883	26	10·833823	76	15·363682
27	11·585156	77	16·166657	27	11·033470	77	15·396800
28	11·786054	78	16·200333	28	11·224801	78	15·428873
29	11·978672	79	16·232948	29	11·408247	79	15·459935
30	12·163433	80	16·264538	30	11·584210	80	15·490020
31	12·340732	81	16·295137	31	11·753066	81	15·519162
32	12·510939	82	16·324778	32	11·915168	82	15·547391
33	12·674400	83	16·353493	33	12·070845	83	15·574739
34	12·831444	84	16·381314	34	12·220410	84	15·601235
35	12·982374	85	16·408269	35	12·364153	85	15·626907
36	13·127480	86	16·434387	36	12·502349	86	15·651780
37	13·267032	87	16·459695	37	12·635256	87	15·675884
38	13·401286	88	16·484221	38	12·763116	88	15·699242
39	13·530480	89	16·507990	39	12·886158	89	15·721879
40	13·654845	90	16·531026	40	13·004601	90	15·743818
41	13·774594·	91	16·553353	41	13·118647	91	15·765081
42	13·889930	92	16·574993	42	13·228491	92	15·785691
43	14·001045	93	16·595970	43	13·334315	93	15·805669
44	14·108119	94	16·616304	44	13·436290	94	15·825035
45	14·211327	95	16·636016	45	13·534583	95	15·843808
46	14·310830	96	16·655127	46	13·629348	96	15·862009
47	14·406783	97	16·673653	47	13·720731	97	15·879653
48	14·499333	98	16·691616	48	13·808875	98	15·896761
49	14·588621	99	16·709033	49	13·893910	99	15·913348
50	14·674777	100	16·725920	50	13·975964	100	15·929431

Present Value (or Years' Purchase) of £1 per Annum in n years, after t years' Deferrence. Redemption of Capital being at 3 per cent., with Interest allowed to a Purchaser at 5 per cent.

n Years	Deferred 5 Years	n Years	Deferred 5 Years	n Years	Deferred 6 Years	n Years	Deferred 6 Years
1	·746215	51	13·385867	1	·710680	51	12·748441
2	1·443992	52	13·458672	2	1·375230	52	12·817779
3	2·097624	53	13·528965	3	1·997736	53	12·884724
4	2·710909	54	13·596842	4	2·581817	54	12·949370
5	3·287228	55	13·662401	5	3·130693	55	13·011806
6	3·829597	56	13·725729	6	3·647234	56	13·072119
7	4·340711	57	13·786913	7	4·134009	57	13·130389
8	4·822993	58	13·846033	8	4·593325	58	13·186694
9	5·278620	59	13·903169	9	5·027256	59	13·241109
10	5·709561	60	13·958395	10	5·437675	60	13·293705
11	6·117595	61	14·011781	11	5·826280	61	13·344549
12	6·504337	62	14·063395	12	6·194605	62	13·393706
13	7·001653	63	14·113308	13	6·544050	63	13·441242
14	7·219685	64	14·161575	14	6·875888	64	13·487210
15	7·550851	65	14·208259	15	7·191284	65	13·531671
16	7·865870	66	14·253417	16	7·491302	66	13·574678
17	8·165767	67	14·297103	17	7·776918	67	13·616284
18	8·451484	68	14·339371	18	8·049029	68	13·656540
19	8·723885	69	14·380270	19	8·308459	69	13·695491
20	8·983772	70	14·419850	20	8·555971	70	13·733186
21	9·231879	71	14·458516	21	8·792263	71	13·769668
22	9·468891	72	14·495233	22	9·017988	72	13·804980
23	9·695435	73	14·531124	23	9·233745	73	13·839162
24	9·912101	74	14·565871	24	9·440094	74	13·872253
25	10·119431	75	14·599510	25	9·637550	75	13·904291
26	10·317930	76	14·632083	26	9·826597	76	13·935313
27	10·508071	77	14·663625	27	10·007683	77	13·965353
28	10·690291	78	14·694170	28	10·181226	78	13·994443
29	10·865001	79	14·723753	29	10·347617	79	14·022617
30	11·032585	80	14·752405	30	10·507221	80	14·049905
31	11·193400	81	14·780159	31	10·660377	81	14·076338
32	11·347783	82	14·807045	32	10·807409	82	14·101943
33	11·496047	83	14·833091	33	10·948613	83	14·126748
34	11·638490	84	14·858325	34	11·084272	84	14·150781
35	11·775388	85	14·882774	35	11·214652	85	14·174066
36	11·907003	86	14·906463	36	11·340000	86	14·196627
37	12·033581	87	14·929419	37	11·460550	87	14·218490
38	12·155353	88	14·951665	38	11·576523	88	14·239676
39	12·272536	89	14·973223	39	11·688126	89	14·260208
40	12·385339	90	14·994118	40	11·795557	90	14·280108
41	12·493954	91	15·014369	41	11·899000	91	14·299394
42	12·598568	92	15·033997	42	11·998632	92	14·318088
43	12·699352	93	15·053024	43	12·094617	93	14·336209
44	12·796471	94	15·071467	44	12·187112	94	14·353774
45	12·890084	95	15·089347	45	12·276267	95	14·370802
46	12·980336	96	15·106681	46	12·362221	96	14·387310
47	13·067368	97	15·123485	47	12·445108	97	14·403314
48	13·151314	98	15·139778	48	12·525057	98	14·418831
49	13·232300	99	15·155575	49	12·602186	99	14·433876
50	13·310446	100	15·170892	50	12·676612	100	14·448464

TABLE X.   cxxiii

Present Value (or Years' Purchase) of £1 per Annum in n years, after t years' Deferrence. Redemption of Capital being 3 per cent., with Interest allowed to a Purchaser at 5 per cent.

n Years	Deferred 7 Years	n Years	Deferred 7 Years	n Years	Deferred 8 Years	n Years	Deferred 8 Years
1	·676839	51	12·141373	1	·644609	51	11·563212
2	1·309743	52	12·207409	2	1·247374	52	11·626103
3	1·902066	53	12·271167	3	1·812006	53	11·686825
4	2·458874	54	12·332734	4	2·341784	54	11·745460
5	2·981612	55	12·392197	5	2·839631	55	11·802092
6	3·473556	56	12·449638	6	3·308149	56	11·856797
7	3·937152	57	12·505133	7	3·749668	57	11·909650
8	4·374595	58	12·558757	8	4·166281	58	11·960720
9	4·787863	59	12·610581	9	4·559869	59	12·010076
10	5·178739	60	12·660673	10	4·932132	60	12·057783
11	5·548838	61	12·709906	11	5·284607	61	12·103900
12	5·899624	62	12·755911	12	5·618689	62	12·148486
13	6·232429	63	12·801183	13	5·935646	63	12·191602
14	6·548466	64	12·844963	14	6·236633	64	12·233297
15	6·848843	65	12·887307	15	6·522707	65	12·273624
16	7·134574	66	12·928266	16	6·794832	66	12·312633
17	7·406589	67	12·967891	17	7·053894	67	12·350371
18	7·665743	68	13·006229	18	7·300707	68	·12·386884
19	7·912819	69	13·043326	19	7·536017	69	12·422214
20	8·148544	70	13·079226	20	7·760518	70	12·456405
21	8·373585	71	13·113970	21	7·974842	71	12·489495
22	8·588561	72	13·147601	22	8·179581	72	12·521524
23	8·794043	73	13·180155	23	8·375279	73	12·552527
24	8·990566	74	13·211671	24	8·562443	74	12·582543
25	9·178620	75	13·242183	25	8·741542	75	12·611602
26	9·358665	76	13·271727	26	8·913013	76	12·639740
27	9·531128	77	13·300337	27	9·077264	77	12·666986
28	9·696406	78	13·328042	28	9·234672	78	12·693372
29	9·854874	79	13·354874	29	9·385594	79	12·718927
30	10·006877	80	13·380863	30	9·530359	80	12·743678
31	10·152741	81	13·406037	31	9·669277	81	12·767653
32	10·292771	82	13·430423	32	9·802638	82	12·790878
33	10·427251	83	13·454047	33	9·930714	83	12·813377
34	10·556451	84	13·476935	34	10·053762	84	12·835175
35	10·680621	85	13·499111	35	10·172020	85	12·856296
36	10·800000	86	13·520598	36	10·285714	86	12·876759
37	10·914810	87	13·541420	37	10·395056	87	12·896589
38	11·025261	88	13·561597	38	10·500247	88	12·915806
39	11·131549	89	13·581152	39	10·601475	89	12·934429
40	11·233865	90	13·600103	40	10·698918	90	12·952478
41	11·332382	91	13·618472	41	10·792744	91	12·969972
42	11·427269	92	13·636275	42	10·883113	92	12·986928
43	11·518684	93	13·653533	43	10·970174	93	13·003364
44	11·606774	94	13·670262	44	11·054070	94	13·019296
45	11·691683	95	13·686479	45	11·134936	95	13·034741
46	11·773544	96	13·702201	46	11·212899	96	13·049714
47	11·852485	97	13·717443	47	11·288080	97	13·064231
48	11·928626	98	13·732221	48	11·360596	98	13·078305
49	12·002083	99	13·746550	49	11·430554	99	13·091951
50	12·072964	100	13·760443	50	11·498060	100	13·105183

Present Value (or Years' Purchase) of £1 per Annum in n years, after t years' Deferrence.   Redemption of Capital being at 3 per cent., with Interest allowed to a Purchaser at 5 per cent.

n Years	Deferred 9 Years	n Years	Deferred 9 Years	n Years	Deferred 10 Years	n Years	Deferred 10 Years
1	·613913	51	11·012590	1	·584679	51	10·488175
2	1·187976	52	11·072487	2	1·131405	52	10·545220
3	1·725721	53	11·130317	3	1·643543	53	10·600296
4	2·230271	54	11·186160	4	2·124068	54	10·653480
5	2·704412	55	11·240095	5	2·575629	55	10·704846
6	3·150620	56	11·292195	6	3·000589	56	10·754466
7	3·571115	57	11·342531	7	3·401060	57	10·802405
8	3·967889	58	11·391170	8	3·778940	58	10·848727
9	4·342735	59	11·438175	9	4·135936	59	10·893495
10	4·697272	60	11·483610	10	4·473590	60	10·936766
11	5·032963	61	11·527531	11	4·793295	61	10·978595
12	5·351136	62	11·569994	12	5·096317	62	11·019036
13	5·653000	63	11·611058	13	5·383806	63	11·058144
14	5·939655	64	11·650767	14	5·656811	64	11·095962
15	6·212106	65	11·689174	15	5·916288	65	11·132541
16	6·471273	66	11·726325	16	6·163114	66	11·167923
17	6·717999	67	11·762266	17	6·398091	67	11·202152
18	6·953059	68	11·797040	18	6·621957	68	11·235270
19	7·177164	69	11·830688	19	6·835391	69	11·267316
20	7·390974	70	11·863251	20	7·039019	70	11·298328
21	7·595093	71	11·894765	21	7·233418	71	11·328341
22	7·790082	72	11·925268	22	7·419122	72	11·357392
23	7·976461	73	11·954796	23	7·596626	73	11·385514
24	8·154713	74	11·983382	24	7·766389	74	11·412738
25	8·325284	75	12·011058	25	7·928837	75	11·439096
26	8·488590	76	12·037855	26	8·084367	76	11·464618
27	8·645019	77	12·063805	27	8·233347	77	11·489331
28	8·794932	78	12·088934	28	8·376121	78	11·513264
29	8·938667	79	12·113272	29	8·513011	79	11·536443
30	9·076538	80	12·136845	30	8·644317	80	11·558893
31	9·208841	81	12·159678	31	8·770320	81	11·580639
32	9·335852	82	12·181796	32	8·871305	82	11·601705
33	9·457830	83	12·203224	33	9·007452	83	11·622112
34	9·575018	84	12·223985	34	9·119059	84	11·641884
35	9·687644	85	12·244099	35	9·226323	85	11·661040
36	9·795924	86	12·263588	36	9·329447	86	11·679602
37	9·900060	87	12·282474	37	9·428624	87	11·697588
38	10·000242	88	12·300776	38	9·524035	88	11·715018
39	10·096649	89	12·318512	39	9·615851	89	11·731910
40	10·189452	90	12·335702	40	9·704235	90	11·748281
41	10·278811	91	12·352363	41	9·789338	91	11·764148
42	10·364876	92	12·368511	42	9·871305	92	11·779528
43	10·447792	93	12·384164	43	9·950272	93	11·794436
44	10·527692	94	12·399338	44	10·026368	94	11·808887
45	10·604708	95	12·414047	45	10·099716	95	11·822896
46	10·678958	96	12·428308	46	10·170431	96	11·836477
47	10·750559	97	12·442133	47	10·238622	97	11·849644
48	10·819622	98	12·455537	48	10·304396	98	11·862409
49	10·886250	99	12·468533	49	10·367851	99	11·874787
50	10·950541	100	12·481135	50	10·429081	100	11·886789

TABLE X. CXXV

Present Value (or Years' Purchase) of £1 per Annum in **n** years, after t years' Deferrence. Redemption of Capital being at **3** per cent., with Interest allowed to a Purchaser at **6** per cent.

n Years	Deferred 1 Year	n Years	Deferred 1 Year	n Years	Deferred 2 Years	n Years	Deferred 2 Years
1	·889996	51	13·765407	1	·839619	51	12·986230
2	1·707162	52	13·829301	2	1·610530	52	13·046507
3	2·459768	53	13·990894	3	2·320536	53	13·104613
4	3·154685	54	13·950283	4	2·976306	54	13·160641
5	3·798585	55	14·007558	5	3·583570	55	13·214674
6	4·396119	56	14·062807	6	4·147281	56	13·266796
7	4·952045	57	14·116114	7	4·671739	57	13·317085
8	5·470345	58	14·167555	8	5·160701	58	13·365615
9	5·954510	59	14·217208	9	5·617461	59	13·412457
10	6·407613	60	14·265141	10	6·044916	60	13·457677
11	6·733368	61	14·311425	11	6·445629	61	13·501341
12	7·231189	62	14·356119	12	6·821875	62	13·543505
13	7·606220	63	14·399295	13	7·175677	63	13·584237
14	7·959378	64	14·441001	14	7·508846	64	13·623583
15	8·292383	65	14·481299	15	7·823001	65	13·661600
16	8·606775	66	14·520241	16	8·119597	66	13·698337
17	8·903950	67	14·557878	17	8·399950	67	13·733843
18	9·185162	68	14·594258	18	8·665245	68	13·768164
19	9·451552	69	14·629430	19	8·916556	69	13·801345
20	9·704152	70	14·663436	20	9·154858	70	13·833427
21	9·943906	71	14·696321	21	9·381041	71	13·864450
22	10·171671	72	14·728126	22	9·595914	72	13·894455
23	10·388233	73	14·758888	23	9·800218	73	13·923476
24	10·594311	74	14·788646	24	9·994630	74	13·951550
25	10·790563	75	14·817436	25	10·179774	75	13·978710
26	10·977598	76	14·845293	26	10·356222	76	14·004990
27	11·155937	77	14·872249	27	10·524500	77	14·030420
28	11·326202	78	14·898337	28	10·685094	78	14·055031
29	11·488763	79	14·923586	29	10·838453	79	14·078851
30	11·644095	80	14·948026	30	10·984993	80	14·101907
31	11·792608	81	14·971685	31	11·125099	81	14·124228
32	11·934679	82	14·994590	32	11·259128	82	14·145836
33	12·070660	83	15·016767	33	11·387412	83	14·166758
34	12·200880	84	15·038242	34	11·510261	84	14·187017
35	12·325643	85	15·059037	35	11·627962	85	14·206635
36	12·445235	86	15·079176	36	11·740785	86	14·225634
37	12·559920	87	15·098680	37	11·848978	87	14·244034
38	12·669949	88	15·117574	38	11·952779	88	14·261858
39	12·775553	89	15·135874	39	12·052405	89	14·279123
40	12·881951	90	15·153602	40	12·148064	90	14·295847
41	12·974347	91	15·170777	41	12·239947	91	14·312051
42	13·067934	92	15·187418	42	12·328236	92	14·327749
43	13·157892	93	15·203541	43	12·413103	93	14·342959
44	13·244393	94	15·219164	44	12·494706	94	14·357698
45	13·327593	95	15·234303	45	12·573198	95	14·371980
46	13·407646	96	15·248975	46	12·648719	96	14·385821
47	13·484694	97	15·263193	47	12·721406	97	14·399235
48	13·558870	98	15·276974	48	12·791384	98	14·412236
49	13·630302	99	15·290330	49	12·858773	99	14·424836
50	13·699110	100	15·303277	50	12·923685	100	14·437050

Present Value (or Years' Purchase) of £1 per Annum in n years, after
t years' Deferrence. Redemption of Capital being at 3 per cent., with Interest
allowed to a Purchaser at 6 per cent.

n Years	Deferred 3 Years	n Years	Deferred 3 Years	n Years	Deferred 4 Years	n Years	Deferred 4 Years
1	·792093	51	12·251162	1	·747258	51	11·557709
2	1·519368	52	12·308028	2	1·433367	52	11·611356
3	2·189185	53	12·362845	3	2·065270	53	11·663070
4	2·807836	54	12·415701	4	2·648904	54	11·712934
5	3·380734	55	12·466676	5	3·189367	55	11·761024
6	3·912530	56	12·515847	6	3·691069	56	11·807412
7	4·407302	57	12·563290	7	4·157835	57	11·852169
8	4·868587	58	12·609072	8	4·593010	58	11·895360
9	5·299492	59	12·653263	9	4·999525	59	11·937040
10	5·702752	60	12·695923	10	5·379959	60	11·977295
11	6·080783	61	12·737116	11	5·736592	61	12·016156
12	6·435732	62	12·776894	12	6·071450	62	12·053683
13	6·769508	63	12·815320	13	6·386333	63	12·089933
14	7·083818	64	12·852438	14	6·682852	64	12·124951
15	7·380191	65	12·888304	15	6·962449	65	12·158786
16	7·659999	66	12·922961	16	7·226419	66	12·191482
17	7·924483	67	12·956458	17	7·475933	67	12·223083
18	8·174761	68	12·988836	18	7·712044	68	12·253628
19	8·411846	69	13·020139	19	7·935710	69	12·283159
20	8·636660	70	13·050405	20	8·147799	70	12·311712
21	8·850040	71	13·079672	21	8·349101	71	12·339323
22	9·052750	72	13·107978	22	8·540337	72	12·366027
23	9·245490	73	13·135356	23	8·722167	73	12·391855
24	9·428898	74	13·161841	24	8·895193	74	12·416841
25	9·603562	75	13·187464	25	9·059971	75	12·441013
26	9·770022	76	13·212256	26	9·217009	76	12·464402
27	9·928775	77	13·236248	27	9·366776	77	12·487036
28	10·080278	78	13·259466	28	9·509704	78	12·508939
29	10·224957	79	13·281937	29	9·646194	79	12·530139
30	10·363202	80	13·303688	30	9·776613	80	12·550659
31	10·495378	81	13·324745	31	9·901307	81	12·570524
32	10·621821	82	13·345130	32	10·020593	82	12·589755
33	10·742844	83	13·364868	33	10·134766	83	12·608376
34	10·858739	84	13·383980	34	10·244101	84	12·626406
35	10·969778	85	13·402488	35	10·348855	85	12·643866
36	11·076214	86	13·420411	36	10·449266	86	12·660775
37	11·178283	87	13·437770	37	10·545558	87	12·677152
38	11·276208	88	13·454585	38	10·637940	88	12·693015
39	11·370195	89	13·470872	39	10·726607	89	12·708380
40	11·460439	90	13·486651	40	10·811743	90	12·723265
41	11·547121	91	13·501937	41	10·893519	91	12·737686
42	11·630413	92	13·516747	42	10·972096	92	12·758658
43	11·710476	93	13·531096	43	11·047627	93	12·765194
44	11·787461	94	13·545001	44	11·120254	94	12·778312
45	11·861509	95	13·558474	45	11·190112	95	12·791023
46	11·932756	96	13·571532	46	11·257325	96	12·803342
47	12·001328	97	13·584186	47	11·322017	97	12·815279
48	12·067345	98	13·596451	48	11·384297	98	12·826850
49	12·130919	99	13·608338	49	11·444272	99	12·838065
50	12·192158	100	13·619860	50	11·502044	100	12·848935

TABLE X.   cxxvii

Present Value (or Years' Purchase) of £1 per Annum in **n** years, after t years' Deferrence.   Redemption of Capital being at 3 per cent., with Interest allowed to a Purchaser at 6 per cent.

n Years	Deferred 5 Years	n Years	Deferred 5 Years	n Years	Deferred 6 Years	n Years	Deferred 6 Years
1	·704960	51	10·903492	1	·665057	51	10·286322
2	1·352232	52	10·954102	2	1·275692	52	10·334068
3	1·948367	53	11·002889	3	1·838084	53	10·380093
4	2·498965	54	11·049930	4	2·357516	54	10·424472
5	3·008835	55	11·095298	5	2·838526	55	10·467272
6	3·482138	56	11·139060	6	3·285039	56	10·508557
7	3·922484	57	11·181284	7	3·700459	57	10·548391
8	4·333026	58	11·222031	8	4·087764	58	10·586831
9	4·716530	59	11·261360	9	4·449560	59	10·623934
10	5·075429	60	11·299328	10	4·788145	60	10·659753
11	5·411876	61	11·335989	11	5·105548	61	10·694339
12	5·727779	62	11·371391	12	5·403570	62	10·727737
13	6·024839	63	11·405590	13	5·683815	63	10·760000
14	6·304573	64	11·438626	14	5·947716	64	10·791166
15	6·568344	65	11·470546	15	6·196556	65	10·821279
16	6·817372	66	11·501391	16	6·431489	66	10·850378
17	7·052762	67	11·531203	17	6·653555	67	10·878503
18	7·275509	68	11·560019	18	6·863693	68	10·905688
19	7·486514	69	11·587879	19	7·062755	69	10·931971
20	7·686598	70	11·614815	20	7·251514	70	10·957382
21	7·876505	71	11·640863	21	7·430672	71	10·981956
22	8·056916	72	11·666055	22	7·600871	72	11·005722
23	8·228454	73	11·690422	23	7·762699	73	11·028709
24	8·391686	74	11·713993	24	7·916692	74	11·050947
25	8·547137	75	11·736797	25	8·063344	75	11·072460
26	8·695286	76	11·758862	26	8·203107	76	11·093276
27	8·836575	77	11·780214	27	8·336399	77	11·113420
28	8·971413	78	11·800878	28	8·463605	78	11·132914
29	9·100177	79	11·820878	29	8·585080	79	11·151782
30	9·223214	80	11·840237	30	8·701153	80	11·170044
31	9·340850	81	11·858977	31	8·812130	81	11·187724
32	9·453384	82	11·877120	32	8·918295	82	11·204840
33	9·561094	83	11·894686	33	9·019908	83	11·221412
34	9·664240	84	11·911696	34	9·117215	84	11·237460
35	9·763064	85	11·928168	35	9·210446	85	11·252998
36	9·857792	86	11·944120	36	9·299812	86	11·268047
37	9·948633	87	11·959569	37	9·385511	87	11·282622
38	10·035786	88	11·974534	38	9·467731	88	11·296741
39	10·119434	89	11·989030	39	9·546645	89	11·310415
40	10·199751	90	12·003072	40	9·622415	90	11·323663
41	10·276898	91	12·016677	41	9·695195	91	11·336498
42	10·351028	92	12·029858	42	9·765129	92	11·348932
43	10·422283	93	12·042628	43	9·832351	93	11·360980
44	10·490799	94	12·055004	44	9·896989	94	11·372655
45	10·556702	95	12·066995	45	9·959162	95	11·383968
46	10·620111	96	12·078616	46	10·018982	96	11·394931
47	10·681141	97	12·089878	47	10·076557	97	11·405556
48	10·739895	98	12·100794	48	10·131986	98	11·415854
49	10·796476	99	12·111374	49	10·185364	99	11·425834
50	10·850978	100	12·121628	50	10·236781	100	11·435509

Present Value (or Years' Purchase) of £1 per Annum in n years, after t years' Deferrence. Redemption of Capital being at 3 per cent., with Interest allowed to a Purchaser at 6 per cent.

n Years	Deferred 7 Years	n Years	Deferred 7 Years	n Years	Deferred 8 Years	n Years	Deferred 8 Years
1	·627412	51	9·704069	1	·591898	51	9·154779
2	1·203482	52	9·749112	2	1·135360	52	9·197272
3	1·734040	53	9·792533	3	1·635886	53	9·238235
4	2·224070	54	9·834399	4	2·098178	54	9·277731
5	2·677853	55	9·874776	5	2·526275	55	9·315823
6	3·099090	56	9·913725	6	2·923669	56	9·352567
7	3·490997	57	9·951304	7	3·293392	57	9·388018
8	3·856378	58	9·987568	8	3·638091	58	9·422230
9	4·197695	59	10·022571	9	3·960088	59	9·455252
10	4·517114	60	10·056362	10	4·261427	60	9·487130
11	4·816550	61	10·088991	11	4·543913	61	·9·517912
12	5·097703	62	10·120498	12	4·809152	62	9·547636
13	5·362085	63	10·150935	13	5·058569	63	9·576350
14	5·611048	64	10·180337	14	5·293439	64	9·604087
15	5·845803	65	10·208745	15	5·514906	65	9·630888
16	6·067437	66	10·236198	16	5·723995	66	9·656786
17	6·276934	67	10·262730	17	5·921633	67	9·681817
18	6·475177	68	10·288377	18	6·108655	68	9·706012
19	6·662971	69	10·313171	19	6·285819	69	9·729403
20	6·841045	70	10·337145	20	6·453813	70	9·752019
21	7·010062	71	10·360327	21	6·613263	71	9·773889
22	7·170627	72	10·382748	22	6·764740	72	9·795041
23	7·323295	73	10·404434	23	6·908766	73	9·815100
24	7·468571	74	10·425413	24	7·045819	74	9·835291
25	7·606922	75	10·445708	25	7·176338	75	9·854438
26	7·738774	76	10·465346	26	7·300727	76	9·872964
27	7·864521	77	10·484350	27	7·419356	77	9·890892
28	7·984526	78	10·502740	28	7·532569	78	9·908242
29	8·099125	79	10·520540	29	7·640681	79	9·925034
30	8·208628	80	10·537769	30	7·743986	80	9·941287
31	8·313324	81	10·554448	31	7·842755	81	9·957022
32	8·413478	82	10·570595	32	7·937241	82	9·972255
33	8·509340	83	10·586229	33	8·027676	83	9·987004
34	8·601140	84	10·601368	34	8·114279	84	10·001286
35	8·689093	85	10·616028	35	8·197254	85	10·015116
36	8·773400	86	10·630225	36	8·276789	86	10·028509
37	8·854248	87	10·643975	37	8·353061	87	10·041481
38	8·931815	88	10·657294	38	8·426237	88	10·054046
39	9·006261	89	10·670194	39	8·496469	89	10·066217
40	9·077743	90	10·682692	40	8·563905	90	10·078007
41	9·146403	91	10·694800	41	8·628679	91	10·089430
42	9·212378	92	10·706531	42	8·690919	92	10·100497
43	9·275795	93	10·717897	43	8·750747	93	10·111219
44	9·336774	94	10·728911	44	8·808274	94	10·121610
45	9·395428	95	10·739583	45	8·863607	95	10·131678
46	9·451862	96	10·749926	46	8·916847	96	10·141435
47	9·506178	97	10·759949	47	8·968088	97	10·150891
48	9·558469	98	10·769664	48	9·017420	98	10·160056
49	9·608826	99	10·779080	49	9·064926	99	10·168939
50	9·657333	100	10·788207	50	9·110687	100	10·177549

TABLE X.   cxxix

Present Value (or Years' Purchase) of £1 per Annum in n years, after t years' Déferrence. Redemption of Capital being at 3 per cent., with Interest allowed to a Purchaser at 6 per cent.

n Years	Deferred 9 Years	n Years	Deferred 9 Years	n Years	Deferred 10 Years	n Years	Deferred 10 Years
1	·558394	51	8·636582	1	·526788	51	8·147729
2	1·071094	52	8·676670	2	1·010467	52	8·185547
3	1·543288	53	8·715314	3	1·455934	53	8·222004
4	1·979413	54	8·752575	4	1·867373	54	8·257156
5	2·383278	55	8·788510	5	2·248378	55	8·291057
6	2·758178	56	8·823174	6	2·602058	56	8·323759
7	3·106973	57	8·856619	7	2·931110	57	8·445311
8	3·432160	58	8·888895	8	3·237891	58	8·385759
9	3·735931	59	8·920047	9	3·524468	59	8·415149
10	4·020213	60	8·950121	10	3·792658	60	8·443520
11	4·286710	61	8·979160	11	4·044071	61	8·470916
12	4·536935	62	9·007202	12	4·280132	62	8·497370
13	4·772234	63	9·034291	13	4·502113	63	8·522836
14	4·993810	64	9·060458	14	4·711147	64	8·547612
15	5·202741	65	9·085741	15	4·908252	65	8·571464
16	5·399994	66	9·110174	16	5·094340	66	8·594514
17	5·586445	67	9·133788	17	5·070238	67	8·616791
18	5·762881	68	9·156613	18	5·436687	68	8·638324
19	5·930017	69	9·178680	19	5·594362	69	8·659142
20	6·088502	70	9·200016	20	5·743877	70	8·679271
21	6·238926	71	9·220649	21	5·885787	71	8·698735
22	6·381829	72	9·240603	22	6·020600	72	8·717561
23	6·517703	73	9·259904	23	6·148783	73	8·735769
24	6·646998	74	9·278575	24	6·270760	74	8·753383
25	6·770129	75	9·296638	25	6·386922	75	8·770423
26	6·887477	76	9·314115	26	6·497628	76	8·786912
27	6·999392	77	9·331028	27	6·603207	77	8·802867
28	7·106195	78	9·347396	28	6·703966	78	8·818308
29	7·208188	79	9·363237	29	6·800186	79	8·833253
30	7·305645	80	9·378571	30	6·892127	80	8·847719
31	7·398824	81	9·393415	31	6·980031	81	8·861723
32	7·487961	82	9·407786	32	7·064123	82	8·875280
33	7·573277	83	9·421700	33	7·144610	83	8·888407
34	7·654979	84	9·435174	34	7·221687	84	8·901118
35	7·733257	85	9·448221	35	7·295534	85	8·913427
36	7·808290	86	9·460856	36	7·366320	86	8·925347
37	7·880245	87	9·473094	37	7·434202	87	8·936891
38	7·949278	88	9·484948	38	7·499328	88	8·948074
39	8·015535	89	9·496429	39	7·561835	89	8·958906
40	8·079154	90	9·507552	40	7·621852	90	8·969399
41	8·140261	91	9·518328	41	7·679501	91	8·979566
42	8·198979	92	9·528769	42	7·734895	92	8·989415
43	8·255420	93	9·538884	43	7·788141	93	8·998958
44	8·309691	94	9·548687	44	7·839340	94	9·008206
45	8·361892	95	9·558185	45	7·888587	95	9·017166
46	8·412118	96	9·567390	46	7·935970	96	9·025850
47	8·460459	97	9·576311	47	7·981575	97	9·034266
48	8·506998	98	9·584957	48	8·025480	98	9·042423
49	8·551816	99	9·593337	49	8·067760	99	9·050329
50	8·594986	100	9·601460	50	8·108487	100	9·057993

Present Value (or Years' Purchase) of £1 per Annum in n years, after t years' Deferrence. Redemption of Capital being at **3** per cent., with Intere[st] allowed to a Purchaser at **8** per cent.

n Years	Deferred 1 Year	n Years	Deferred 1 Year	n Years	Deferred 2 Years	n Years	Deferred 2 Yea[rs]
1	·857339	51	10·458443	1	·793832	51	9·683745
2	1·617025	52	10·495982	2	1·497246	52	9·718503
3	2·294563	53	10·532094	3	2·124596	53	9·751940
4	2·902343	54	10·566844	4	2·687355	54	9·784116
5	3·450383	55	10·600296	5	3·194799	55	9·815090
6	3·946871	56	10·632504	6	3·654510	56	9·844912
7	4·398565	57	10·663524	7	4·072746	57	9·873635
8	4·811095	58	10·693406	8	4·454718	58	9·901304
9	5·189184	59	10·722205	9	4·804800	59	9·927969
10	· 5·536825	60	10·749962	10	5·126690	60	9·953670
11	5·857420	61	10·776721	11	5·423538	61	9·978447
12	6·153883	62	10·802524	12	5·698040	62	10·002338
13	6·428723	63	10·827414	13	5·952522	63	10·025384
14	6·684115	64	10·851424	14	6·188996	64	10·047616
15	6·921953	65	10·874593	15	6·409217	65	10·069068
16	7·143892	66	10·896952	16	6·614716	66	10·089771
17	7·351389	67	10·918536	17	6·806842	67	10·109757
18	7·545725	68	10·939374	18	6·986783	68	10·129051
19	7·728036	69	10·959497	19	7·155590	69	10·147684
20	7·899334	70	10·978931	20	7·314199	70	10·165679
21	8·060518	71	10·997705	21	7·463444	71	10·183061
22	8·212394	72	11·015841	22	7·604070	72	10·199854
23	8·355684	73	11·033367	23	7·736745	73	10·216082
24	8·491040	74	11·050303	24	7·862076	74	10·231763
25	8·619048	75	11·066673	25	7·980601	75	10·246921
26	8·740240	76	11·082498	26	8·092815	76	10·261574
27	8·855095	77	11·097798	27	8·199163	77	10·275740
28	8·964053	78	11·112592	28	8·300050	78	10·289438
29	9·067512	79	11·126898	29	8·395845	79	10·302685
30	9·165836	80	11·140736	30	8·486886	80	10·315498
31	9·259359	81	11·154120	31	8·573482	81	10·327891
32	9·348386	82	11·167069	32	8·655914	82	10·339880
33	9·433197	83	11·179597	33	8·734443	83	10·351480
34	9·514050	84	11·191719	34	8·809307	84	10·362705
35	9·591183	85	11·203450	35	8·880726	85	10·373566
36	9·664816	86	11·214804	36	8·948904	86	10·384079
37	9·735151	87	11·225793	37	9·014030	87	10·394254
38	9·802377	88	11·236430	38	9·076276	88	10·404103
39	9·866667	89	11·246728	39	9·135804	89	10·413638
40	9·9:8183	90	11·256698	40	9·192763	90	10·422870
41	9·987078	91	11·266352	41	9·247295	91	10·431809
42	10·043489	92	11·275700	42	9·299528	92	10·440464
43	10·097547	93	11·284752	43	9·349582	93	10·448846
44	10·149376	94	11·293519	44	9·397571	94	10·456964
45	10·199088	95	11·302011	45	9·443601	95	10·464826
46	10·246791	96	11·310236	46	9·487770	96	10·472442
47	10·292583	97	11·318:05	47	9·530171	97	10·479820
48	10·336560	98	11·325923	48	9·570890	98	10·486967
49	10·378807	99	11·333401	49	9·610008	99	10·4938 1
50	10·419410	100	11·340647	50	9·647603	100	10·500601

TABLE X.   cxxxi

Present Value (or Years' Purchase) of £1 per Annum in n years, after t years' Deferrence. Redemption of Capital being at 3 per cent., with Interest allowed to a Purchaser at 8 per cent.

n Years	Deferred 3 Years	Years n	Deferred 3 Years	n Years	Deferred 4 Years	n Years	Deferred 4 Years
1	·735030	51	8·966426	1	·680583	51	8·302251
2	1·386338	52	8·998610	2	1·283647	52	8·332050
3	1·967217	53	9·029570	3	1·821498	53	8·360717
4	2·488290	54	9·059362	4	2·303974	54	8·388303
5	2·958146	55	9·088042	5	2·739025	55	8·414858
6	3·383804	56	9·115655	6	3·133154	56	8·440425
7	3·771059	57	9·142250	7	3·491723	57	8·465050
8	4·124737	58	9·167869	8	3·819203	58	8·488772
9	4·448887	59	9·192559	9	4·119342	59	8·511633
10	4·746933	60	9·216356	10	4·395310	60	8·533667
11	5·021792	61	9·239298	11	4·649809	61	8·554910
12	5·275961	62	9·261420	12	4·885151	62	8·575393
13	5·511591	63	9·282759	13	5·103328	63	8·595151
14	5·730549	64	9·303343	14	5·306067	64	8·614211
15	5·934457	65	9·323207	15	5·494870	65	8·632603
16	6·124734	66	9·342376	16	5·671053	66	8·650353
17	6·302628	67	9·360881	17	5·835770	67	8·667487
18	6·469240	68	9·378746	18	5·990040	68	8·684029
19	6·625543	69	9·395999	19	6·134765	69	8·700003
20	6·772403	70	9·412661	20	6·270747	70	8·715431
21	6·910593	71	9·428755	21	6·398700	71	8·730334
22	7·040802	72	9·444304	22	6·519264	72	8·744731
23	7·163650	73	9·459330	23	6·633012	73	8·758643
24	7·279696	74	9·473850	24	6·740463	74	8·772088
25	7·389442	75	9·487885	25	6·842079	75	8·785083
26	7·493344	76	9·501452	26	6·938285	76	8·797645
27	7·591814	77	9·514569	27	7·029461	77	8·809791
28	7·685227	78	9·527253	28	7·115955	78	8·821535
29	7·773927	79	9·539518	29	7·198084	79	8·832892
30	7·858224	80	9·551382	30	7·276137	80	8·843877
31	7·938405	81	9·562857	31	7·350379	81	8·854501
32	8·014731	82	9·573958	32	7·421051	82	8·864781
33	8·087443	83	9·584699	33	7·488377	83	8·874726
34	8·156761	84	9·595092	34	7·552560	84	8·884349
35	8·222891	85	9·605149	35	7·613791	85	8·893661
36	8·286018	86	9·614883	36	7·672243	86	8·902674
37	8·346319	87	9·624304	37	7·728077	87	8·911397
38	8·403955	88	9·633424	38	7·781444	88	8·919841
39	8·459073	89	9·642253	39	7·832479	89	8·928016
40	8·511814	90	9·650801	40	7·881313	90	8·935931
41	8·562306	91	9·659077	41	7·928065	91	8·943594
42	8·610670	92	9·667092	42	7·972846	92	8·951015
43	8·657016	93	9·674852	43	8·015760	93	8·958201
44	8·701451	94	9·682369	44	8·056903	94	8·965161
45	8·744071	95	9·689649	45	8·096366	95	8·971902
46	8·784968	96	9·696701	46	8·134234	96	8·978431
47	8·824228	97	9·703532	47	8·170585	97	8·984757
48	8·861931	98	9·710150	48	8·205496	98	8·990884
49	8·898151	99	9·716561	49	8·239033	99	8·996820
50	8·932961	100	9·722773	50	8·271265	100	9·002572

Present Value (or Years' Purchase) of £1 per Annum in **n** years, after
t years' Deferrence.   Redemption of Capital being at 3 per cent., with Interest
allowed to a Purchaser at 8 per cent.

n Years	Deferred 5 Years	n Years	Deferred 5 Years	n Years	Deferred 6 Years	n Years	Deferrrd 6 Years
1	·630169	51	7·687265	1	·583491	51	7·117845
2	1·188561	52	7·714857	2	1·100521	52	7·143393
3	1·686572	53	7·741401	3	1·561642	53	7·167970
4	2·133308	54	7·766943	4	1·975287	54	7·191620
5	2·536133	55	7·791531	5	2·348274	55	7·214387
6	2·901067	56	7·815205	6	2·686175	56	7·236307
7	3·233076	57	7·838006	7	2·993591	57	7·257419
8	3·536297	58	7·859970	8	3·274352	58	7·277756
9	3·814204	59	7·881138	9	3·531673	59	7·297357
10	4·069730	60	7·901540	10	3·768272	60	7·316247
11	4·099730	61	7·921209	11	3·986464	61	7·334459
12	4·523286	62	7·940175	12	4·188231	62	7·352020
13	4·725301	63	7·958469	13	4·375283	63	7·368960
14	4·913023	64	7·976118	14	4·549099	64	7·385301
15	5·087840	65	7·993147	15	4·710968	65	7·401069
16	5·250972	66	8·009582	16	4·862016	66	7·416286
17	5·403488	67	8·025447	17	5·003234	67	7·430976
18	5·546331	68	8·040764	18	5·135496	68	7·445158
19	5·680335	69	8·055555	19	5·259574	69	7·458853
20	5·806244	70	8·069839	20	5·376157	70	7·472080
21	5·924719	71	8·083638	21	5·485856	71	7·484857
22	6·036353	72	8·096969	22	5·589220	72	7·497200
23	6·141675	73	8·109851	23	5·686741	73	7·509128
24	6·241166	74	8·122299	24	5·778862	74	7·520654
25	6·335255	75	8·134332	25	5·865982	75	7·531796
26	6·424335	76	8·145964	26	5·948463	76	·7·542546
27	6·508757	77	8·157210	27	6·026632	77	7·552979
28	6·588844	78	8·168084	28	6·100787	78	7·563047
29	6·664889	79	8·178599	29	6·171199	79	7·572784
30	6·737161	80	8·188771	30	6·238117	80	7·582202
31	6·805903	81	8·198608	31	6·301767	81	7·591311
32	6·871340	82	8·208126	32	6·362358	82	7·600124
33	6·933679	83	8·217335 ·	33	6·420079	83	7·608650
34	6·993108	84	8·226245	34	6·475106	84	7·616900
35	7·049803	85	8·234867	35	6·527601	85	7·624884
36	7·103925	86	8·243212	36	6·577714	86	7·632611
37	7·155624	87	8·251290	37	6·625583	87	7·640090
38	7·205037	88	8·259108	38	6·671336	88	7·647329
39	7·252292	89	8·266678	39	6·715091	89	7·654338
40	7·297508	90	8·274006	40	6·756958	90	7·661124
41	7·340797	91	8·281102	41	6·797041	91	7·667694
42	7·382261	92	8·287973	42	6·835433	92	7·674056
43	7·421996	93	8·294626	43	6·872224	93	7·680216
44	7·460091	94	8·301071	44	6·907498	94	7·686183
45	7·496631	95	8·307312	45	6·941331	95	7·691963
46	7·531694	96	8·313358	46	6·973797	96	7·697561
47	7·565355	97	8·319215	47	7·004963	97	7·702984
48	7·597677	98	8·324889	48	7·034893	98	7·708237
49	7·628730	99	8·330385	49	7·063645	99	7·713326
50	7·658575	100	8·335711	50	7·091279	100	7·718258

TABLE X. cxxxiii

Present Value (or Years' Purchase) of £1 per Annum in **n** years, after t years' Deferrence. Redemption of Capital being at **3** per cent., with Interest allowed to a Purchaser at **8** per cent.

n Years	Deferred 7 Years	n Years	Deferred 7 Years	n Years	Deferred 8 Years	n Years	Deferred 8 Years
1	·540269	51	6·590588	1	·500249	51	6·102402
2	1·018999	52	6·614244	2	·943519	52	6·124306
3	1·445963	53	6·637001	3	1·338856	53	6·145377
4	1·828967	54	6·658899	4	1·693489	54	6·165653
5	2·174325	55	6·679980	5	2·013265	55	6·185172
6	2·487196	56	6·700276	6	2·302961	56	6·203965
7	2·771840	57	6·719824	7	2·566521	57	6·222065
8	3·031804	58	6·738655	8	2·807228	58	6·239501
9	3·270054	59	6·756803	9	3·027839	59	6·256305
10	3·489136	60	6·774294	10	3·230684	60	6·272500
11	3·691165	61	6·791157	11	3·417749	61	6·288114
12	3·877987	62	6·807417	12	3·590732	62	6·303170
13	4·051183	63	6·823102	13	3·751098	63	6·317693
14	4·212123	64	6·838233	14	3·900118	64	6·331703
15	4·362001	65	6·852833	15	4·038894	65	6·345221
16	4·501860	66	6·866923	16	4·168393	66	6·358268
17	4·632618	67	6·880525	17	4·289465	67	6·370862
18	4·755083	68	6·893656	18	4·402858	68	6·383021
19	4·869970	69	6·906337	19	4·509235	69	6·394762
20	4·977917	70	6·918584	20	4·609186	70	6·406102
21	5·079490	71	6·930414	21	4·703236	71	6·417056
22	5·175197	72	6·941843	22	4·791854	72	6·427638
23	5·265494	73	6·952887	23	4·875462	73	6·437864
24	5·350792	74	6·963560	24	4·954441	74	6·447746
25	5·431458	75	6·973876	25	5·029132	75	6·457298
26	5·507829	76	6·983848	26	5·099847	76	6·466532
27	5·580208	77	6·993490	27	5·166864	77	6·475459
28	5·648869	78	7·002812	28	5·230439	78	6·484091
29	5·714066	79	7·011828	29	5·290807	79	6·492439
30	5·776027	80	7·020548	30	5·348178	80	6·500513
31	5·834962	81	7·028983	31	5·402748	81	6·508323
32	5·891064	82	7·037143	32	5·454694	82	6·515879
33	5·944510	83	7·045037	33	5·504181	83	6·523188
34	5·995461	84	7·052676	34	5·551357	84	6·530262
35	6·044068	85	7·060069	35	5·596364	85	6·537106
36	6·090469	86	7·067223	36	5·639328	86	6·543731
37	6·134792	87	7·074148	37	5·680368	87	6·550143
38	6·177155	88	7·080851	38	5·719593	88	6·556350
39	6·217669	89	7·087341	39	5·757106	89	6·562358
40	6·256435	90	7·093624	40	5·793000	90	6·568176
41	6·293548	91	7·099707	41	5·827365	91	6·572809
42	6·329097	92	7·105598	42	5·860280	92	6·579254
43	6·363163	93	7·111302	43	5·891823	93	6·584545
44	6·395824	94	7·116828	44	5·922064	94	6·589661
45	6·427151	95	7·122179	45	5·951071	95	6·594616
46	6·457211	96	7·127362	46	5·978905	96	6·599415
47	6·486068	97	7·132383	47	6·005624	97	6·604065
48	6·513781	98	7·137247	48	6·031284	98	6·608568
49	6·540404	99	7·141960	49	6·055935	99	6·612931
50	6·565991	100	7·146526	50	6·079627	100	6·617160

Present Value (or Years' Purchase) of £1 per Annum in n years, after t years' Deferrence. Redemption of Capital being at 3 per cent., with Interest allowed to a Purchaser at 8 per cent.

n Years	Deferred 9 Years	n Years	Deferred 9 Years	n Years	Deferred 10 Years	n Years	Deferred 10 Years
1	·463194	51	5·650371	1	·428882	51	5·231820
2	·873628	52	5·670653	2	·808914	52	5·250598
3	1·239681	53	5·690163	3	1·147852	53	5·268663
4	1·568045	54	5·708937	4	1·451892	54	5·286047
5	1·864134	55	5·727010	5	1·726048	55	5·302781
6	2·132371	56	5·744411	6	1·974416	56	5·318893
7	2·376408	57	5·761170	7	2·200375	57	5·334411
8	2·599285	58	5·777315	8	2·406742	58	5·349359
9	2·803554	59	5·792874	9	2·595881	59	5·363766
10	2·991374	60	5·807870	10	2·769788	60	5·377651
11	3·164582	61	5·822327	11	2·930165	61	5·391038
12	3·324751	62	5·836268	12	3·078470	62	5·403945
13	3·473239	63	5·849715	13	3·215958	63	5·416396
14	3·611219	64	5·862687	14	3·343718	64	5·428408
15	3·739716	65	5·875204	15	3·462696	65	5·439998
16	3·859623	66	5·887284	16	3·573721	66	5·451183
17	3·971726	67	5·898945	17	3·677520	67	5·461980
18	4·076720	68	5·910203	18	3·774737	68	5·472404
19	4·175217	69	5·921075	19	3·865938	69	5·482471
20	4·267764	70	5·931575	20	3·951629	70	5·492193
21	4·354847	71	5·941718	21	4·032261	71	5·501584
22	4·436901	72	5·951516	22	4·108237	72	5·510657
23	4·514316	73	5·960985	23	4·179918	73	5·519424
24	4·587445	74	5·970135	24	4·247629	74	5·527896
25	4·656603	75	5·978979	25	4·311665	75	5·536086
26	4·722079	76	5·987529	26	4·372291	76	5·544002
27	4·784132	77	5·995795	27	4·429747	77	5·551656
28	4·842999	78	6·003787	28	4·484253	78	5·559056
29	4·898894	79	6·011517	29	4·536008	79	5·566213
30	4·952016	80	6·018993	30	4·585195	80	5·573135
31	5·002544	81	6·026224	31	4·631980	81	5·579831
32	5·050642	82	6·033220	32	4·676515	82	5·586309
33	5·096463	83	6·039988	33	4·718942	83	5·592576
34	5·140145	84	6·046538	34	4·759388	84	5·598640
35	5·181818	85	6·052875	35	4·797974	85	5·604508
36	5·221599	86	6·059009	36	4·834809	86	5·610188
37	5·259599	87	6·064946	37	4·869994	87	5·615685
38	5·295919	88	6·070693	38	4·903623	88	5·621006
39	5·330653	89	6·076257	39	4·935784	89	5·626158
40	5·363888	90	6·081644	40	4·966558	90	5·631145
41	5·395707	91	6·086859	41	4·996020	91	5·635974
42	5·426184	92	6·091910	42	5·024239	92	5·640651
43	5·455390	93	6·096800	43	5·051282	93	5·645179
44	5·483392	94	6·101537	44	5·077209	94	5·649565
45	5·510250	95	6·106125	45	5·102077	95	5·653813
46	5·536022	96	6·110569	46	5·125941	96	5·657928
47	5·560762	97	6·114874	47	5·148848	97	5·661914
48	5·584522	98	6·119044	48	5·170848	98	5·665775
49	5·607347	99	6·123084	49	5·191982	99	5·669516
50	5·629283	100	6·126999	50	5·212293	100	5·673141

TABLE X. CXXXV

Present Value (or Years' Purchase) of £1 per Annum in n years, after t years' Deferrence. Redemption of Capital being at **3** per cent., with Interest allowed to a Purchaser at **10** per cent.

n Years	Deferred 1 Year	n Years	Deferred 1 Year	n Years	Deferred 2 Years	n Years	Deferred 2 Years
1	·826446	51	8·376107	1	·751315	51	7·614643
2	1·534044	52	8·400615	2	1·394585	52	7·636923
3	2·146460	53	8·424161	3	1·951327	53	7·658329
4	2·681470	54	8·446792	4	2·437700	54	7·678902
5	3·152684	55	8·468549	5	2·866077	55	7·698681
6	3·570698	56	8·489475	6	3·246090	56	7·717693
7	3·943887	57	8·509606	7	3·585351	57	7·736005
8	4·278953	58	8·528979	8	3·889957	58	7·753617
9	4·581330	59	8·547628	9	4·164845	59	7·770571
10	4·855464	60	8·565585	10	4·414058	60	7·786896
11	5·105031	61	8·582881	11	4·640937	61	7·802619
12	5·333097	62	8·599543	12	4·848270	62	7·817767
13	5·542239	63	8·615602	13	5·038399	63	7·832365
14	5·734636	64	8·631079	14	5·213306	64	7·846436
15	5·912149	65	8·646002	15	5·374681	65	7·860002
16	6·076370	66	8·660392	16	5·523973	66	7·873084
17	6·228675	67	8·674273	17	5·662432	67	7·885702
18	6·370256	68	8·687664	18	5·791142	68	7·897876
19	6·502151	69	8·700585	19	5·911046	69	7·909623
20	6·625268	70	8·713056	20	6·022971	70	7·920960
21	6·740409	71	8·725095	21	6·127644	71	7·931905
22	6·848277	72	8·736719	22	6·225706	72	7·942471
23	6·949498	73	8·747943	23	6·317725	73	7·952675
24	7·044627	74	8·758784	24	6·404206	74	7·962531
25	7·134161	75	8·769256	25	6·485601	75	7·972051
26	7·218542	76	8·779374	26	6·562311	76	7·981249
27	7·298171	77	8·789150	27	6·634701	77	7·990137
28	7·373405	78	8·798599	28	6·703095	78	7·998726
29	7·444567	79	8·807731	29	6·767789	79	8·007029
30	7·511952	80	8·816560	30	6·829048	80	8·015055
31	7·575825	81	8·825096	31	6·887114	81	8·022815
32	7·636428	82	8·833350	32	6·942207	82	8·030318
33	7·693981	83	8·841332	33	6·994528	83	8·037575
34	7·748685	84	8·849053	34	7·044259	84	8·044593
35	7·800724	85	8·856521	35	7·091568	85	8·051383
36	7·850269	86	8·863746	36	7·136608	86	8·057951
37	7·897473	87	8·870736	37	7·179521	87	8·064306
38	7·942479	88	8·877500	38	7·220435	88	8·070455
39	7·985421	89	8·884046	39	7·259474	89	8·076405
40	8·026415	90	8·890381	40	7·296471	90	8·082165
41	8·065580	91	8·896513	41	7·332345	91	8·087739
42	8·103016	92	8·902449	42	7·366378	92	8·093136
43	8·138821	93	8·908196	43	7·398928	93	8·098360
44	8·173084	94	8·913759	44	7·430077	94	8·103418
45	8·205889	95	8·919146	45	7·459900	95	8·108315
46	8·237314	96	8·924363	46	7·488468	96	8·113057
47	8·267431	97	8·929415	47	7·515846	97	8·117650
48	8·296307	98	8·934308	48	7·542098	98	8·122098
49	8·324007	99	8·939047	49	7·567279	99	8·126406
50	8·350588	100	8·943637	50	7·591444	100	8·130579

Present Value (or Years' Purchase) of £1 per Annum in n years, after t years' Deferrence. Redemption of Capital being at 3 per cent., with Interest allowed to a Purchaser at 10 per cent.

n Years	Deferred 3 Years	n Years	Deferred 3 Years	n Years	Deferred 4 Years	n Years	Deferred 4 Years
1	·683013	51	6·922402	1	·620921	51	6·293093
2	1·267805	52	6·942657	2	1·152550	52	6·311507
3	1·773934	53	6·962117	3	1·612667	53	6·329197
4	2·216091	54	6·980820	4	2·014628	54	6·346200
5	2·605524	55	6·998801	5	2·368658	55	6·362547
6	2·950990	56	7·016095	6	2·682719	56	6·378268
7	3·259410	57	7·032732	7	2·963100	57	6·393393
8	3·536325	58	7·048743	8	3·214841	58	6·407948
9	3·786223	59	7·064155	9	3·442021	59	6·421959
10	4·012780	60	7·078996	10	3·647982	60	6·435451
11	4·219034	61	7·093290	11	3·835485	61	6·448446
12	4·407519	62	7·107061	12	4·006835	62	6·460964
13	4·580363	63	7·120332	13	4·163966	63	6·473029
14	4·739369	64	7·133124	14	4·308517	64	6·484658
15	4·886073	65	7·145456	15	4·441885	65	6·495869
16	5·021794	66	7·157349	16	4·565267	66	6·506681
17	5·147666	67	7·168820	17	4·679696	67	6·517109
18	5·264674	68	7·179887	18	4·786068	68	6·527170
19	5·373678	69	7·190566	19	4·885162	69	6·536879
20	5·475429	70	7·200873	20	4·977662	70	6·546248
21	5·570586	71	7·210823	21	5·064169	71	6·555293
22	5·659733	72	7·220429	22	5·145212	72	6·564026
23	5·743386	73	7·229705	23	5·221260	73	6·572459
24	5·822006	74	7·238664	24	5·292733	74	6·580604
25	5·896001	75	7·247319	25	5·360001	75	6·588472
26	5·965738	76	7·255681	26	5·423398	76	6·596073
27	6·031546	77	7·263761	27	5·483224	77	6·603419
28	6·093723	78	7·271569	28	5·539748	78	6·610518
29	6·152535	79	7·279117	29	5·593214	79	6·617379
30	6·208225	80	7·286413	30	5·643841	80	6·624012
31	6·261013	81	7·293468	31	5·691830	81	6·630425
32	6·311097	82	7·300289	32	5·737361	82	6·636627
33	6·358662	83	7·306886	33	5·780602	83	6·642624
34	6·403872	84	7·313267	34	5·821702	84	6·648424
35	6·446880	85	7·319439	35	5·860800	85	6·654035
36	6·487825	86	7·325410	36	5·898023	86	6·659463
37	6·526837	87	7·331187	37	5·933488	87	6·664715
38	6·564032	88	7·336777	38	5·967302	88	6·669797
39	6·599522	89	7·342187	39	5·999565	89	6·674715
40	6·633401	90	7·347422	40	6·030365	90	6·679475
41	6·665768	91	7·352490	41	6·059789	91	6·684082
42	6·696707	92	7·357596	42	6·087916	92	6·688542
43	6·726298	93	7·362145	43	6·114817	93	6·692859
44	6·754615	94	7·366743	44	6·140559	94	6·697039
45	6·781727	95	7·371196	45	6·165206	95	6·701087
46	6·807698	96	7·375507	46	6·188816	96	6·705006
47	6·832588	97	7·379682	47	6·211443	97	6·708802
48	6·856452	98	7·383726	48	6·233139	98	6·712478
49	6·879344	99	7·387642	49	6·253950	99	6·716038
50	6·901313	100	7·391436	50	6·273920	100	6·719487

Present Value (or Years' Purchase) of £1 per Annum in n years after t years' Deferrence. Redemption of Capital being at 3 per cent., with Interest allowed to a Purchaser at 10 per cent.

n Years	Deferred 5 Years	n Years	Deferred 5 Years	n Years	Deferred 6 Years	n Years	Deferred 6 Years
1	·564474	51	5·720994	1	·513158	51	5·200903
2	1·047772	52	5·737733	2	·952520	52	5·216121
3	1·466061	53	5·753816	3	1·332783	53	5·230741
4	1·831480	54	5·769272	4	1·664982	54	5·244793
5	2·153326	55	5·784133	5	1·957569	55	5·258303
6	2·438835	56	5·798425	6	2·217123	56	5·271296
7	2·693728	57	5·812175	7	2·448843	57	5·283796
8	2·922582	58	5·825407	8	2·656893	58	5·295825
9	3·129110	59	5·838145	9	2·844645	59	5·307404
10	3·316347	60	5·850410	10	3·014861	60	5·318555
11	3·486805	61	5·862223	11	3·169823	61	5·329294
12	3·642577	62	5·873604	12	3·311434	62	5·339640
13	3·785424	63	5·884572	13	3·441294	63	5·349611
14	3·916834	64	5·895143	14	3·560758	64	5·359221
15	4·038077	65	5·905336	15	3·670979	65	5·368487
16	4·150243	66	5·915164	16	3·772948	66	5·377422
17	4·254269	67	5·924645	17	3·867517	67	5·386041
18	4·350970	68	5·933791	18	3·955428	68	5·394356
19	4·441056	69	5·942617	19	4·037324	69	5·402379
20	4·525148	70	5·951135	20	4·113771	70	5·410123
21	4·603790	71	5·959357	21	4·185264	71	5·417598
22	4·677465	72	5·967296	22	4·252241	72	5·424815
23	4·746600	73	5·974963	23	4·315091	73	5·431784
24	4·811575	74	5·982367	24	4·374159	74	5·438516
25	4·872728	75	5·989520	25	4·429753	75	5·445018
26	4·930362	76	5·996430	26	4·482147	76	5·451300
27	4·984749	77	6·003108	27	4·531590	77	5·457371
28	5·036135	78	6·009561	28	4·578304	78	5·463238
29	5·084740	79	6·015799	29	4·622491	79	5·468908
30	5·130765	80	6·021829	30	4·664331	80	5·474390
31	5·174391	81	6·027659	31	4·703991	81	5·479690
32	5·215783	82	6·033297	32	4·741621	82	5·484815
33	5·255092	83	6·038749	33	4·777357	83	5·489772
34	5·292456	84	6·044022	34	4·811324	84	5·494566
35	5·328000	85	6·049123	35	4·843636	85	5·499203
36	5·361839	86	6·054058	36	4·874399	86	5·503689
37	5·394080	87	6·058832	37	4·903709	87	5·508029
38	5·424820	88	6·063452	38	4·931655	88	5·512229
39	5·454150	89	6·067923	39	4·958318	89	5·516294
40	5·482150	90	6·072250	40	4·983773	90	5·520227
41	5·508899	91	6·076438	41	5·008090	91	5·524035
42	5·534469	92	6·080493	42	5·031335	92	5·527721
43	5·558924	93	6·084417	43	5·053567	93	5·531289
44	5·582326	94	6·088218	44	5·074842	94	5·534743
45	5·604733	95	6·091897	45	5·095212	95	5·538088
46	5·626196	96	6·095460	46	5·114724	96	5·541327
47	5·646767	97	6·098911	47	5·133424	97	5·544464
48	5·666490	98	6·102252	48	5·151354	98	5·547502
49	5·685409	99	6·105489	49	·5·168553	99	5·550445
50	5·703564	100	6·108625	50	5·185058	100	5·553295

Present Value (or Years' Purchase) of £1 per Annum in n years, after t years' Deferrence. Redemption of Capital being at 3 per cent., with Interest allowed to a Purchaser at 12 per cent.

n Years	Deferred 1 Year	n Years	Deferred 1 Year	n Years	Deferred 2 Years	n Years	Deferred 2 Years
1	·797194	51	6·946476	1	·711780	51	6·202211
2	1·457462	52	6·963631	2	1·301306	52	6·217528
3	2·013069	53	6·980098	3	1·797383	53	6·232230
4	2·486880	54	6·995910	4	2·220428	54	6·246349
5	2·895553	55	7·011101	5	2·585316	55	6·259911
6	3·251512	56	7·025698	6	2·903136	56	6·272945
7	3·564210	57	7·039731	7	3·182330	57	6·285474
8	3·840966	58	7·053226	8	3·429434	58	6·297523
9	4·087540	59	7·066207	9	3·649589	59	6·309113
10	4·308522	60	7·078698	10	3·846894	60	6·320266
11	4·507616	61	7·090721	11	4·024657	61	6·331001
12	4·687847	62	7·102297	12	4·185578	62	6·341337
13	4·851705	63	7·113446	13	4·331880	63	6·351291
14	5·001263	64	7·124186	14	4·465413	64	6·360880
15	5·138256	65	7·134535	15	4·587728	65	6·370120
16	5·264151	66	7·144509	16	4·700135	66	6·379026
17	5·380196	67	7·154125	17	4·803746	67	6·387611
18	5·487458	68	7·163397	18	4·899516	68	6·395890
19	5·586856	69	7·172340	19	4·988264	69	6·403875
20	5·679185	70	7·180967	20	5·070701	70	6·411577
21	5·765138	71	7·189291	21	5·147445	71	6·419009
22	5·845318	72	7·197324	22	5·219034	72	6·426182
23	5·923829	73	7·205078	23	5·289133	73	6·433106
24	5·990423	74	7·212564	24	5·348592	74	6·439790
25	6·056230	75	7·219793	25	5·407349	75	6·446244
26	6·118046	76	7·226775	26	5·462541	76	6·452478
27	6·176198	77	7·233519	27	5·514462	77	6·458499
28	6·230979	78	7·240034	28	5·563374	78	6·464316
29	6·282653	79	7·246323	29	5·609511	79	6·469932
30	6·331455	80	7·252412	30	5·653085	80	6·475368
31	6·377600	81	7·258292	31	5·694286	81	6·480618
32	6·421279	82	7·263976	32	5·733285	82	6·485693
33	6·462668	83	7·269471	33	5·770239	83	6·490600
34	6·501925	84	7·274785	34	5·805290	84	6·495344
35	6·539196	85	7·279923	35	5·838568	85	6·499932
36	6·574611	86	7·284893	36	5·870189	86	6·504369
37	6·608293	87	7·289700	37	5·900261	87	6·508661
38	6·640350	88	7·294350	38	5·928884	88	6·512813
39	6·670886	89	7·298849	39	5·956149	89	6·516830
40	6·699994	90	7·303203	40	5·982138	90	6·520717
41	6·727759	91	7·307416	41	6·006928	91	6·524478
42	6·754262	92	7·311493	42	6·030591	92	6·528118
43	6·779574	93	7·315439	43	6·053191	93	6·531642
44	6·803766	94	7·319259	44	6·074791	94	6·535052
45	6·826899	95	7·322957	45	6·095445	95	6·538354
46	6·849032	96	7·326537	46	6·115207	96	6·541551
47	6·870219	97	7·330003	47	6·134124	97	6·544646
48	6·890512	98	7·333360	48	6·152242	98	6·547643
49	6·909956	99	7·336610	49	6·169604	99	6·550545
50	6·928597	100	7·339758	50	6·186248	100	6·553356

TABLE X.   cxli

Present Value (or Years' Purchase) of £1 per Annum in **n** years, after
t years' Deferrence. Redemption of Capital being at **3** per cent., with Interest
allowed to a Purchaser at **12** per cent.

n Years	Deferred 3 Years	n Years	Deferred 3 Years	n Years	Deferred 4 Years	n Years	Deferred 4 Years
1	·635518	51	5·537688	1	·567427	51	4·944365
2	1·161880	52	5·551364	2	1·037393	52	4·956575
3	1·604806	53	5·564491	3	1·432863	53	4·968296
4	1·982525	54	5·577097	4	1·770112	54	4·979551
5	2·308317	55	5·589206	5	2·060998	55	4·990363
6	2·592086	56	5·600844	6	2·314362	56	5·000753
7	2·841366	57	5·612031	7	2·536934	57	5·010742
8	·3·061995	58	5·622788	8	2·733924	58	5·020347
9	3·258562	59	5·633137	9	2·909430	59	5·029587
10	3·434727	60	5·643095	10	3·066721	60	5·038478
11	3·593444	61	5·652680	11	3·208432	61	5·047035
12	3·737123	62	5·661908	12	3·336717	62	5·055275
13	3·867750	63	5·670796	13	3·453348	63	5·063210
14	3·986976	64	5·679357	14	3·559800	64	5·070855
15	4·096186	65	5·687607	15	3·657309 ·	65	5·078221
16	4·196549	66	5·695559	16	3·746919	66	5·085320
17	4·289059	67	5·703224	17	3·829517	67	5·092165
18	4·374568	68	5·710616	18	3·905864	68	5·098764
19	4·453807	69	5·717745	19	3·976614	69	5·105130
20	4·527412	70	5·724623	20	4·042332	70	5·111270
21	4·595933	71	5·731258	21	4·103511	71	5·117195
22	4·659852	72	5·737663	22	4·160582	72	5·122913
23	4·722440	73	5·743844	23	4·216464	73	5·128432
24	4·775529	74	5·749812	24	4·263865	74	5·133761
25	4·827990	75	5·755575	25	4·310705	75	5·138906
26	4·877269	76	5·761141	26	4·354704	76	5·143876
27	4·923627	77	5·766517	27	4·396096	77	5·148676
28	4·967298	78	5·771711	28	4·435088	78	5·153313
29	5·008492	79	5·776725	29	4·471868	79	5·157790
30	5·047397	80	5·781579	30	4·506605	80	5·162124
31	5·084184	81	5·786266	31	4·539450	81	5·166309
32	5·119004	82	5·790797	32	4·570540	82	5·170355
33	5·151999	83	5·795178	33	4·599999	83	5·174266
34	5·183295	84	5·799414	34	4·627942	84	5·178048
35	5·213007	85	5·803510	35	4·654470	85	5·181706
36	5·241240	86	5·807472	36	4·679678	86	5·185243
37	5·268090	87	5·811304	37	4·703652	87	5·188664
38	5·293647	88	5·815011	38	4·726470	88	5·191974
39	5·317990	89	5·818598	39	4·748205	89	5·195177
40	5·341194	90	5·822069	40	4·768923	90	5·198275
41	5·363329	91	5·825427	41	4·788686	91	5·201274
42	5·384456	92	5·828677	42	4·807550	92	5·204176
43	5·404635	93	5·831823	43	4·825567	93	5·206985
44	5·423920	94	5·834868	44	4·842786	94	5·209704
45	5·442362	95	5·837816	45	4·859252	95	5·212336
46	5·460006	96	5·840670	46	4·875005	96	5·214884
47	5·476897	97	5·843434	47	4·890086	97	5·217352
48	5·493074	98	5·846110	48	4·904530	98	5·219741
49	5·508575	99	5·848701	49	4·918370	99	5·222054
50	5·523436	100	5·851211	50	4·931639	100	5·224295 .

Present Value (or Years' Purchase) of £1 per Annum in n years, after t years' Deferrence. Redemption of Capital being at 3 per cent., with Interest allowed to a Purchaser at 12 per cent.

n Years	Deferred 5 Years	n Years	Deferred 5 Years	n Years	Deferred 6 Years	n Years	Deferred 6 Years
1	·506631	51	4·414611	1	·452349	51	3·941617
2	·926244	52	4·425513	2	·827003	52	3·951351
3	1·279342	53	4·435978	3	1·142269	53	3·960695
4	1·580457	54	4·446028	4	1·411122	54	3·969667
5	1·840177	55	4·455681	5	1·643015	55	3·978287
6	2·066395	56	4·464958	6	1·844995	56	3·986570
7	2·265120	57	4·473876	7	2·022428	57	3·994532
8	2·441004	58	4·482453	8	2·179467	58	4·002190
9	2·597706	59	4·490702	9	2·319380	59	4·009556
10	2·738143	60	4·498641	10	2·444771	60	4·016643
11	2·864672	61	4·506282	11	2·557743	61	4·023466
12	2·979212	62	4·513638	12	2·660010	62	4·030034
13	3·083347	63	4·520724	13	2·752988	63	4·036360
14	3·178393	64	4·527549	14	2·837851	64	4·042454
15	3·265454	65	4·534126	15	2·915584	65	4·048327
16	3·345463	66	4·540465	16	2·987021	66	4·053986
17	3·419212	67	4·546576	17	3·052868	67	4·059442
18	3·487379	68	4·552468	18	3·113731	68	4·064704
19	3·550548	69	4·558151	19	3·170132	69	4·069778
20	3·609225	70	4·563634	20	3·222522	70	4·074673
21	3·663849	71	4·568924	21	3·271294	71	4·079397
22	3·714805	72	4·574030	22	3·316791	72	4·083955
23	3·764700	73	4·578957	23	3·361340	73	4·088355
24	3·807022	74	4·583715	24	3·399127	74	4·092603
25	3·848844	75	4·588309	25	3·436468	75	4·096705
26	3·888129	76	4·592746	26	3·471543	76	4·100666
27	3·925085	77	4·597032	27	3·504540	77	4·104493
28	3·959900	78	4·601172	28	3·535625	78	4·108190
29	3·992740	79	4·605170	29	3·564946	79	4·111759
30	4·023754	80	4·609039	30	3·592638	80	4·115214
31	4·053080	81	4·612776	31	3·618821	81	4·118550
32	4·080839	82	4·616388	32	3·643606	82	4·121775
33	4·107142	83	4·619881 ·	33	3·667091	83	4·124893
34	4·132091	84	4·623257	34	3·689367	84	4·127908
35	4·155777	85	4·626523	35	3·710515	85	4·130824
36	4·178284	86	4·629681	36	3·730611	86	4·133644
37	4·199689	87	4·632736	37	3·749723	87	4·136372
38	4·220063	88	4·635691	38	3·767913	88	4·139010
39	4·239469	89	4·638551	39	3·785240	89	4·141563
40	4·257967	90	4·641317	40	3·801757	90	4·144033
41	4·275613	91	4·643995	41	3·817511	91	4·146424
42	4·292455	92	4·646586	42	3·832549	92	4·148737
43	4·308542	93	4·649094	43	3·846913	93	4·150976
44	4·323916	94	4·651521	44	3·860639	94	4·153144
45	4·338618	95	4·653871	45	3·873766	95	4·155242
46	4·352683	96	4·656147	46	3·886324	96	4·157274
47	4·366148	97	4·658350	47	3·898347	97	4·159241
48	4·379045	98	4·660483	48	3·909861	98	4·161145
49	4·391402	99	4·662548	49	3·920895	99	4·162990
50	4·403249	100	4·664549	50	3·931472	100	4·164776

TABLE X. cxliii

Present Value (or Years' Purchase) of £1 per Annum in **n** years, after t years' Deferrence. Redemption of Capital being at 3 per cent., with Interest allowed to a Purchaser at 12 per cent.

n Years	Deferred 7 Years	n Years	Deferred 7 Years	n Years	Deferred 8 Years	n Years	Deferred 8 Years
1	·403883	51	3·519301	1	·360610	51	3·142233
2	·738396	52	3·527992	2	·659282	52	3·149993
3	1·019883	53	3·536335	3	·910610	53	3·157442
4	1·259931	54	3·544346	4	1·124938	54	3·164595
5	1·466978	55	3·552042	5	1·309801	55	3·171466
6	1·647317	56	3·559437	6	1·470819	56	3·178069
7	1·805740	57	3·566547	7	1·612267	57	3·184417
8	1·945953	58	3·573384	8	1·737458	58	3·190521
9	2·070875	59	3·579960	9	1·848996	59	3·196393
10	2·182831	60	3·586289	10	1·948956	60	3·202044
11	2·283699	61	3·592380	11	2·039017	61	3·207482
12	2·375009	62	3·598245	12	2·120544	62	3·212718
13	2·458025	63	3·603893	13	2·194665	63	3·217762
14	2·533795	64	3·609334	14	2·262317	64	3·222620
15	2·603200	65	3·614577	15	2·324286	65	3·227301
16	2·666983	66	3·619631	16	2·381235	66	3·231813
17	2·724775	67	3·624502	17	2·433727	67	3·236163
18	2·780017	68	3·629200	18	2·482247	68	3·240357
19	2·830475	69	3·633730	19	2·527210	69	3·244402
20	2·877252	70	3·638101	20	2·568975	70	3·248305
21	2·920798	71	3·642318	21	2·607856	71	3·252070
22	2·961420	72	3·646388	22	2·644125	72	3·255704
23	3·001196	73	3·650317	23	2·679639	73	3·259211
24	3·034935	74	3·654110	24	2·709763	74	3·262598
25	3·068275	75	3·657772	25	2·739531	75	3·265868
26	3·099592	76	3·661309	26	2·767493	76	3·269026
27	3·129054	77	3·664726	27	2·793798	77	3·272076
28	3·156808	78	3·668026	28	2·818579	78	3·275024
29	3·182987	79	3·671213	29	2·841953	79	3·277869
30	3·207712	80	3·674298	30	2·864029	80	3·280623
31	3·231091	81	3·677277	31	2·884902	81	3·283283
32	3·253220	82	3·680156	32	2·904661	82	3·285854
33	3·274189	83	3·682941	33	2·923383	83	3·288340
34	3·294078	84	3·685632	34	2·941141	84	3·290743
35	3·312960	85	3·688236	35	2·958000	85	3·293068
36	3·330903	86	3·690754	36	2·974020	86	3·295316
37	3·347967	87	3·693189	37	2·989256	87	3·297490
38	3·364208	88	3·695545	38	3·003757	88	3·299594
39	3·379679	89	3·697824	39	3·017570	89	3·301629
40	3·394426	90	3·700030	40	3·030737	90	3·303598
41	3·408492	91	3·702164	41	3·043297	91	3·305504
42	3·421919	92	3·704230	42	3·055285	92	3·307348
43	3·434743	93	3·706229	43	3·066735	93	3·309133
44	3·447000	94	3·708164	44	3·077678	94	3·310861
45	3·458719	95	3·710038	45	3·088142	95	3·312534
46	3·469933	96	3·711852	46	3·098154	96	3·314153
47	3·480667	97	3·713608	47	3·107738	97	3·315721
48	3·490948	98	3·715308	48	3·116917	98	3·317240
49	3·500799	99	3·716955	49	3·125713	99	3·318710
50	3·510243	100	3·718550	50	3·134146	100	3·320134

Present Value (or Years' Purchase) of £1 per Annum in n years after t years' Deferrence. Redemption of Capital being at 3 per cent., with Interest allowed to a Purchaser at 12 per cent.

n Years	Deferred 9 Years	n Years	Deferred 9 Years	n Years	Deferred 10 Years	n Years	Deferred 10 Years
1	·321973	51	2·805565	1	·287476	51	2·504969
2	·588645	52	2·812494	2	·525575	52	2·511155
3	·813045	53	2·819144	3	·725933	53	2·517093
4	1·004409	54	2·825531	4	·896794	54	2·522795
5	1·169465	55	2·831666	5	1·044166	55	2·528273
6	1·313231	56	2·837562	6	1·172528	56	2·533537
7	1·439524	57	2·843229	7	1·285290	57	2·538598
8	1·551302	58	2·848680	8	1·385091	58	2·543464
9	1·650889	59	2·853923	9	1·474008	59	2·548145
10	1·740140	60	2·858968	10	1·553696	60	2·552650
11	1·820551	61	2·863823	11	1·625492	61	2·556985
12	1·893343	62	2·868499	12	1·690485	62	2·561160
13	1·959522	63	2·873002	13	1·749574	63	2·565180
14	2·019926	64	2·877339	14	1·803506	64	2·569053
15	2·075255	65	2·881519	15	1·852907	65	2·572785
16	2·126102	66	2·885547	16	1·898306	66	2·576382
17	2·172971	67	2·889431	17	1·940153	67	2·579849
18	2·216292	68	2·893176	18	1·978832	68	2·583193
19	2·256437	69	2·896788	19	2·014676	69	2·586418
20	2·293728	70	2·900272	20	2·047971	70	2·589529
21	2·328443	71	2·903633	21	2·078967	71	2·592530
22	2·360826	72	2·906878	22	2·107880	72	2·595427
23	2·392535	73	2·910010	23	2·136192	73	2·598223
24	2·419432	74	2·913034	24	2·160206	74	2·600923
25	2·446010	75	2·915954	25	2·183937	75	2·603530
26	2·470976	76	2·918773	26	2·206229	76	2·606047
27	2·494463	77	2·921497	27	2·227199	77	2·608479
28	2·516588	78	2·924128	28	2·246954	78	2·610829
29	2·537458	79	2·926669	29	2·265588	79	2·613097
30	2·557169	80	2·929128	30	2·283186	80	2·615293
31	2·575806	81	2·931503	31	2·299826	81	2·617413
32	2·593447	82	2·933798	32	2·315578	82	2·619463
33	2·610163	83	2·936018	33	2·330503	83	2·621444
34	2·626019	84	2·938164	34	2·344659	84	2·623360
35	2·641071	85	2·940239	35	2·358100	85	2·625213
36	2·655375	86	2·942246	36	2·370871	86	2·627005
37	2·668979	87	2·944188	37	2·383017	87	2·628739
38	2·681926	88	2·946066	38	2·394577	88	2·630416
39	2·694259	89	2·947883	39	2·405589	89	2·632038
40	2·706015	90	2·949641	40	2·416085	90	2·633608
41	2·717229	91	2·951343	41	2·426097	91	2·635127
42	2·727933	92	2·952989	42	2·435654	92	2·636598
43	2·738156	93	2·954583	43	2·444783	93	2·638021
44	2·747927	94	2·956126	44	2·453506	94	2·639398
45	2·757270	95	2·957619	45	2·461848	95	2·640732
46	2·766209	96	2·959065	46	2·469829	96	3·642023
47	2·774766	97	2·960465	47	2·477470	97	2·643273
48	2·782962	98	2·961821	48	2·484788	98	2·644483
49	2·790816	99	2·963134	49	2·491800	99	2·645655
50	2·798344	100	2·964405	50	2·498522	100	2·646791

TABLE X.   cxlv

Present Value (or Years' Purchase) of £1 per Annum in n years, after t years' Deferrence. Redemption of Capital being at 3 per cent., with Interest allowed to a Purchaser at 15 per cent.

n Years	Deferred 1 Year	n Years	Deferred 1 Year	n Years	Deferred 2 Years	n Years	Deferred 2 Years
1	·756144	51	5·485046	1	·657516	51	4·769605
2	1·353176	52	5·496023	2	1·176674	52	4·779150
3	1·836345	53	5·506550	3	1·596822	53	4·788304
4	2·235231	54	5·516651	4	1·943679	54	4·797087
5	2·569982	55	5·526345	5	2·234767	55	4·805517
6	2·854801	56	5·535654	6	2·482436	56	4·813612
7	3·099984	57	5·544596	7	2·695638	57	4·821387
8	3·313180	58	5·553188	8	2·881026	58	4·828859
9	3·500188	59	5·561448	9	3·043642	59	4·836041
10	3·665487	60	5·569390	10	3·187380	60	4·842947
11	3·812588	61	5·577029	11	3·315293	61	4·849590
12	3·944285	62	5·584380	12	3·429813	62	4·855982
13	4·062828	63	5·591455	13	3·532894	63	4·862134
14	4·170050	64	5·598266	14	3·626130	64	4·868057
15	4·267458	65	5·604826	15	3·710833	65	4·873761
16	4·356303	66	5·611145	16	3·788089	66	4·879256
17	4·437632	67	5·617233	17	3·858810	67	4·884550
18	4·512330	68	5·623101	18	3·923765	68	4·889653
19	4·581147	69	5·628758	19	3·983606	69	4·894572
20	4·644724	70	5·634212	20	4·038890	70	4·899315
21	4·703613	71	5·639473	21	4·090098	71	4·903889
22	4·758291	72	5·644547	22	4·137644	72	4·908301
23	4·809173	73	5·649443	23	4·181889	73	4·912559
24	4·856620	74	5·654168	24	4·223148	74	4·916667
25	4·900951	75	5·658729	25	4·261696	75	4·920633
26	4·942445	76	5·663131	26	4·297778	76	4·924462
27	4·981349	77	5·667383	27	4·331608	77	4·928158
28	5·017884	78	5·671488	28	4·363377	78	4·931729
29	5·052244	79	5·675454	29	4·393255	79	4·935177
30	5·084605	80	5·679285	30	4·421395	80	4·938509
31	5·115122	81	5·682987	31	4·447932	81	4·941728
32	5·143938	82	5·686564	32	4·472989	82	4·944838
33	5·171178	83	5·690022	33	4·496676	83	4·947845
34	5·196959	84	5·693364	34	4·519094	84	4·950751
35	5·221382	85	5·696595	35	4·540332	85	4·953560
36	5·244544	86	5·699719	36	4·560473	86	4·956277
37	5·266530	87	5·702740	37	4·579591	87	4·958904
38	5·287418	88	5·705662	38	4·597755	88	4·961445
39	5·307281	89	5·708488	39	4·615026	89	4·963902
40	5·326183	90	5·711222	40	4·631463	90	4·966280
41	5·344186	91	5·713867	41	4·647117	91	4·968580
42	5·361344	92	5·716426	42	4·662038	92	4·970805
43	5·377708	93	5·718903	43	4·676268	93	4·972959
44	5·393327	94	5·721300	44	4·689849	94	4·975043
45	5·408243	95	5·723619	45	4·702820	95	4·977060
46	5·422497	96	5·725865	46	4·715214	96	4·979012
47	5·436125	97	5·728039	47	4·727065	97	4·980903
48	5·449163	98	5·730143	48	4·738402	98	4·982733
49	5·461643	99	5·732181	49	4·749254	99	4·984505
50	5·473595	100	5·734154	50	4·759647	100	4·986220

t

Present Value (or Years' Purchase) of £1 per Annum in n years, after
t years' Deferrence. Redemption of Capital being at 3 per cent., with Interest
allowed to a Purchaser at 15 per cent.

n Years	Deferred 3 Years	n Years	Deferred 3 Years	n Years	Deferred 4 Years	n Years	Deferred 4 Years
1	˙571753	51	4˙147482	1	˙497177	51	3˙606506
2	1˙023195	52	4˙155783	2	˙889735	52	3˙613724
3	1˙388541	53	4˙163743	3	1˙207427	53	3˙620646
4	1˙690156	54	4˙171380	4	1˙469701	54	3˙627287
5	1˙943276	55	4˙178711	5	1˙689805	55	3˙633661
6	2˙158640	56	4˙185749	6	1˙877078	56	3˙639782
7	2˙344033	57	4˙192511	7	2˙038290	57	3˙645661
8	2˙505240	58	4˙199008	8	2˙178470	58	3˙651311
9	2˙646645	59	4˙205253	9	2˙301430	59	3˙656742
10	2˙771634	60	4˙211258	10	2˙410117	60	3˙661964
11	2˙882864	61	4˙217035	11	2˙506838	61	3˙666987
12	2˙982446	62	4˙222593	12	2˙593431	62	3˙671820
13	3˙072082	63	4˙227943	13	2˙671375	63	3˙676472
14	3˙153157	64	4˙233093	14	2˙741875	64	3˙680951
15	3˙226811	65	4˙238053	15	2˙805923	65	3˙685264
16	3˙293990	66	4˙242831	16	2˙864340	66	3˙689419
17	3˙355487	67	4˙247435	17	2˙917815	67	3˙693422
18	3˙411970	68	4˙251872	18	2˙966930	68	3˙697280
19	3˙464005	69	4˙256149	19	3˙012178	69	3˙700999
20	3˙512078	70	4˙260273	20	3˙053981	70	3˙704586
21	3˙556607	71	4˙264251	21	3˙092702	71	3˙708044
22	3˙597951	72	4˙268088	22	3˙128653	72	3˙711380
23	3˙636425	73	4˙271790	23	3˙162109	73	3˙714600
24	3˙672302	74	4˙275363	24	3˙193306	74	3˙717707
25	3˙705823	75	4˙278811	25	3˙222455	75	3˙720706
26	3˙737198	76	4˙282141	26	3˙249737	76	3˙723601
27	3˙766615	77	4˙285355	27	3˙275318	77	3˙726396
28	3˙794241	78	4˙288460	28	3˙299340	78	3˙729095
29	3˙820222	79	4˙291458	29	3˙321932	79	3˙731703
30	3˙844691	80	4˙294355	30	3˙343210	80	3˙734222
31	3˙867767	81	4˙297154	31	3˙363276	81	3˙736656
32	3˙889556	82	4˙299859	32	3˙382222	82	3˙739008
33	3˙910153	83	4˙302474	33	3˙400133	83	3˙741281
34	3˙929647	84	4˙305001	34	3˙417084	84	3˙743479
35	3˙948115	85	4˙307444	35	3˙433143	85	3˙745603
36	3˙965628	86	4˙309806	36	3˙448373	86	3˙747657
37	3˙982253	87	4˙312090	37	3˙462829	87	3˙749644
38	3˙998047	88	4˙314300	38	3˙476563	88	3˙751565
39	4˙013066	89	4˙316437	39	3˙489623	89	3˙753423
40	4˙027359	90	4˙318504	40	3˙502051	90	3˙755221
41	4˙040972	91	4˙320504	41	3˙513888	91	3˙756960
42	4˙053946	92	4˙322439	42	3˙525170	92	3˙758643
43	4˙066320	93	4˙324312	43	3˙535930	93	3˙760271
44	4˙078130	94	4˙326124	44	3˙546200	94	3˙761847
45	4˙089408	95	4˙327878	45	3˙556007	95	3˙763372
46	4˙100186	96	4˙329576	46	3˙565379	96	3˙764849
47	4˙110491	97	4˙331220	47	3˙574340	97	3˙766278
48	4˙120350	98	4˙332811	48	3˙582913	98	3˙767662
49	4˙129786	99	4˙334352	49	3˙591119	99	3˙769002
50	4˙138823	100	4˙335844	50	3˙598977	100	3˙770299

TABLE X. cxlvii

Present Value (or Years' Purchase) of £1 per Annum in n years, after t years' Deferrence. Redemption of Capital being at 3 per cent., with Interest allowed to a Purchaser at 15 per cent.

n Years	Deferred 5 Years	n Years	Deferred 5 Years	n Years	Deferred 6 Years	n Years	Deferred 6 Years
1	·432328	51	3·136093	1	·375937	51	2·727037
2	·773682	52	3·142369	2	·672767	52	2·732495
3	1·049936	53	3·148388	3	·912988	53	2·737729
4	1·278000	54	3·154163	4	1·111305	54	2·742750
5	1·469396	55	3·159706	5	1·277735	55	2·747570
6	1·632242	56	3·165028	6	1·419341	56	2·752198
7	1·772426	57	3·170140	7	1·541240	57	2·756644
8	1·894321	58	3·175053	8	1·647236	58	2·760916
9	2·001244	59	3·179775	9	1·740212	59	2·765022
10	2·095754	60	3·184316	10	1·822395	60	2·768971
11	2·179859	61	3·188684	11	1·895530	61	2·772769
12	2·255158	62	3·192887	12	1·961007	62	2·776423
13	2·322935	63	3·196932	13	2·019944	63	2·779941
14	2·384240	64	3·200827	14	2·073252	64	2·783328
15	2·439933	65	3·204577	15	2·121681	65	2·786589
16	2·490730	66	3·208190	16	2·165852	66	2·789730
17	2·537231	67	3·211671	17	2·206287	67	2·792758
18	2·579939	68	3·215026	18	2·243425	68	2·795675
19	2·619285	69	3·218260	19	2·277640	69	2·798487
20	2·655636	70	3·221379	20	2·309249	70	·2·801199
21	2·689306	71	3·224386	21	2·338527	71	2·803814
22	2·720568	72	3·227288	22	2·365711	72	2·806337
23	2·749660	73	3·230087	23	2·391009	73	2·808771
24	2·776788	74	3·232789	24	2·414598	74	2·811121
25	2·802134	75	3·235396	25	2·436639	75	2·813388
26	2·825859	76	3·237914	26	2·457268	76	2·815577
27	2·848102	77	3·240344	27	2·476611	77	2·817691
28	2·868991	78	3·242692	28	2·494775	78	2·819732
29	2·888637	79	3·244959	29	2·511858	79	2·821704
30	2·907139	80	3·247150	30	2·527947	80	2·823608
31	2·924587	81	3·249266	31	2·543120	81	2·825449
32	2·941063	82	3·251311	32	2·557446	82	2·827227
33	2·956638	83	3·253288	33	2·570989	83	2·828946
34	2·971378	84	3·255199	34	2·583807	84	2·830608
35	2·985342	85	3·257046	35	2·595950	85	2·832214
36	2·998585	86	3·258833	36	2·607465	86	2·833767
37	3·011155	87	3·260560	37	2·618396	87	2·835269
38	3·023098	88	3·262230	38	2·628781	88	2·836722
39	3·034455	89	3·263846	39	2·638656	89	2·838127
40	3·045262	90	3·265409	40	2·648054	90	2·839486
41	3·055555	91	3·266922	41	2·657004	91	2·840802
42	3·065365	92	3·268385	42	2·665535	92	2·842074
43	3·074722	93	3·269801	43	2·673671	93	2·843305
44	3·083652	94	3·271171	44	2·681437	94	2·844497
45	3·092180	95	3·272498	45	2·688852	95	2·845650
46	3·100330	96	3·273782	46	2·695939	96	2·846767
47	3·108122	97	3·275024	47	2·702715	97	2·847847
48	3·115576	98	3·276228	48	2·709197	98	2·848894
49	3·122712	99	3·277393	49	2·715402	99	2·849907
50	3·129545	100	3·278521	50	2·721344	100	2·850888

Present Value (or Years' Purchase) of £1 per Annum in n years, after t years' Deferrence.   Redemption of Capital being at 3 per cent., with Interest allowed to a Purchaser at 15 per cent.

n Years	Deferred 7 Years	n Years	Deferred 7 Years	n Years	Deferred 8 Years	n Years	Deferred 8 Years
1	·326902	51	2·371336	1	·284262	51	2·062032
2	·585015	52	2·376082	2	·508709	52	2·066158
3	·793903	53	2·380634	3	·690350	53	2·070116
4	·966352	54	2·385000	4	·840306	54	2·073913
5	1·111075	55	2·389191	5	·966151	55	2·077558
6	1·234209	56	2·393216	6	1·073225	56	2·081057
7	1·340209	57	2·397082	7	1·165399	57	2·084419
8	1·432379	58	2·400796	8	1·245547	58	2·087649
9	1·513228	59	2·404367	9	1·315850	59	2·090754
10	1·584691	60	2·407801	10	1·377992	60	2·093740
11	1·648287	61	2·411106	11	1·433293	61	2·096612
12	1·705223	62	2·414281	12	1·482803	62	2·099375
13	1·756473	63	2·417340	13	1·527368	63	2·102035
14	1·802828	64	2·420285	14	1·567676	64	2·104595
15	1·844940	65	2·423121	15	1·604295	65	2·107062
16	1·883350	66	2·425853	16	1·637695	66	2·109437
17	1·918511	67	2·428485	17	1·668270	67	2·111726
18	1·950805	68	2·431022	18	1·696352	68	2·113932
19	1·980556	69	2·433467	19	1·722223	69	2·116058
20	2·008042	70	2·435825	20	1·746124	70	2·118109
21	2·033502	71	2·438099	21	1·768262	71	2·120086
22	2·057140	72	2·440293	22	1·788818	72	2·121994
23	2·079138	73	2·442410	23	1·807946	73	2·123835
24	2·099651	74	2·444453	24	1·825783	74	2·125611
25	2·118816	75	2·446424	25	1·842449	75	2·127325
26	2·136755	76	2·448328	26	1·858048	76	2·128981
27	2·153575	77	2·450166	27	1·872674	77	2·130579
28	2·169369	78	2·451941	28	1·886408	78	2·132122
29	2·184224	79	2·453655	29	1·899325	79	2·133613
30	2·198215	80	2·455312	30	1·911491	80	2·135054
31	2·211408	81	2·456912	31	1·922964	81	2·136445
32	2·223866	82	2·458458	32	1·933797	82	2·137790
33	2·235643	83	2·459953	33	1·944037	83	2·139090
34	2·246788	84	2·461398	34	1·953729	84	2·140346
35	2·257347	85	2·462795	35	1·962911	85	2·141561
36	2·267361	86	2·464146	36	1·971618	86	2·142735
37	2·276866	87	2·465452	37	1·979883	87	2·143871
38	2·285897	88	2·466715	38	1·987736	88	2·144969
39	2·294484	89	2·467937	39	1·995203	89	2·146032
40	2·302656	90	2·469119	40	2·002309	90	2·147060
41	2·310439	91	2·470262	41	2·009077	91	2·148054
42	2·317857	92	2·471369	42	2·015527	92	2·149016
43	2·324932	93	2·472439	43	2·021680	93	2·149947
44	2·331684	94	2·473475	44	2·027551	94	2·150848
45	2·338132	95	2·474478	45	2·033159	95	2·151720
46	2·344295	96	2·475449	46	2·038517	96	2·152564
47	2·350187	97	2·476389	47	2·043641	97	2·153382
48	2·355823	98	2·477299	48	2·048542	98	2·154173
49	2·361219	99	2·478180	49	2·053234	99	2·154939
50	2·366386	100	2·479033	50	2·057727	100	2·155681

Present Value (or Years' Purchase) of £1 per Annum in **n** years, after t years' Deferrence. Redemption of Capital being at **3** per cent., with Interest allowed to a Purchaser at **15** per cent.

n Years	Deferred 9 Years	n Years	Deferred 9 Years	n Years	Deferred 10 Years	n Years	Deferred 10 Years
1	·247185	51	1·793071	1	·214943	51	1·559192
2	·442355	52	1·796660	2	·384657	52	1·562313
3	·600305	53	1·800101	3	·522004	53	1·565305
4	·730701	54	1·803403	4	·635392	54	1·568176
5	·840132	55	1·806572	5	·730549	55	1·570932
6	·933239	56	1·809615	6	·811513	56	1·573578
7	1·013390	57	1·812538	7	·881209	57	1·576120
8	1·083084	58	1·815347	8	·941813	58	1·578563
9	1·144218	59	1·818046	9	·994972	59	1·580910
10	1·198254	60	1·820643	10	1·041960	60	1·583168
11	1·246342	61	1·823141	11	1·083775	61	1·585340
12	1·289394	62	1·825543	12	1·121212	62	1·587429
13	1·328146	63	1·827856	13	1·154909	63	1·589440
14	1·363197	64	1·830083	14	1·185388	64	1·591377
15	1·395039	65	1·832227	15	1·213078	65	1·593241
16	1·424083	66	1·834293	16	1·238333	66	1·595037
17	1·450670	67	1·836283	17	1·261452	67	1·596768
18	1·475089	68	1·838202	18	1·282686	68	1·598436
19	1·497585	69	1·840051	19	1·302248	69	1·600044
20	1·518368	70	1·841834	20	1·320320	70	1·601595
21	1·537619	71	1·843553	21	1·337060	71	1·603090
22	1·555494	72	1·845212	22	1·352603	72	1·604532
23	1·572128	73	1·846813	23	1·367067	73	1·605924
24	1·587638	74	1·848357	24	1·380554	74	1·607267
25	1·602129	75	1·849848	25	1·393156	75	1·608564
26	1·615694	76	1·851288	26	1·404951	76	1·609815
27	1·628412	77	1·852677	27	1·416010	77	1·611024
28	1·640355	78	1·854019	28	1·426396	78	1·612191
29	1·651587	79	1·855316	29	1·436163	79	1·613318
30	1·662166	80	1·856568	30	1·445362	80	1·614407
31	1·672142	81	1·857778	31	1·454037	81	1·615460
32	1·681562	82	1·858948	32	1·462228	82	1·616476
33	1·690467	83	1·860078	33	1·469972	83	1·617459
34	1·698895	84	1·861171	34	1·477300	84	1·618409
35	1·706879	85	1·862227	35	1·484243	85	1·619328
36	1·714451	86	1·863248	36	1·490827	86	1·620216
37	1·721638	87	1·864236	37	1·497076	87	1·621075
38	1·728466	88	1·865191	38	1·503014	88	1·621905
39	1·734959	89	1·866115	39	1·508660	89	1·622708
40	1·741138	90	1·867008	40	1·514034	90	1·623486
41	1·747024	91	1·867873	41	1·519151	91	1·624237
42	1·752633	92	1·868710	42	1·524028	92	1·624965
43	1·757983	93	1·869519	43	1·528680	93	1·625669
44	1·763088	94	1·870303	44	1·533120	94	1·626350
45	1·767964	95	1·871061	45	1·537360	95	1·627010
46	1·772624	96	1·871795	46	1·541412	96	1·627648
47	1·777079	97	1·872506	47	1·545286	97	1·628266
48	1·781341	98	1·873194	48	1·548992	98	1·628864
49	1·785421	99	1·873860	49	1·552540	99	1·629443
50	1·789328	100	1·874505	50	1·555937	100	1·630004

Present Value (or Years' Purchase) of £1 per Annum in n years, after t years' Deferrence. Redemption of Capital being at 3 per cent., with Interest allowed to a Purchaser at 18 per cent.

n Years	Deferred 1 Year	n Years	Deferred 1 Year	n Years	Deferred 2 Years	n Years	Deferred 2 Year
1	·718184	51	4·494990	1	·608631	51	3·809313
2	1·259952	52	4·502552	2	1·067756	52	3·815722
3	1·683032	53	4·509799	3	1·426298	53	3·821864
4	2·022441	54	4·516749	4	1·713933	54	3·827753
5	2·300657	55	4·523415	5	1·949710	55	3·833403
6	2·532767	56	4·529813	6	2·146413	56	3·838824
7	2·729276	57	4·535955	7	2·312946	57	3·844030
8	2·897723	58	4·541854	8	2·455698	58	3·849029
9	3·043659	59	4·547522	9	2·579372	59	3·853832
10	3·171261	60	4·552969	10	2·687509	60	3·858448
11	3·283734	61	4·558207	11	2·782825	61	3·862887
12	3·383577	62	4·563244	12	2·867438	62	3·867156
13	3·472777	63	4·568091	13	2·943022	63	3·871263
14	3·552889	64	4·572755	14	3·010923	64	3·875216
15	3·625230	65	4·577244	15	3·072229	65	3·879021
16	3·690843	66	4·581568 ·	16	3·127833	66	3·882685
17	3·750600	67	4·585732	17	3·178475	67	3·886214
18	3·805229	68	4·589744	18	3·224770	68	3·889614
19	3·855342	69	4·593610	19	3·267239	69	3·892890
20	3·901456	70	4·597337	20	3·306319	70	3·896048
21	3·944016	71	4·600930	21	3·342386	71	3·899094
22	3·983398	72	4·604396	22	3·375761	72	3·902030
23	4·019932	73	4·607738	23	3·406722	73	3·904863
24	4·053901	74	4·610963	24	3·435510	74	3·907595
25	4·085553	75	4·614074	25	3·462333	75	3·910232
26	4·115105	76	4·617078	26	3·487377	76	3·912778
27	4·142746	77	4·619977	27	3·510802	77	3·915235
28	4·168647	78	4·622776	28	3·532751	78	3·917607
29	4·192955	79	4·625479	29	3·553351	79	3·919898
30	4·215803	80	4·628090	30	3·572715	80	3·922110
31	4·237311	81	4·630612	31	3·590941	81	3·924247
32	4·257584	82	4·633049	32	3·608122	82	3·926313
33	4·276717	83	4·635403	33	3·624336	83	3·928308
34	4·294796	84	4·637679	34	3·639658·	84	3·930237 ·
35	4·311899	85	4·639879	35	3·654152	85	3·932101
36	4·328096	86	4·642005	36	3·667878	86	3·933903
37	4·343450	87	4·644061	37	3·680890	87	3·935645
38	4·358020	88	4·646049	38	3·693237	88	3·937330
39	4·371857	89	4·647972	39	3·704964	89	3·938959
40	4·385011	90	4·649831	40	3·716111	90	3·940535
41	4·397525	91	4·651630	41	3·726716	91	3·942060
42	4·409440	92	4·653371	42	3·736813	92	3·943534
43	4·420793	93	4·655054	43	3·746435	93	3·944961
44	4·431618	94	4·656684	44	3·755608	94	3·946342
45	4·441947	95	4·658260	45	3·764362	95	3·947678
46	4·451809	96	4·659786	46	3·772719	96	3·948972
47	4·461231	97	4·661264	47	3·780704	97	3·950223
48	4·470237	98	4·662694	48	3·788337	98	3·951435
49	4·478852	99	4·664078	49	3·795637.	99	3·952608
50	4·487096	100	4·665418	50	3·803624	100	3·953744

TABLE X. cli

Present Value (or Years' Purchase) of £1 per Annum in n years, after
t years' Deferrence. Redemption of Capital being at 3 per cent., with Interest
allowed to a Purchaser at 18 per cent.

n Years	Deferred 3 Years	n Years	Deferred 3 Years	n Years	Deferred 4 Years	n Years	Deferred 4 Years
1	·515789	51	3·228232	1	·437109	51	2·735790
2	·904878	52	3·233663	2	·766846	52	2·740392
3	1·208727	53	3·238868	3	1·024345	53	2·744803
4	1·452486	54	3·243859	4	1·230920	54	2·749033
5	1·652296	55	3·248646	5	1·400251	55	2·753090
6	1·818994	56	3·253241	6	1·541520	56	2·756984
7	1·961024	57	3·257652	7	1·661122	57	2·760722
8	2·081100	58	3·261889	8	1·763644	58	2·764313
9	2·185908	59	3·265959	9	1·852465	59	2·767762
10	2·277550	60	3·269872	10	1·930127	60	2·771078
11	2·358326	61	3·273633	11	1·998582	61	2·774265
12	2·430032	62	3·277251	12	2·059349	62	2·777331
13	2·494087	63	3·280731	13	2·113633	63	2·780281
14	2·551630	64	3·284081	14	2·162398	64	2·783120
15	2·603584	65	3·287306	15	2·206427	65	2·785852
16	2·650706	66	3·290411	16	2·246361	66	2·788484
17	2·693623	67	3·293401	17	2·282731	67	2·791018
18	2·732856	68	3·296283	18	2·315980	68	2·793460
19	2·768846	69	3·299059	19	2·346480	69	2·795813
20	2·801965	70	3·301736	20	2·374547	70	2·798081
21	2·832531	71	3·304317	21	2·400450	71	2·800268
22	2·860815	72	3·306805	22	2·424419	72	2·802377
23	2·887053	73	3·309206	23	2·446655	73	2·804412
24	2·911449	74	3·311522	24	2·467330	74	2·806374
25	2·934181	75	3·313756	25	2·486594	75	2·808268
26	2·955404	76	3·315913	26	2·504580	76	2·810096
27	2·975256	77	3·317995	27	2·521403	77	2·811861
28	2·993857	78	3·320006	28	2·537167	78	2·813564
29	3·011315	79	3·321947	29	2·551962	79	2·815209
30	3·027724	80	3·323822	30	2·565868	80	2·816798
31	3·043171	81	3·325633	31	2·578958	81	2·818333
32	3·057730	82	3·327383	32	2·591297	82	2·819817
33	3·071471	83	3·329075	33	2·602942	83	2·821250
34	3·084456	84	3·330709	34	2·613946	84	2·822635
35	3·096739	85	3·332289	35	2·624355	85	2·823973
36	3·108371	86	3·333816	36	2·634213	86	2·825268
37	3·119398	87	3·335292	37	2·643558	87	2·826519
38	3·129862	88	3·336720	38	2·652425	88	2·827729
39	3·139800	89	3·338101	39	2·660847	89	2·828899
40	3·149247	90	3·339437	40	2·668853	90	2·830031
41	3·158234	91	3·340728	41	2·676469	91	2·831126
42	3·166791	92	3·341978	42	2·683721	92	2·832185
43	3·174945	93	3·343188	43	2·690631	93	2·833210
44	3·182719	94	3·344358	44	2·697220	94	2·834201
45	3·190137	95	3·345490	45	2·703506	95	2·835161
46	3·197220	96	3·346586	46	2·709508	96	2·836090
47	3·203986	97	3·347647	47	2·715243	97	2·836989
48	3·210455	98	3·348674	48	2·720724	98	2·837859
49	3·216642	99	3·349668	49	2·725968	99	2·838702
50	3·222563	100	3·350631	50	2·730985	100	2·839517

Present Value (or Years' Purchase) of £1 per Annum in n years, after t years' Deferrence. Redemption of Capital being at 3 per cent., with Interest allowed to a Purchaser at 18 per cent.

n Years	Deferred 5 Years	n Years	Deferred 5 Years	n Years	Deferred 6 Years	n Years	Deferred 6 Years
1	·370432	51	2·318466	1	·313925	51	1·964802
2	·649869	52	2·322366	2	·550737	52	1·968107
3	·868089	53	2·326104	3	·735669	53	1·971275
4	1·043153	54	2·329689	4	·884028	54	1·974313
5	1·186653	55	2·333127	5	1·005639	55	1·977226
6	1·306373	56	2·336427	6	1·107096	56	1·980023
7	1·407730	57	2·339595	7	1·192992	57	1·982708
8	1·494613	58	2·342638	8	1·266621	58	1·985286
9	1·569885	59	2·345561	9	1·330411	59	1·987764
10	1·635701	60	2·348371	10	1·386187	60	1·990145
11	1·693713	61	2·351072	11	1·435350	61	1·992434
12	1·745211	62	2·299824	12	1·478992	62	1·994636
13	1·791214	63	2·356170	13	1·517978	63	1·996754
14	1·832541	64	2·358576	14	1·553001	64	1·998793
15	1·869853	65	2·360892	15	1·584621	65	2·000756
16	1·903696	66	2·363122	16	1·613302	66	2·002646
17	1·934518	67	2·365270	17	1·639422	67	2·004466
18	1·962695	68	2·367339	18	1·663301	68	2·006219
19	1·988542	69	2·369333	19	1·685205	69	2·007909
20	2·012328	70	2·371255	20	1·705363	70	2·009538
21	2·034279	71	2·373109	21	1·723966	71	2·011109
22	2·054593	72	2·374896	22	1·741180	72	2·012624
23	2·073436	73	2·376620	23	1·757149	73	2·014085
24	2·090957	74	2·378283	24	1·771998	74	2·015494
25	2·107283	75	2·379888	25	1·785833	75	2·016854
26	2·122525	76	2·381437	26	1·798750	76	2·018167
27	2·136782	77	2·382933	27	1·810833	77	2·019434
28	2·150142	78	2·384376	28	1·822154	78	2·020658
29	2·162679	79	2·385771	29	1·832779	79	2·021840
30	2·174464	80	2·387117	30	1·842766	80	2·022981
31	2·185558	81	2·388418	31	1·852168	81	2·024083
32	2·196014	82	2·389675	32	1·861029	82	2·025148
33	2·205883	83	2·390890	33	1·869392	83	2·026178
34	2·215208	84	2·392063	34	1·877295	84	2·027172
35	2·224030	85	2·393198	35	1·884771	85	2·028134
36	2·232384	86	2·394295	36	1·891851	86	2·029063
37	2·240303	87	2·395355	37	1·898562	87	2·029962
38	2·247818	88	2·396380	38	1·904931	88	2·030831
39	2·254955	89	2·397372	39	1·910979	89	2·031671
40	2·261740	90	2·398331	40	1·916729	90	2·032484
41	2·268194	91	2·399259	41	1·922199	91	2·033270
42	2·274340	92	2·400157	42	1·927407	92	2·034031
43	2·280196	93	2·401025	43	1·932369	93	2·034767
44	2·285779	94	2·401866	44	1·937101	94	2·035479
45	2·291107	95	2·402679	45	1·941616	95	2·036169
46	2·296193	96	2·403466	46	1·945927	96	2·036836
47	2·301053	97	2·404228	47	1·950045	97	2·037481
48	2·305699	98	2·404965	48	1·953982	98	2·038106
49	2·310142	99	2·405679	49	1·957747	99	2·038711
50	2·314394	100	2·406371	50	1·961351	100	2·039297

TABLE X.

Present Value (or Years' Purchase) of £1 per Annum in **n** years, after
t years' Deferrence.   Redemption of Capital being at **3** per cent., with Interest
allowed to a Purchaser at **18** per cent.

n Years	Deferred 7 Years	n Years	Deferred 7 Years	n Years	Deferred 8 Years	n Years	Deferred 8 Years
1	·266038	51	1·665086	1	·225456	51	1·411090
2	·466726	52	1·667887	2	·395531	52	1·413464
3	·623448	53	1·670572	3	·528346	53	1·415739
4	·749176	54	1·673146	4	·634895	54	1·417920
5	·852236	55	1·675616	5	·722234	55	1·420013
6	·938217	56	1·677985	6	·795099	56	1·422022
7	1·011010	57	1·680261	7	·856788	57	1·423950
8	1·073408	58	1·682446	8	·909668	58	1·425802
9	1·127467	59	1·684545	9	·955481	59	1·427581
10	1·174735	60	1·686563	10	·995538	60	1·429291
11	1·216399	61	1·688504	11	1·030846	61	1·430935
12	1·253383	62	1·690369	12	1·062189	62	1·432516
13	1·286422	63	1·692165	13	1·090188	63	1·434038
14	1·316102	64	1·693892	14	1·115341	64	1·435502
15	1·342900	65	1·695556	15	1·138050	65	1·436912
16	1·367205	66	1·697157	16	1·158648	66	1·438269
17	1·389341	67	1·698700	17	1·177407	67	1·439576
18	1·409577	68	1·700186	18	1·194557	68	1·440836
19	1·428140	69	1·701618	19	1·210288	69	1·442049
20	1·445223	70	1·702999	20	1·224765	70	1·443219
21	1·460988	71	1·704330	21	1·238125	71	1·444347
22	1·475576	72	1·705613	22	1·250488	72	1·445435
23	1·489110	73	1·706851	23	1·261957	73	1·446484
24	1·501693	74	1·708046	24	1·272621	74	1·447497
25	1·513418	75	1·709199	25	1·282557	75	1·448473
26	1·524365	76	1·710311	26	1·291834	76	1·449416
27	1·534604	77	1·711385	27	1·300512	77	1·450326
28	1·544198	78	1·712422	28	1·308642	78	1·451205
29	1·553193	79	1·713423	29	1·316273	79	1·452054
30	1·561666	80	1·714390	30	1·323446	80	1·452873
31	1·569634	81	1·715325	31	1·330198	81	1·453665
32	1·577143	82	1·716227	32	1·336562	82	1·454430
33	1·584231	83	1·717100	33	1·342568	83	1·455169
34	1·590928	84	1·717943	34	1·348244	84	1·455884
35	1·597263	85	1·718757	35	1·353613	85	1·456574
36	1·603263	86	1·719545	36	1·358698	86	1·457242
37	1·608951	87	1·720307	37	1·363518	87	1·457887
38	1·614348	88	1·721043	38	1·368091	88	1·458511
39	1·619474	89	1·721755	39	1·372435	89	1·459115
40	1·624346	90	1·722444	40	1·376565	90	1·459698
41	1·628982	91	1·723111	41	1·380493	91	1·460263
42	1·633396	92	1·723755	42	1·384234	92	1·460809
43	1·637601	93	1·724379	43	1·387798	93	1·461338
44	1·641611	94	1·724982	44	1·391196	94	1·461850
45	1·645437	95	1·725567	45	1·394438	95	1·462345
46	1·649090	96	1·726132	46	1·397534	96	1·462824
47	1·652581	97	1·726679	47	1·400492	97	1·463287
48	1·655917	98	1·727209	48	1·403319	98	1·463736
49	1·659108	99	1·727721	49	1·406024	99	1·464171
50	1·662162	100	1·728218	50	1·408612	100	1·464591

Present Value (or Years' Purchase) of £1 per Annum in n years, after t years' Deferrence. Redemption of Capital being at **3** per cent., with Interest allowed to a Purchaser at **18** per cent.

n Years	Deferred 9 Years	n Years	Deferred 9 Years	n Years	Deferred 10 Years	n Years	Deferred 10 Years
1	·191064	51	1·195839	1	·161919	51	1·013423
2	·335195	52	1·197851	2	·284064	52	1·015128
3	·447751	53	1·199779	3	·379450	53	1·016762
4	·538047	54	1·201628	4	·455972	54	1·018328
5	·612063	55	1·203401	5	·518697	55	1·019831
6	·673813	56	1·205103	6	·571028	56	1·021274
7	·726092	57	1·206737	7	·615332	57	1·022659
8	·770905	58	1·208306	8	·653309	58	1·023989
9	·809729	59	1·209814	9	·686211	59	1·025266
10	·843676	60	1·211264	10	·714980	60	1·026495
11	·873598	61	1·212657	11	·740338	61	1·027675
12	·900160	62	1·213997	12	·762848	62	1·028811
13	·923888	63	1·215286	13	·782956	63	1·029904
14	·945204	64	1·216527	14	·801020	64	1·030955
15	·964450	65	1·217722	15	·817330	65	1·031968
16	·981905	66	1·218872	16	·832123	66	1·032942
17	·997803	67	1·219980	17	·845596	67	1·033881
18	1·012336	68	1·221047	18	·857912	68	1·034786
19	1·025668	69	1·222076	19	·869210	69	1·035657
20	1·037936	70	1·223067	20	·879607	70	1·036498
21	1·049259	71	1·224023	21	·889202	71	1·037308
22	1·059736	72	1·224945	22	·898081	72	1·038089
23	1·069455	73	1·225834	23	·906318	73	1·038843
24	1·078493	74	1·226692	24	·913977	74	1·039570
25	1·086913	75	1·227520	25	·921113	75	1·040271
26	1·094775	76	1·228319	26	·927775	76	1·040948
27	1·102129	77	1·229090	27	·934007	77	1·041602
28	1·109019	78	1·229835	28	·939847	78	1·042233
29	1·115486	79	1·230554	29	·945327	79	1·042842
30	1·121565	80	1·231249	30	·950478	80	1·043431
31	1·127286	81	1·231920	31	·955327	81	1·044000
32	1·132680	82	1·232568	32	·959898	82	1·044549
33	1·137770	83	1·233194	33	·964212	83	1·045080
34	1·142580	84	1·233800	34	·968288	84	1·045593
35	1·147130	85	1·234385	35	·972144	85	1·046089
36	1·151439	86	1·234951	36	·975796	86	1·046568
37	1·155524	87	1·235497	37	·979257	87	1·047032
38	1·159400	88	1·236026	38	·982542	88	1·047480
39	1·163081	89	1·236538	39	·985662	89	1·047913
40	1·166580	90	1·237033	40	·988627	90	1·048333
41	1·169909	91	1·237511	41	·991449	91	1·048738
42	1·173079	92	1·237974	42	·994135	92	1·049131
43	1·176100	93	1·238422	43	·996695	93	1·049510
44	1·178980	94	1·238856	44	·999135	94	1·049878
45	1·181727	95	1·239275	45	1·001464	95	1·050233
46	1·184351	96	1·239681	46	1·003687	96	1·050577
47	1·186858	97	1·240074	47	1·005812	97	1·050910
48	1·189254	98	1·240454	48	1·007842	98	1·051233
49	1·191546	99	1·240823	49	1·009784	99	1·051545
50	1·193739	100	1·241179	50	1·011643	100	1·051847

TABLE X.   clv

Present Value (or Years' Purchase) of £1 per Annum in n years, after t years' Deferrence. Redemption of Capital being at 3 per cent., with Interest allowed to a Purchaser at 20 per cent.

n Years	Deferred 1 Year	n Years	Deferred 1 Year	n Years	Deferred 2 Years	n Years	Deferred 2 Years
1	·694444	51	3·996154	1	·578704	51	3·330129
2	1·203177	52	4·002231	2	1·002647	52	3·335193
3	1·591757	53	4·008054	3	1·326465	53	3·340045
4	1·898137	54	4·013635	4	1·581781	54	3·344696
5	2·145805	55	4·018988	5	1·788171	55	3·349156
6	2·350082	56	4·024123	6	1·958402	56	3·353436
7	2·521384	57	4·029052	7	2·101153	57	3·357543
8	2·667039	58	4·033784	8	2·222532	58	3·361487
9	2·792355	59	4·038330	9	2·326963	59	3·365275
10	2·901270	60	4·042698	10	2·417725	60	3·368915
11	2·996767	61	4·046897	11	2·497306	61	3·372414
12	3·081147	62	4·050934	12	2·567622	62	3·375779
13	3·156212	63	4·054818	13	2·630177	63	3·379015
14	3·223398	64	4·058555	14	2·686165	64	3·382129
15	3·283858	65	4·062151	15	2·736548	65	3·385126
16	3·338530	66	4·065614	16	2·782108	66	3·388011
17	3·388188	67	4·068948	17	2·823490	67	3·390790
18	3·433471	68	4·072160	18	2·861226	68	3·393467
19	3·474917	69	4·075255	19	2·895764	69	3·396046
20	3·512977	70	4·078238	20	2·927481	70	3·398531
21	3·548035	71	4·081113	21	2·956696	71	3·400927
22	3·580420	72	4·083885	22	2·983683	72	3·403238
23	3·610412	73	4·086559	23	3·008677	73	3·405366
24	3·638257	74	4·089138	24	3·031881	74	3·407615
25	3·664165	75	4·091627	25	3·053471	75	3·409689
26	3·688323	76	4·094028	26	3·073602	76	3·411690
27	3·710891	77	4·096347	27	3·092409	77	3·413622
28	3·732013	78	4·098584	28	3·110011	78	3·415487
29	3·751816	79	4·100745	29	3·126513	79	3·417288
30	3·770410	80	4·102832	30	3·142009	80	3·419026
31	3·787897	81	4·104847	31	3·156581	81	3·420706
32	3·804365	82	4·106795	32	3·170304	82	3·422329
33	3·819894	83	4·108676	33	3·183245	83	3·423897
34	3·834556	84	4·110494	34	3·195464	84	3·425412
35	3·848416	85	4·112251	35	3·207014	85	3·426876
36	3·861532	86	4·113950	36	3·217944	86	3·428291
37	3·873958	87	4·115592	37	3·228298	87	3·429660
38	3·885740	88	4·117180	38	3·238117	88	3·430983
39	3·896924	89	4·118715	39	3·247437	89	3·432262
40	3·907550	90	4·120200	40	3·256291	90	3·433500
41	3·917653	91	4·121636	41	3·264711	91	3·434697
42	3·927267	92	4·123026	42	3·272722	92	3·435855
43	3·936423	93	4·124370	43	3·280353	93	3·436975
44	3·945150	94	4·125670	44	3·287625	94	3·438059
45	3·953472	95	4·126929	45	3·294560	95	3·439108
46	3·961415	96	4·128147	46	3·301179	96	3·440123
47	3·969000	97	4·129326	47	3·307500	97	3·441105
48	3·976249	98	4·130467	48	3·313540	98	3·442056
49	3·983179	99	4·131572	49	3·319316	99	3·442976
50	3·989809	100	4·132641	50	3·324840	100	3·443868

Present Value (or Years' Purchase) of £1 per Annum in n years, after t years' Deferrence. Redemption of Capital being at 3 per cent., with Interest allowed to a Purchaser at 20 per cent.

n Years	Deferred 3 Years	n Years	Deferred 3 Years	n Years	Deferred 4 Years	n Years	Deferred 4 Years
1	·482253	51	2·775107	1	·401878	51	2·312589
2	·835539	52	2·779327	2	·696283	52	2·316106
3	1·105387	53	2·783371	3	·921156	53	2·319476
4	1·318150	54	2·787247	4	1·098459	54	2·322706
5	1·490143	55	2·790964	5	1·241786	55	2·325803
6	1·632002	56	2·794530	6	1·360001	56	2·328775
7	1·750961	57	2·797952	7	1·459134	57	2·331627
8	1·852110	58	2·801239	8	1·543425	58	2·334366
9	1·939136	59	2·804396	9	1·615946	59	2·336997
10	2·014771	60	2·807429	10	1·678976	60	2·339524
11	2·081088	61	2·810345	11	1·734240	61	2·341954
12	2·139685	62	2·813149	12	1·783071	62	2·344291
13	2·191814	63	2·815846	13	1·826512	63	2·346538
14	2·238471	64	2·818441	14	1·865392	64	2·348701
15	2·280457	65	2·820938	15	1·900381	65	2·350782
16	2·318424	66	2·823343	16	1·932020	66	2·352786
17	2·352908	67	2·825658	17	1·960757	67	2·354715
18	2·384355	68	2·827889	18	1·986963	68	2·356574
19	2·413137	69	2·830038	19	2·010947	69	2·358365
20	2·439567	70	2·832110	20	2·032973	70	2·360091
21	2·463913	71	2·834106	21	2·053261	71	2·361755
22	2·486402	72	2·836032	22	2·072002	72	2·363360
23	2·507231	73	2·837888	23	2·089359	73	2·364907
24	2·526567	74	2·839679	24	2·105473	74	2·366400
25	2·544559	75	2·841408	25	2·120466	75	2·367840
26	2·561335	76	2·843075	26	2·134446	76	2·369229
27	2·577008	77	2·844685	27	2·147506	77	2·370571
28	2·591676	78	2·846239	28	2·159730	78	2·371866
29	2·605428	79	2·847740	29	2·171190	79	2·373116
30	2·618341	80	2·849189	30	2·181951	80	2·374324
31	2·630484	81	2·850588	31	2·192070	81	2·375490
32	2·641920	82	2·851941	32	2·201600	82	2·376617
33	2·652704	83	2·853247	33	2·210587	83	2·377706
34	2·662886	84	2·854510	34	2·219072	84	2·378758
35	2·672511	85	2·855730	35	2·227093	85	2·379775
36	2·681620	86	2·856910	36	2·234683	86	2·380758
37	2·690248	87	2·858050	37	2·241874	87	2·381708
38	2·698431	88	2·859152	38	2·248692	88	2·382627
39	2·706198	89	2·860219	39	2·255165	89	2·383516
40	2·713576	90	2·861250	40	2·261313	90	2·384375
41	2·720592	91	2·862247	41	2·267160	91	2·385206
42	2·727269	92	2·863212	42	2·272724	92	2·386010
43	2·733627	93	2·864146	43	2·278023	93	2·386788
44	2·739687	94	2·865049	44	2·283073	94	2·387541
45	2·745467	95	2·865923	45	2·287889	95	2·388269
46	2·750983	96	2·866769	46	2·292486	96	2·388974
47	2·756250	97	2·867587	47	2·296875	97	2·389656
48	2·761284	98	2·868380	48	2·301070	98	2·390317
49	2·766096	99	2·869147	49	2·305080	99	2·390956
50	2·770702	100	2·869890	50	2·308917	100	2·391575

TABLE X.   clvii

Present Value (or Years' Purchase) of £1 per Annum in n years, after t years' Deferrence. Redemption of Capital being at 3 per cent., with Interest allowed to a Purchaser at 20 per cent.

n Years	Deferred 5 Years	n Years	Deferred 5 Years	n Years	Deferred 6 Years	n Years	Deferred 6 Years
1	·334898	51	1·927158	1	·279082	51	1·605965
2	·580236	52	1·930088	2	·483530	52	1·608407
3	·767630	53	1·932896	3	·639692	53	1·610747
4	·915382	54	1·935588	4	·762819	54	1·612990
5	1·034821	55	1·938169	5	·862351	55	1·615141
6	1·133334	56	1·940646	6	·944445	56	1·617205
7	1·215945	57	1·943023	7	1·013288	57	1·619185
8	1·286188	58	1·945305	8	1·071823	58	1·621087
9	1·346622	59	1·947497	9	1·122185	59	1·622914
10	1·399147	60	1·949604	10	1·165955	60	1·624670
11	1·445200	61	1·951629	11	1·204333	61	1·626357
12	1·485892	62	1·953576	12	1·238244	62	1·627980
13	1·522093	63	1·955449	13	1·268411	63	1·629540
14	1·554494	64	1·957251	14	1·295411	64	1·631042
15	1·583650	65	1·958985	15	1·319709	65	1·632488
16	1·610016	66	1·960655	16	1·341680	66	1·633879
17	1·633964	67	1·962263	17	1·361637	67	1·635219
18	1·655802	68	1·963812	18	1·379835	68	1·636510
19	1·675789	69	1·965304	19	1·396491	69	1·637754
20	1·694144	70	1·966743	20	1·411787	70	1·638952
21	1·711051	71	1·968129	21	1·425876	71	1·640108
22	1·726668	72	1·969466	22	1·438890	72	1·641222
23	1·741132	73	1·970756	23	1·450944	73	1·642297
24	1·754561	74	1·972000	24	1·462134	74	1·643333
25	1·767055	75	1·973200	25	1·472546	75	1·644333
26	1·778705	76	1·974358	26	1·482254	76	1·645298
27	1·789589	77	1·975476	27	1·491324	77	1·646230
28	1·799775	78	1·976555	28	1·499812	78	1·647129
29	1·809325	79	1·977597	29	1·507771	79	1·647997
30	1·818292	80	1·978603	30	1·515243	80	1·648836
31	1·826725	81	1·979575	31	1·522271	81	1·649646
32	1·834667	82	1·980514	32	1·528889	82	1·650429
33	1·842156	83	1·981422	33	1·535130	83	1·651185
34	1·849227	84	1·982298	34	1·541022	84	1·651915
35	1·855911	85	1·983146	35	1·546592	85	1·652622
36	1·862236	86	1·983965	36	1·551863	86	1·653304
37	1·868228	87	1·984757	37	1·556857	87	1·653964
38	1·873910	88	1·985523	38	1·561592	88	1·654602
39	1·879304	89	1·986263	39	1·566087	89	1·655219
40	1·884428	90	1·986979	40	1·570357	90	1·655816
41	1·889300	91	1·987672	41	1·574417	91	1·656393
42	1·893937	92	1·988342	42	1·578281	92	1·656952
43	1·898352	93	1·988990	43	1·581960	93	1·657492
44	1·902561	94	1·989617	44	1·585467	94	1·658014
45	1·906574	95	1·990224	45	1·588812	95	1·658520
46	1·910405	96	1·990812	46	1·592004	96	1·659010
47	1·914063	97	1·991380	47	1·595052	97	1·659484
48	1·917558	98	1·991931	48	1·597965	98	1·659942
49	1·920900	99	1·992463	49	1·600750	99	1·660386
50	1·924097	100	1·992979	50	1·603415	100	1·660816

Present Value (or Years' Purchase) of £1 per Annum in n years, after t years' Deferrence. Redemption of Capital being at 3 per cent., with Interest allowed to a Purchaser at 20 per cent.

n Years	Deferred 7 Years	n Years	Deferred 7 Years	n Years	Deferred 8 Years	n Years	Deferred 8 Years
1	·232568	51	1·338304	1	·193807	51	1·115253
2	·402942	52	1·340339	2	·335785	52	1·116949
3	·533076	53	1·342289	3	·444230	53	1·118574
4	·635682	54	1·344158	4	·529735	54	1·120132
5	·718626	55	1·345951	5	·598855	55	1·121626
6	·787038	56	1·347671	6	·655865	56	1·123059
7	·844406	57	1·349321	7	·703672	57	1·124434
8	·893186	58	1·350906	8	·744322	58	1·125755
9	·935154	59	1·352429	9	·779295	59	1·127024
10	·971630	60	1·353891	10	·809691	60	1·128243
11	1·003611	61	1·355298	11	·836343	61	1·129415
12	1·031870	62	1·356650	12	·859891	62	1·130541
13	1·057009	63	1·357950	13	·880841	63	1·131625
14	1·079510	64	1·359202	14	·899591	64	1·132668
15	1·099757	65	1·360406	15	·916464	65	1·133672
16	1·118067	66	1·361566	16	·931722	66	1·134638
17	1·134697	67	1·362683	17	·945581	67	1·135569
18	1·149863	68	1·363758	18	·958219	68	1·136465
19	1·163743	69	1·364795	19	·969786	69	1·137329
20	1·176489	70	1·365794	20	·980407	70	1·138161
21	1·188230	71	1·366757	21	·990192	71	1·138964
22	1·199075	72	1·367685	22	·999229	72	1·139737
23	1·209120	73	1·368580	23	1·007600	73	1·140484
24	1·218445	74	1·369444	24	1·015371	74	1·141203
25	1·227122	75	1·370278	25	1·022601	75	1·141898
26	1·235212	76	1·371082	26	1·029343	76	1·142568
27	1·242770	77	1·371858	27	1·035642	77	1·143215
28	1·249844	78	1·372608	28	1·041536	78	1·143840
29	1·256476	79	1·373331	29	1.047063	79	1·144443
30	1·262703	80	1·374030	30	1·052252	80	1·145025
31	1·268559	81	1·374705	31	1·057133	81	1·145588
32	1·274074	82	1·375357	32	1·061729	82	1·146131
33	1·279275	83	1·375987	33	1·066062	83	1·146656
34	1·284185	84	1·376596	34	1·070154	84	1·147163
35	1·288827	85	1·377185	35	1·074022	85	1·147654
36	1·293219	86	1·377753	36	1·077683	86	1·148128
37	1·297381	87	1·378303	37	1·081151	87	1·148586
38	1·301327	88	1·378835	38	1·084439	88	1·149029
39	1·305072	89	1·379349	39	1·087560	89	1·149458
40	1·308630	90	1·379847	40	1·090525	90	1·149872
41	1·312014	91	1·380328	41	1·093345	91	1·150273
42	1·315234	92	1·380793	42	1·096028	92	1·150661
43	1·318300	93	1·381243	43	1·098583	93	1·151036
44	1·321223	94	1·381679	44	1·101019	94	1·151399
45	1·324010	95	1·382100	45	1·103342	95	1·151750
46	1·326670	96	1·382508	46	1·105558	96	1·152090
47	1·329210	97	1·382903	47	1·107675	97	1·152419
48	1·331638	98	1·383285	48	1·109698	98	1·152738
49	1·333959	99	1·383655	49	1·111632	99	1·153046
50	1·336179	100	1·384013	50	1·113482	100	1·153344

TABLE X. clix

Present Value (or Years' Purchase) of £1 per Annum in n years, after t years' Deferrence. Redemption of Capital being at 3 per cent., with Interest allowed to a Purchaser at 20 per cent.

n Years	Deferred 9 Years	n Years	Deferred 9 Years	n Years	Deferred 10 Years	n Years	Deferred 10 Years
1	·161506	51	·929378	1	·134588	51	·774481
2	·279820	52	·930791	2	·233184	52	·775659
3	·370192	53	·932145	3	·308493	53	·776788
4	·441446	54	·933443	4	·367872	54	·777869
5	·499046	55	·934688	5	·415871	55	·778907
6	·546554	56	·935882	6	·455462	56	·779902
7	·586393	57	·937029	7	·488661	57	·780857
8	·620268	58	·938129	8	·516890	58	·781774
9	·649413	59	·939186	9	·541177	59	·782655
10	·674743	60	·940202	10	·562286	60	·783502
11	·696952	61	·941179	11	·580794	61	·784316
12	·716576	62	·942118	12	·597147	62	·785098
13	·734034	63	·943021	13	·611695	63	·785851
14	·749659	64	·943890	14	·624716	64	·786575
15	·763720	65	·944727	15	·636434	65	·787272
16	·776435	66	·945532	16	·647029	66	·787943
17	·787984	67	·946307	17	·656653	67	·788589
18	·798519	68	·947054	18	·665430	68	·789212
19	·808155	69	·947774	19	·673462	69	·789812
20	·817006	70	·948468	20	·680838	70	·790390
21	·825160	71	·949136	21	·687633	71	·790947
22	·832691	72	·949781	22	·693909	72	·791484
23	·839666	73	·950403	23	·699722	73	·792003
24	·846142	74	·951003	24	·705119	74	·792502
25	·852168	75	·951582	25	·710140	75	·792933
26	·857786	76	·952140	26	·714822	76	·793450
27	·863035	77	·952679	27	·719196	77	·793899
28	·867947	78	·953200	28	·723289	78	·794333
29	·872552	79	·953702	29	·727127	79	·794752
30	·876877	80	·954188	30	·730731	80	·795156
31	·880944	81	·954656	31	·734120	81	·795547
32	·884774	82	·955109	32	·737311	82	·795924
33	·888385	83	·955547	33	·740321	83	·796289
34	·891795	84	·955970	34	·743163	84	·796641
35	·895019	85	·956378	35	·745849	85	·796982
36	·898069	86	·956773	36	·748391	86	·797311
37	·900959	87	·957155	37	·750799	87	·797652
38	·903699	88	·957524	38	·753083	88	·797937
39	·906300	89	·957881	39	·755250	89	·798235
40	·908771	90	·958227	40	·757309	90	·798522
41	·911121	91	·958561	41	·759267	91	·798801
42	·913357	92	·958884	42	·761131	92	·799070
43	·915486	93	·959197	43	·762905	93	·799330
44	·917516	94	·959499	44	·764596	94	·799583
45	·919451	95	·959792	45	·766209	95	·799826
46	·921299	96	·960075	46	·767749	96	·800063
47	·923063	97	·960349	47	·769219	97	·800291
48	·924748	98	·960615	48	·770624	98	·800512
49	·926360	99	·960872	49	·771967	99	·800726
50	·927902	100	·961120	50	·773252	100	·800934

TABLE XI.

FOR

VALUING MINERAL AND OTHER PROPERTIES,

OR

The Present Value (or Years' Purchase) of £1 per annum in n years, deferred 1, 2, 3, 4, 5, 6, 7, 8, 9, *and* 10 *years, allowing interest to a present purchaser upon his purchase money, or capital invested, at the rate of* **20** *per cent. per annum, and to redeem the capital so invested, by an Annual Redemption Fund, at the rates of* **3½** *and* **4** *per cent. per annum.*

Calculated to **6** *places of decimals, and to* **100** *years for each percentage.*

TABLE XI.   clxiii

Present Value (or Years' Purchase) of £1 per Annum in **n** years, after
t years' Deferrence.   Redemption of Capital being at **3½** and **4** per cent., with
Interest allowed to a Purchaser at **20** per cent.

DEFERRED 1 YEAR.

n Years	Redemption 3½ per cent.	n Years	Redemption 3½ per cent.	n Years	Redemption 4 per cent.	n Years	Redemption 4 per cent.
1	·694444	51	4·019519	1	·694444	51	4·040229
2	1·205283	52	4·025291	2	1·207386	52	4·045657
3	1·596625	53	4·030807	3	1·601490	53	4·050828
4	1·905846	54	4·036080	4	1·913553	54	4·055758
5	2·156206	55	4·041123	5	2·166600	55	4·060458
6	2·362939	56	4·045948	6	2·375780	56	4·064943
7	2·536440	57	4·050717	7	2·551465	57	4·069221
8	2·684045	58	4·054988	8	2·700998	58	4·073305
9	2·811080	59	4·059224	9	2·829721	59	4·077204
10	2·921502	60	4·063283	10	2·941615	60	4·080928
11	3·018318	61	4·067172	11	3·039708	61	4·084486
12	3·103848	62	4·070901	12	3·126341	62	4·087885
13	3·179914	63	4·074478	13	3·203356	63	4·091135
14	3·247967	64	4·077908	14	3·272220	64	4·094242
15	3·309175	65	4·081199	15	3·334118	65	4·097213
16	3·364491	66	4·084358	16	3·390013	66	4·100055
17	3·414697	67	4·087392	17	3·440702	67	4·102774
18	3·460445	68	4·090304	18	3·486846	68	4·105376
19	3·502279	69	4·093101	19	3·528999	69	4·108069
20	3·540659	70	4·095788	20	3·567629	70	4·110251
21	3·575976	71	4·098370	21	3·603133	71	4·112533
22	3·608565	72	4·100852	22	3·635852	72	4·114719
23	3·638711	73	4·103237	23	3·666079	73	4·116813
24	3·666665	74	4·105530	24	3·694068	74	4·118818
25	3·692642	75	4·107736	25	3·720038	75	4·120739
26	3·716830	76	4·109857	26	3·744183	76	4·122581
27	3·739396	77	4·111897	27	3·766673	77	4·124345
28	3·760485	78	4·113859	28	3·787653	78	4·126036
29	3·780226	79	4·115565	29	3·807261	79	4·127658
30	3·798734	80	4·117565	30	3·825610	80	4·129207
31	3·816111	81	4·119315	31	3·842806	81	4·130702
32	3·832448	82	4·120998	32	3·858942	82	4·132131
33	3·847828	83	4·122620	33	3·874101	83	4·133502
34	3·862322	84	4·124181	34	3·888360	84	4·134816
35	3·875999	85	4·125684	35	3·901786	85	4·136077
36	3·888917	86	4·127132	36	3·914440	86	4·137287
37	3·901131	87	4·128527	37	3·926379	87	4·138447
38	3·912691	88	4·129870	38	3·937653	88	4·139561
39	3·923641	89	4·131164	39	3·948308	89	4·140629
40	3·934023	90	4·132411	40	3·958385	90	4·141654
41	3·943873	91	4·133613	41	3·967924	91	4·142638
42	3·953226	92	4·134771	42	3·976960	92	4·143583
43	3·962115	93	4·135887	43	3·985525	93	4·144490
44	3·970567	94	4·136963	44	3·993649	94	4·145360
45	3·978609	95	4·138000	45	4·001360	95	4·146196
46	3·986267	96	4·138999	46	4·008682	96	4·146998
47	3·993562	97	4·139963	47	4·015639	97	4·147768
48	4·000516	98	4·140893	48	4·022253	98	4·148508
49	4·007149	99	4·141788	49	4·028543	99	4·149218
50	4·013478	100	4·142653	50	4·034530	100	4·149899

Present Value (or Years' Purchase) of £1 per Annum in n years, after t years' Deferrence. Redemption of Capital being at 3½ and 4 per cent., with Interest allowed to a Purchaser at 20 per cent.

DEFERRED 2 YEARS.

n Years	Redemption 3½ per cent.	n Years	Redemption 3½ per cent.	n Years	Redemption 4 per cent.	n Years	Redemption 4 per cent.
1	·578704	51	3·349599	1	·578703	51	3·366857
2	1·004403	52	3·354408	2	1·006155	52	3·371380
3	1·330521	53	3·359005	3	1·334575	53	3·375690
4	1·588205	54	3·363399	4	1·594627	54	3·379798
5	1·796838	55	3·367601	5	1·805500	55	3·383715
6	1·969116	56	3·371623	6	1·979816	56	3·387451
7	2·113700	57	3·375596	7	2·126220	57	3·391017
8	2·236704	58	3·379156	8	2·250831	58	3·394420
9	2·342566	59	3·382686	9	2·358101	59	3·397669
10	2·434585	60	3·386068	10	2·451346	60	3·400772
11	2·515265	61	3·389309	11	2·533089	61	3·403737
12	2·586540	62	3·392417	12	2·605284	62	3·406570
13	2·649929	63	3·395397	13	2·669463	63	3·409278
14	2·706639	64	3·398255	14	2·726850	64	3·411867
15	2·757646	65	3·400999	15	2·778431	65	3·414344
16	2·803742	66	3·403631	16	2·825010	66	3·416712
17	2·845581	67	3·406159	17	2·867251	67	3·418978
18	2·883704	68	3·408586	18	2·905704	68	3·421146
19	2·918566	69	3·410917	19	2·940832	69	3·423390
20	2·950549	70	3·413156	20	2·973023	70	3·425208
21	2·979980	71	3·415308	21	3·002610	71	3·427110
22	3·007137	72	3·417376	22	3·029876	72	3·428932
23	3·032259	73	3·419363	23	3·055065	73	3·430676
24	3·055554	74	3·421274	24	3·078389	74	3·432347
25	3·077201	75	3·423112	25	3·100031	75	3·433948
26	3·097358	76	3·424880	26	3·120151	76	3·435483
27	3·116163	77	3·426580	27	3·138893	77	3·436953
28	3·133737	78	3·428215	28	3·156377	78	3·438362
29	3·150188	79	3·429637	29	3·172717	79	3·439714
30	3·165612	80	3·431303	30	3·188008	80	3·441005
31	3·180093	81	3·432762	31	3·202338	81	3·442251
32	3·193707	82	3·434164	32	3·215784	82	3·443442
33	3·206523	83	3·435516	33	3·228417	83	3·444584
34	3·218602	84	3·436817	34	3·240299	84	3·445679
35	3·229999	85	3·438069	35	3·251488	85	3·446730
36	3·240764	86	3·439276	36	3·262033	86	3·447738
37	3·250943	87	3·440438	37	3·271982	87	3·448705
38	3·260576	88	3·441558	38	3·281376	88	3·449633
39	3·269701	89	3·442636	39	3·290256	89	3·450523
40	3·278352	90	3·443675	40	3·298653	90	3·451378
41	3·286561	91	3·444676	41	3·306603	91	3·452198
42	3·294355	92	3·445642	42	3·314133	92	3·452985
43	3·301762	93	3·446571	43	3·321270	93	3·453741
44	3·308806	94	3·447468	44	3·328040	94	3·454466
45	3·315508	95	3·448333	45	3·334466	95	3·455162
46	3·321889	96	3·449165	46	3·340567	96	3·455830
47	3·327968	97	3·449969	47	3·346365	97	3·456473
48	3·333763	98	3·450743	48	3·351877	98	3·457089
49	3·339291	99	3·451489	49	3·357119	99	3·457680
50	3·344565	100	3·452210	50	3·362108	100	3·458248

TABLE XI. clxv

Present Value (or Years' Purchase) of £1 per Annum in **n** years, after
t years' Deferrence. Redemption of Capital being at 3½ and 4 per cent., with
Interest allowed to a Purchaser at **20** per cent.

DEFERRED **3** YEARS.

n Years	Redemption 3½ per cent.	n Years	Redemption 3½ per cent.	n Years	Redemption 4 per cent.	n Years	Redemption 4 per cent.
1	·482253	51	2·791335	1	·482253	51	2·805717
2	·837003	52	2·795343	2	·838463	52	2·809486
3	1·108768	53	2·799174	3	1·112147	53	2·813078
4	1·323505	54	2·802836	4	1·328857	54	2·816501
5	1·497366	55	2·806338	5	1·504585	55	2·819765
6	1·640930	56	2·809689	6	1·649849	56	2·822879
7	1·761418	57	2·813000	7	1·771852	57	2·825850
8	1·863921	58	2·815967	8	1·875695	58	2·828687
9	1·952140	59	2·818908	9	1·965086	59	2·831394
10	2·028822	60	2·821727	10	2·042790	60	2·833980
11	2·096055	61	2·824427	11	2·110910	61	2·836451
12	2·155451	62	2·827017	12	2·171072	62	2·838812
13	2·208275	63	2·829501	13	2·224555	63	2·841069
14	2·255534	64	2·831883	14	2·272377	64	2·843226
15	2·298040	65	2·834169	15	2·315362	65	2·845290
16	2·336453	66	2·836363	16	2·354178	66	2·847263
17	2·371318	67	2·838469	17	2·389379	67	2·849151
18	2·403088	68	2·840492	18	2·421423	68	2·850958
19	2·432139	69	2·842434	19	2·450696	69	2·852828
20	2·458792	70	2·844300	20	2·477522	70	2·854343
21	2·483318	71	2·846093	21	2·502178	71	2·855929
22	2·505949	72	2·847816	22	2·524899	72	2·857446
23	2·526884	73	2·849473	23	2·545891	73	2·858900
24	2·546296	74	2·851065	24	2·565327	74	2·860293
25	2·564336	75	2·852597	25	2·583362	75	2·861627
26	2·581133	76	2·854070	26	2·600129	76	2·862906
27	2·596804	77	2·855486	27	2·615747	77	2·864131
28	2·611449	78	2·856849	28	2·630317	78	2·865305
29	2·625158	79	2·858034	29	2·643934	79	2·866431
30	2·638011	80	2·859423	30	2·656676	80	2·867507
31	2·650078	81	2·860638	31	2·668618	81	2·868545
32	2·661424	82	2·861807	32	2·679823	82	2·869538
33	2·672104	83	2·862933	33	2·690350	83	2·870490
34	2·682170	84	2·864017	34	2·700253	84	2·871403
35	2·691667	85	2·865061	35	2·709576	85	2·872278
36	2·700638	86	2·866066	36	2·718364	86	2·873118
37	2·709121	87	2·867035	37	2·726655	87	2·873924
38	2·717148	88	2·867968	38	2·734483	88	2·874698
39	2·724752	89	2·868867	39	2·741883	89	2·875440
40	2·731961	90	2·869732	40	2·748881	90	2·876151
41	2·738802	91	2·870567	41	2·755505	91	2·876835
42	2·745298	92	2·871371	42	2·761780	92	2·877491
43	2·751470	93	2·872146	43	2·767728	93	2·878121
44	2·757339	94	2·872893	44	2·773370	94	2·878725·
45	2·762924	95	2·873614	45	2·778725	95	2·879305
46	2·768242	96	2·874308	46	2·783809	96	2·879862
47	2·773308	97	2·874977	47	2·788641	97	2·880397
48	2·778137	98	2·875622	48	2·793234	98	2·880911
49	2·782744	99	2·876245	49	2·797602	99	2·881404
50	2·787139	100	2·876845	50	2·801760	100	2·881877

Present Value (or Years' Purchase) of £1 per Annum in n years, after
t years' Deferrence. Redemption of Capital being at 3½ and 4 per cent., with
Interest allowed to a Purchaser at 20 per cent.

DEFERRED 4 YEARS.

n Years	Redemption 3¼ per cent.	n Years	Redemption 3½ per cent.	n Years	Redemption 4 per cent.	n Years	Redemption 4 per cent.
1	·401877	51	2·326111	1	·401877	51	2·338096
2	·697502	52	2·329451	2	·698719	52	2·341237
3	·923973	53	2·332643	3	·926789	53	2·344230
4	1·102920	54	2·335695	4	1·107380	54	2·347083
5	1·247804	55	2·338613	5	1·253820	55	2·349803
6	1·367441	56	2·341406	6	1·374873	56	2·352398
7	1·467847	57	2·344165	7	1·476542	57	2·354874
8	1·553267	58	2·346637	8	1·563078	58	2·357237
9	1·626782	59	2·349089	9	1·637571	59	2·359494
10	1·690684	60	2·351437	10	1·702324	60	2·361648
11	1·746712	61	2·353688	11	1·759091	61	2·363708
12	1·796208	62	2·355846	12	1·809226	62	2·365675
13	1·840228	63	2·357916	13	1·853795	63	2·367556
14	1·879610	64	2·359901	14	1·893646	64	2·369353
15	1·915032	65	2·361806	15	1·929467	65	2·371073
16	1·947043	66	2·363634	16	1·961813	66	2·372718
17	1·976097	67	2·365389	17	1·991147	67	2·374291
18	2·002572	68	2·367075	18	2·017851	68	2·375797
19	2·026781	69	2·368693	19	2·042245	69	2·377355
20	2·048992	70	2·370248	20	2·064600	70	2·378618
21	2·069431	71	2·371742	21	2·085147	71	2·379939
22	2·088290	72	2·373179	22	2·104081	72	2·381204
23	2·105736	73	2·374559	23	2·121574	73	2·382415
24	2·121912	74	2·375886	24	2·137771	74	2·383575
25	2·136945	75	2·377162	25	2·152800	75	2·384688
26	2·150943	76	2·378390	26	2·166773	76	2·385753
27	2·164002	77	2·379570	27	2·179788	77	2·386774
28	2·176206	78	2·380706	28	2·191930	78	2·387753
29	2·187630	79	2·381693	29	2·203277	79	2·388691
30	2·198341	80	2·382851	30	2·213895	80	2·389588
31	2·208397	81	2·383863	31	2·223847	81	2·390453
32	2·217852	82	2·384838	32	2·233184	82	2·391280
33	2·226752	83	2·385776	33	2·241957	83	2·392073
34	2·235140	84	2·386679	34	2·250209	84	2·392834
35	2·243055	85	2·387549	35	2·257978	85	2·393563
36	2·250530	86	2·388387	36	2·265301	86	2·394264
37	2·257599	87	2·389194	37	2·272211	87	2·394935
38	2·264288	88	2·389972	38	2·278735	88	2·395580
39	2·270625	89	2·390721	39	2·284901	89	2·396198
40	2·276633	90	2·391442	40	2·290733	90	2·396791
41	2·282334	91	2·392137	41	2·296253	91	2·397361
42	2·287747	92	2·392808	42	2·301482	92	2·397908
43	2·292890	93	2·393453	43	2·306439	93	2·398432
44	2·297781	94	2·394076	44	2·311140	94	2·398936
45	2·302435	95	2·394676	45	2·315602	95	2·399419
46	2·306867	96	2·395255	46	2·319839	96	2·399883
47	2·311089	97	2·395813	47	2·323866	97	2·400329
48	2·315113	98	2·396350	48	2·327693	98	2·400757
49	2·318952	99	2·396869	49	2·331333	99	2·401168
50	2·322614	100	2·397369	50	2·334798	100	2·401563

Present Value (or Years' Purchase) of £1 per Annum in n years, after
t years' Deferrence. Redemption of Capital being at 3½ and 4 per cent, with
Interest allowed to a Purchaser at 20 per cent.

DEFERRED 5 YEARS.

n Years	Redemption 3½ per cent.	n Years	Redemption 3½ per cent.	n Years	Redemption 4 per cent.	n Years	Redemption 4 per cent.
1	·334898	51	1·938428	1	·334898	51	1·948416
2	·581252	52	1·941212	2	·582267	52	1·951033
3	·769978	53	1·943872	3	·772325	53	1·953527
4	·919101	54	1·946415	4	·922818	54	1·955905
5	1·039838	55	1·948847	5	1·044851	55	1·958171
6	1·139536	56	1·951174	6	1·145729	56	1·960334
7	1·223207	57	1·953473	7	1·230454	57	1·962397
8	1·294390	58	1·955534	8	1·302567	58	1·964367
9	1·355653	59	1·957576	9	1·364644	59	1·966247
10	1·408905	60	1·959533	10	1·418605	60	1·968043
11	1·455595	61	1·961409	11	1·465911	61	1·969759
12	1·496842	62	1·963207	12	1·507690	62	1·971398
13	1·533525	63	1·964932	13	1·544831	63	1·972965
14	1·566344	64	1·966586	14	1·578041	64	1·974464
15	1·595862	65	1·968174	15	1·607891	65	1·975897
16	1·622538	66	1·969697	16	1·634847	66	1·977267
17	1·646750	67	1·971160	17	1·659292	67	1·978578
18	1·668812	68	1·972565	18	1·681545	68	1·979833
19	1·688986	69	1·973913	19	1·701873	69	1·981132
20	1·707496	70	1·975209	20	1·720502	70	1·982184
21	1·724528	71	1·976454	21	1·737624	71	1·983285
22	1·740243	72	1·977651	22	1·753403	72	1·984339
23	1·754782	73	1·978801	23	1·767981	73	1·985348
24	1·768262	74	1·979907	24	1·781478	74	1·986315
25	1·780790	75	1·980971	25	1·794002	75	1·987242
26	1·792455	76	1·981994	26	1·805646	76	1·988130
27	1·803337	77	1·982978	27	1·816492	77	1·988981
28	1·813507	78	1·983924	28	1·826610	78	1·989796
29	1·823028	79	1·984747	29	1·836066	79	1·990578
30	1·831953	80	1·985711	30	1·844915	80	1·991326
31	1·840333	81	1·986555	31	1·853208	81	1·992047
32	1·848212	82	1·987367	32	1·860989	82	1·992736
33	1·855629	83	1·988149	33	1·868300	83	1·993397
34	1·862619	84	1·988902	34	1·875176	84	1·994031
35	1·869215	85	1·989627	35	1·881651	85	1·994639
36	1·875444	86	1·990325	36	1·887754	86	1·995222
37	1·881335	87	1·990998	37	1·893511	87	1·995782
38	1·886909	88	1·991645	38	1·898948	88	1·996319
39	1·892190	89	1·992270	39	1·904086	89	1·996834
40	1·897197	90	1·992871	40	1·908946	90	1·997328
41	1·901947	91	1·993450	41	1·913547	91	1·997803
42	1·906458	92	1·994009	42	1·917904	92	1·998259
43	1·910744	93	1·944547	43	1·922035	93	1·998696
44	1·914820	94	1·995066	44	1·925952	94	1·999116
45	1·918699	95	1·995566	45	1·929671	95	1·999519
46	1·922391	96	1·996048	46	1·933202	96	1·999905
47	1·925909	97	1·996513	47	1·936557	97	2·000277
48	1·929263	98	1·996961	48	1·939747	98	2·000633
49	1·932462	99	1·997393	49	1·942780	99	2·000976
50	1·935514	100	1·997810	50	1·945667	100	2·001305

Present Value (or Years' Purchase) of £1 per Annum in n years, after t years' Deferrence. Redemption of Capital being at 3½ and 4 per cent., with Interest allowed to a Purchaser at 20 per cent.

DEFERRED 6 YEARS.

n Years	Redemption 3½ per cent.	n Years	Redemption 3½ per cent.	n Years	Redemption 4 per cent.	n Years	Redemption 4 per cent.
1	·279082	51	1·615355	1	·279082	51	1·623678
2	·484376	52	1·617675	2	·485222	52	1·625859
3	·641648	53	1·619892	3	·643603	53	1·627938
4	·765917	54	1·622011	4	·769014	54	1·629919
5	·866531	55	1·624037	5	·870708	55	1·631808
6	·949612	56	1·625977	6	·954773	56	1·633610
7	1·019338	57	1·627893	7	1·025377	57	1·635329
8	1·078658	58	1·629610	8	1·085471	58	1·636971
9	1·129710	59	1·631312	9	1·137202	59	1·638538
10	1·174086	60	1·632943	10	1·182170	60	1·640034
11	1·212995	61	1·634506	11	1·221591	61	1·641464
12	1·247367	62	1·636005	12	1·256407	62	1·642830
13	1·277936	63	1·637442	13	1·287358	63	1·644136
14	1·305285	64	1·638820	14	1·315033	64	1·645385
15	1·329884	65	1·640143	15	1·339908	65	1·646579
16	1·352114	66	1·641413	16	1·362371	66	1.647721
17	1·372290	67	1·642632	17	1·382742	67	1·648814
18	1·390675	68	1·643802	18	1·401286	68	1·649859
19	1·407487	69	1·644926	19	1·418226	69	1·650942
20	1·422912	70	1·646006	20	1·433751	70	1·651809
21	1·437104	71	1·647044	21	1·448019	71	1·652736
22	1·450201	72	1·648041	22	1·461168	72	1·653614
23	1·462317	73	1·649000	23	1·473316	73	1·654455
24	1·473551	74	1·649921	24	1·484564	74	1·655261
25	1·483990	75	1·650808	25	1·495000	75	1·656033
26	1·493711	76	1·651660	26	1·504704	76	1·656774
27	1·502779	77	1·652480	27	1·513742	77	1·657483
28	1·511255	78	1·653269	28	1·522174	78	1·658162
29	1·519188	79	1·653954	29	1·530054	79	1·658814
30	1·526626	80	1·654758	30	1·537428	80	1·659436
31	1·533609	81	1·655461	31	1·544338	81	1·660037
32	1·540175	82	1·656138	32	1·550823	82	1·660611
33	1·546356	83	1·656789	33	1·556915	83	1·661162
34	1·552181	84	1·657417	34	1·562645	84	1·661691
35	1·557677	85	1·658021	35	1·568041	85	1·662197
36	1·562868	86	1·658602	36	1·573126	86	1·662684
37	1·567777	87	1·659163	37	1·577925	87	1·663150
38	1·572423	88	1·659703	38	1·582455	88	1·663597
39	1·576823	89	1·660223	39	1·586737	89	1·664027
40	1·580996	90	1·660724	40	1·590787	90	1·664439
41	1·584954	91	1·661207	41	1·594621	91	1·664834
42	1·588713	92	1·661672	42	1·598252	92	1·665214
43	1·592285	93	1·662121	43	1·601694	93	1·665578
44	1·595682	94	1·662553	44	1·604959	94	1·665928
45	1·598914	95	1·662970	45	1·608058	95	1·666264
46	1·601991	96	1·663372	46	1·611000	96	1·666586
47	1·604923	97	1·663759	47	1·613796	97	1·666896
48	1·607718	98	1·664133	48	1·616454	98	1·667193
49	1·610383	99	1·664493	49	1·618982	99	1·667478
50	1·612927	100	1·664840	50	1·621388	100	1·667752

TABLE XI.   clxix

Present Value (or Years' Purchase) of £1 per Annum in **n** years, after
t years' Deferrence. Redemption of Capital being at **3½** and **4** per cent., with
Interest allowed to a Purchaser at **20** per cent.

DEFERRED **7** YEARS.

n Years	Redemption 3½ per cent.	n Years	Redemption 3½ per cent.	n Years	Redemption 4 per cent.	n Years	Redemption 4 per cent.
1	·232568	51	1·346131	1	·232568	51	1·353067
2	·403647	52	1·348064	2	·404352	52	1·354885
3	·534707	53	1·349911	3	·536337	53	1·356616
4	·638265	54	1·351677	4	·640846	54	1·358268
5	·722110	55	1·353366	5	·725591	55	1·359843
6	·791344	56	1·354982	6	·795645	56	1·361342
7	·849450	57	1·356579	7	·854482	57	1·362776
8	·898882	58	1·358010	8	·904560	58	1·364144
9	·941426	59	1·359428	9	·947670	59	1·365450
10	·978407	60	1·360787	10	·985143	60	1·366697
11	1·010830	61	1·362090	11	1·017994	61	1·367888
12	1·039474	62	1·363339	12	1·047007	62	1·369027
13	1·064948	63	1·364537	13	1·072799	63	1·370115
14	1·087739	64	1·365685	14	1·095862	64	1·371156
15	1·108238	65	1·366788	15	1·116591	65	1·372151
16	1·126763	66	1·367846	16	1·135310	66	1·373102
17	1·143576	67	1·368861	17	1·152286	67	1·374013
18	1·158997	68	1·369837	18	1·167740	68	1·374884
19	1·172907	69	1·370773	19	1·181857	69	1·375786
20	1·185761	70	1·371674	20	1·194794	70	1·376517
21	1·197589	71	1·372538	21	1·206684	71	1·377281
22	1·208503	72	1·373369	22	1·217641	72	1·378013
23	1·218599	73	1·374168	23	1·227765	73	1·378714
24	1·227960	74	1·374936	24	1·237138	74	1·379386
25	1·236660	75	1·375675	25	1·245835	75	1·380030
26	1·244760	76	1·376385	26	1·253921	76	1·380646
27	1·252318	77	1·377068	27	1·261453	77	1·381237
28	1·259380	78	1·377725	28	1·268480	78	1·381803
29	1·265992	79	1·378297	29	1·275046	79	1·382346
30	1·272190	80	1·378966	30	1·281191	80	1·382865
31	1·278009	81	1·379553	31	1·286950	81	1·383366
32	1·283481	82	1·380116	32	1·292354	82	1·383845
33	1·288631	83	1·380659	33	1·297431	83	1·384304
34	1·293486	84	1·381182	34	1·302206	84	1·384744
35	1·298066	85	1·381686	35	1·306702	85	1·385166
36	1·302392	86	1·382170	36	1·310940	86	1·385571
37	1·306483	87	1·382638	37	1·314939	87	1·385960
38	1·310354	88	1·383087	38	1·318714	88	1·386333
39	1·314021	89	1·383521	39	1·322282	89	1·386691
40	1·317498	90	1·383938	40	1·325657	90	1·387034
41	1·320797	91	1·384341	41	1·328852	91	1·387364
42	1·323929	92	1·384729	42	1·331878	92	1·387680
43	1·326906	93	1·385102	43	1·334747	93	1·387984
44	1·329736	94	1·385463	44	1·337467	94	1·388275
45	1·332430	95	1·385810	45	1·340050	95	1·388555
46	1·334994	96	1·386145	46	1·342502	96	1·388823
47	1·337437	97	1·386568	47	1·344832	97	1·389082
48	1·339766	98	1·386779	48	1·347047	98	1·389329
49	1·341988	99	1·387079	49	1·349153	99	1·389567
50	1·344107	100	1·387368	50	1·351158	100	1·389795

Present Value (or Years' Purchase) of £1 per Annum in n years, after
t years' Deferrence. Redemption of Capital being at 3½ and 4 per cent., with
Interest allowed to a Purchaser at 20 per cent.

DEFERRED 8 YEARS.

n Years	Redemption 3½ per cent.	n Years	Redemption 3½ per cent.	n Years	Redemption 4 per cent.	n Years	Redemption 4 per cent.
1	·193807	51	1·121774	1	·193807	51	1·127544
2	·336372	52	1·123385	2	·336959	52	1·129069
3	·445589	53	1·124924	3	·446947	53	1·130512
4	·531887	54	1·126396	4	·534038	54	1·131888
5	·601757	55	1·127803	5	·604658	55	1·133200
6	·659453	56	1·129150	6	·663037	56	1·134451
7	·707874	57	1·130481	7	·712067	57	1·135645
8	·749068	58	1·131673	8	·753799	58	1·136785
9	·784521	59	1·132855	9	·789723	59	1·137873
10	·815338	60	1·133988	10	·820951	60	1·138912
11	·842357	61	1·135073	11	·848327	61	1·139905
12	·866227	62	1·136114	12	·872505	62	1·140854
13	·887456	63	1·137112	13	·893998	63	1·141761
14	·906448	64	1·138069	14	·913217	64	1·142628
15	·923530	65	1·138988	15	·930491	65	1·143457
16	·938967	66	1·139870	16	·946091	66	1·144250
17	·952979	67	1·140716	17	·960237	67	1·145009
18	·965747	68	1·141528	18	·973115	68	1·145735
19	·977421	69	1·142310	19	·984879	69	1·146487
20	·988133	70	1·143060	20	·995660	70	1·147096
21	·997989	71	1·143780	21	1·005568	71	1·147733
22	1·007084	72	1·144473	22	1·014700	72	1·148343
23	1·015497	73	1·145138	23	1·023136	73	1·148927
24	1·023299	74	1·145778	24	1·030947	74	1·149487
25	1·030548	75	1·146394	25	1·038194	75	1·150023
26	1·037299	76	1·146986	26	1·044933	76	1·150537
27	1·043597	77	1·147555	27	1·051209	77	1·151029
28	1·049482	78	1·148103	28	1·057065	78	1·151501
29	1·054991	79	1·148579	29	1·062537	79	1·151954
30	1·060157	80	1·149137	30	1·067658	80	1·152386
31	1·065006	81	1·149625	31	1·072457	81	1·152803
32	1·069566	82	1·150095	32	1·076960	82	1·153202
33	1·073858	83	1·150548	33	1·081191	83	1·153585
34	1·077903	84	1·150983	34	1·085170	84	1·153951
35	1·081720	85	1·151403	35	1·088917	85	1·154303
36	1·085325	86	1·151807	36	1·092449	86	1·154641
37	1·088734	87	1·152196	37	1·095781	87	1·154965
38	1·091960	88	1·152571	38	1·098927	88	1·155276
39	1·095016	89	1·152932	39	1·101900	89	1·155574
40	1·097913	90	1·153280	40	1·104713	90	1·155860
41	1·100662	91	1·153616	41	1·107375	91	1·156135
42	1·103273	92	1·153939	42	1·109897	92	1·156398
43	1·105753	93	1·154250	43	1·112287	93	1·156651
44	1·108112	94	1·154551	44	1·114554	94	1·156894
45	1·110357	95	1·154840	45	1·116706	95	1·157127
46	1·112494	96	1·155119	46	1·118750	96	1·157351
47	1·114530	97	1·155388	47	1·120691	97	1·157566
48	1·116470	98	1·155647	48	1·122537	98	1·157773
49	1·118322	99	1·155897	49	1·124293	99	1·157971
50	1·120088	100	1·156139	50	1·125964	100	1·158161

TABLE XI. clxxi

Present Value (or Years' Purchase) of £1 per Annum in n years, after t years' Deferrence. Redemption of Capital being at 3½ and 4 per cent., with Interest allowed to a Purchaser at 20 per cent.

DEFERRED 9 YEARS.

n Years	Redemption 3½ per cent.	n Years	Redemption 3½ per cent.	n Years	Redemption 4 per cent.	n Years	Redemption 4 per cent.
1	·161506	51	·934814	1	·161506	51	·939630
2	·280311	52	·936156	2	·280800	52	·940892
3	·371325	53	·937439	3	·372456	53	·942095
4	·443240	54	·938665	4	·445032	54	·943242
5	·501465	55	·939838	5	·503883	55	·944335
6	·549545	56	·940960	6	·552532	56	·945378
7	·589896	57	·942069	7	·593390	57	·946373
8	·624224	58	·943163	8	·628167	58	·947322
9	·653768	59	·944048	9	·658104	59	·948229
10	·679449	60	·944991	10	·684127	60	·949095
11	·701965	61	·945896	11	·706940	61	·949923
12	·721857	62	·946763	12	·727088	62	·950713
13	·739548	63	·947595	13	·745000	63	·951469
14	·755375	64	·948393	14	·761015	64	·952192
15	·769610	65	·949158	15	·775411	65	·952883
16	·782474	66	·949893	16	·788410	66	·953544
17	·794151	67	·950599	17	·800199	67	·954176
18	·804790	68	·951276	18	·810931	68	·954781
19	·814519	69	·951926	19	·820734	69	·955407
20	·823445	70	·952551	20	·829718	70	·955915
21	·831659	71	·953152	21	·837975	71	·956446
22	·839238	72	·953729	22	·845585	72	·956954
23	·846249	73	·954284	23	·852615	73	·957441
24	·852750	74	·954817	24	·859124	74	·957907
25	·858792	75	·955330	25	·865164	75	·958354
26	·864417	76	·955823	26	·870779	76	·958782
27	·869665	77	·956298	27	·876009	77	·959193
28	·874570	78	·956754	28	·880889	78	·959586
29	·879161	79	·957151	29	·885449	79	·959963
30	·883466	80	·957616	30	·889716	80	·960323
31	·887507	81	·958023	31	·893716	81	·960671
32	·891306	82	·958414	32	·897468	82	·961003
33	·894883	83	·958792	33	·900994	83	·961322
34	·898254	84	·959155	34	·904310	84	·961628
35	·901435	85	·959504	35	·907432	85	·961921
36	·904439	86	·959841	36	·910375	86	·962203
37	·907280	87	·960165	37	·913152	87	·962472
38	·909968	88	·960478	38	·915774	88	·962731
39	·912515	89	·960779	39	·918252	89	·962980
40	·914929	90	·961069	40	·920596	90	·963218
41	·917220	91	·961348	41	·922814	91	·963447
42	·919396	92	·961617	42	·924916	92	·963667
43	·921463	93	·961877	43	·926908	93	·963878
44	·923428	94	·962127	44	·928797	94	·964080
45	·925299	95	·962368	45	·930590	95	·964275
46	·927080	96	·962601	46	·932293	96	·964461
47	·928776	97	·962825	47	·933911	97	·964640
48	·930394	98	·963041	48	·935449	98	·964812
49	·931936	99	·963249	49	·936912	99	·964977
50	·933408	100	·963450	50	·938305	100	·965136

Present Value (or Years' Purchase) of £1 per Annum in **n** years, after t years' Deferrence.   Redemption of Capital being at 3½ and 4 per cent., with Interest allowed to a Purchaser at 20 per cent.

DEFERRED 10 YEARS.

n Years	Redemption 3½ per cent.	n Years	Redemption 3½ per cent.	n Years	Redemption 4 per cent.	n Years	Redemption 4 per cent.
1	·134588	51	·779012	1	·134588	51	·783026
2	·233593	52	·780131	2	·234000	52	·784078
3	·309437	53	·781200	3	·310380	53	·785080
4	·369367	54	·782222	4	·370860	54	·786035
5	·417888	55	·783199	5	·419903	55	·786946
6	·457954	56	·784134	6	·460443	56	·787815
7	·491580	57	·785058	7	·494492	57	·788645
8	·520187	58	·785886	8	·523473	58	·789436
9	·544808	59	·786707	9	·548421	59	·790192
10	·566208	60	·787494	10	·570106	60	·790913
11	·584972	61	·788247	11	·589118	61	·791603
12	·601548	62	·788970	12	·605908	62	·792262
13	·616290	63	·789663	13	·620834	63	·792892
14	·629479	64	·790328	14	·634180	64	·793494
15	·641342	65	·790966	15	·646176	65	·794070
16	·652063	66	·791578	16	·657009	66	·794621
17	·661793	67	·792166	17	·666833	67	·795147
18	·670659	68	·792731	18	·675776	68	·795652
19	·678767	69	·793273	19	·683946	69	·796174
20	·686205	70	·793794	20	·691432	70	·796597
21	·693050	71	·794294	21	·698313	71	·797039
22	·609366	72	·794775	22	·704655	72	·797463
23	·705209	73	·795237	23	·710513	73	·797868
24	·710626	74	·795682	24	·715937	74	·798257
25	·715661	75	·796109	25	·720970	75	·798629
26	·720348	76	·796520	26	·725650	76	·798986
27	·724722	77	·796916	27	·730009	77	·799328
28	·728809	78	·797296	28	·734075	78	·799656
29	·732635	79	·797626	29	·737875	79	·799970
30	·736222	80	·798014	30	·741431	80	·800270
31	·739590	81	·798353	31	·744764	81	·800560
32	·742756	82	·798679	32	·747891	82	·800837
33	·745737	83	·798994	33	·750829	83	·801103
34	·748546	84	·799296	34	·753592	84	·801357
35	·751197	85	·799588	35	·756195	85	·801602
36	·753700	86	·799868	36	·758647	86	·801836
37	·756067	87	·800139	37	·760961	87	·802061
38	·758308	88	·800399	38	·763146	88	·802277
39	·760430	89	·800650	39	·765211	89	·802484
40	·762442	90	·800891	40	·767164	90	·802683
41	·764351	91	·801124	41	·769013	91	·802873
42	·766164	92	·801349	42	·770764	92	·803057
43	·767886	93	·801565	43	·772424	93	·803232
44	·769524	94	·801773	44	·773998	94	·803401
45	·771083	95	·801975	45	·775493	95	·803563
46	·772567	96	·802168	46	·776912	96	·803718
47	·773981	97	·802355	47	·778260	97	·803868
48	·775329	98	·802535	48	·779542	98	·804011
49	·776614	99	·802709	49	·780761	99	·804149
50	·777841	100	·802876	50	·781921	100	·804281

TABLE XII.

Comparison of the Difference in Value between the OLD *or ordinary* TABLES *of the Present Value of £1 per annum, and a portion of the* NEW TABLES *in this work, which allow a purchaser interest on his purchase money, or capital invested, at one rate per cent., and redeem the capital so invested at another rate per cent.*

The Difference in Value is shewn in decimals of a £, and in £. s. d., for the rates of **4, 5, 8, 10, 12, 15, 18,** *and* **20** *per cent. per annum. Also the rate per cent. lost on the purchase of every £1 Annuity by the use of the* OLD TABLES.

NOTICE.

The words 'every £1 Annuity,' used above, and in the headings of the third, fourth, and fifth columns all through the Table, must be taken to mean *every £1 of the first year's income purchased.*

The reference made to the same Table, and to the rate per cent. lost on every £1 Annuity, on pp. 52, 54, 56, and 57, must also be taken in the same sense; the rate per cent. lost on the *Capital* being a distinct question. For example: an Annuity of £1 for 21 years, at 15 per cent. on the Capital, and the same rate for redemption, is worth £6·31246. But if we can redeem the *Capital* at only 3 *per cent.* the value is £5·40915, showing a difference of ·90331; and this is the loss, or 90·331 per cent., on the first year's Annuity, or income. For the loss on Capital we have $\frac{·90331 \times 100}{6·31246} = 14·31$ per cent. lost by the use of the OLD TABLES; and $\frac{·90331 \times 100}{5·40915} = 16·69$ per cent. gained by the use of the NEW TABLES.

Comparison of the Difference in Value between the old or ordinary Table of Present Values, and those calculated at one rate of interest on Capital, and at another rate for its Redemption, at the following rates.

Years	The ordinary or old Table of Present Values, Interest on Capital and for Redemption being 4 per cent.	The new Table of Present Values, Interest for Redemption being 3 per cent. and on Capital 4 per cent.	Difference in Excess of True Value on every £1 Annuity purchased by old Table, in Decimals of a Pound	Difference in Excess of True Value on every £1 Annuity purchased by old Table, in Pounds, Shillings, and Pence £	s.	d.	Rate per Cent. lost on the Purchase of every £1 Annuity by the old Table	Years
1	·96154	·96154	·00000	0	0	0	·000	1
2	1·88610	1·87754	·00856	0	0	2	·856	2
3	2·77509	2·75080	·02429	0	0	5¾	2·429	3
4	3·62990	3·58388	·04602	0	0	11	4·602	4
5	4·45182	4·37915	·07267	0	1	5¼	7·267	5
6	5·24214	5·13881	·10333	0	2	0¾	10·333	6
7	6·00206	5·86488	·13718	0	2	8¾	13·718	7
8	6·73275	6·55925	·17350	0	3	5½	17·350	8
9	7·43533	7·22367	·21166	0	4	2¾	21·166	9
10	8·11090	7·85975	·25115	0	5	0¾	25·115	**10**
11	8·76048	8·46902	·29146	0	5	9¾	29·146	11
12	9·38507	9·05288	·33219	0	6	7½	33·219	12
13	9·98565	9·61265	·37300	0	7	5½	37·300	13
14	10·56312	10·14957	·41355	0	8	3¼	41·355	14
15	11·11839	10·66478	·45361	0	9	0¾	45·361	15
16	11·65230	11·15936	·49294	0	9	10¼	49·294	16
17	12·16567	11·63433	·53134	0	10	7½	53·134	17
18	12·65930	12·09063	·56867	0	11	4¼	56·867	18
19	13·13394	12·52915	·60479	0	12	1	60·479	19
20	13·59033	12·95073	·63960	0	12	9½	63·960	**20**
21	14·02916	13·35617	·67299	0	13	5½	67·299	21
22	14·45112	13·74620	·70492	0	14	1	70·492	22
23	14·85684	14·12152	·73532	0	14	8¼	73·532	23
24	15·24696	14·48280	·76416	0	15	3¼	76·416	24
25	15·62208	14·83066	·79142	0	15	9¾	79·142	25
26	15·98277	15·16570	·81707	0	16	4	81·707	26
27	16·32959	15·48846	·84113	0	16	9¾	84·113	27
28	16·66306	15·79948	·86358	0	17	3¼	86·358	28
29	16·98371	16·09926	·88445	0	17	8¼	88·445	29
30	17·29203	16·38827	·90376	0	18	0¾	90·376	**30**
31	17·58849	16·66696	·92153	0	18	5	92·153	31
32	17·87355	16·93577	·93778	0	18	9	93·778	32
33	18·14765	17·19509	·95256	0	19	0½	95·256	33
34	18·41120	17·44532	·96588	0	19	3¾	96·588	34
35	18·66461	17·68682	·97779	0	19	6½	97·779	35
36	18·90828	17·91993	·98835	0	19	9	98·835	36
37	19·14258	18·14499	·99759	0	19	11¼	99·759	37
38	19·36786	18·36232	1·00554	1	0	1¼	100·554	38
39	19·58448	18·57222	1·01226	1	0	2¾	101·226	39
40	19·79277	18·77498	1·01779	1	0	4¼	101·779	**40**
41	19·99305	18·97087	1·02218	1	0	5¼	102·218	41
42	20·18563	19·16014	1·02549	1	0	6	102·549	42
43	20·37079	19·34307	1·02772	1	0	6½	102·772	43
44	20·54884	19·51987	1·02897	1	0	6¾	102·897	44
45	20·72004	19·69079	1·02925	1	0	7	102·925	45
46	20·88465	19·85603	1·02862	1	0	6¾	102·862	46
47	21·04294	20·01581	1·02713	1	0	6½	102·713	47
48	21·19513	20·17033	1·02480	1	0	5¾	102·480	48
49	21·34147	20·31978	1·02169	1	0	5	102·169	49
50	21·48218	20·46434	1·01784	1	0	4¼	101·784	**50**

Comparison of the Difference in Value between the old or ordinary Table of Present Values, and those calculated at one rate of interest on Capital, and at another rate for its Redemption, at the following rates.

Years	The ordinary or old Table of Present Values, Interest on Capital and for Redemption being 5 per cent.	The new Table of Present Values, Interest for Redemption being 3 per cent. and on Capital 5 per cent.	Difference in Excess of True Value on every £1 Annuity purchased by old Table, in Decimals of a Pound	Difference in Excess of True Value on every £1 Annuity purchased by old Table, in Pounds, Shillings, and Pence £	s.	d.	Rate per Cent. lost on the Purchase of every £1 Annuity by the old Table	Years
1	·95238	·95238	·00000	0	0	0	·000	1
2	1·85941	1·84294	·01647	0	0	4	1·647	2
3	2·72325	2·67716	·04609	0	0	11¼	4·609	3
4	3·54595	3·45988	·08607	0	1	8¾	8·607	4
5	4·32948	4·19543	·13405	0	2	8	13·405	5
6	5·07569	4·88765	·18804	0	3	9¼	18·804	6
7	5·78637	5·53997	·24640	0	4	11¼	24·640	7
8	6·46321	6·15550	·30771	0	6	1¾	30·771	8
9	7·10782	6·73701	·37081	0	7	5	37·081	9
10	7·72174	7·28701	·43473	0	8	8¼	43·473	**10**
11	8·30641	7·80778	·49863	0	9	11½	49·863	11
12	8·86325	8·30137	·56188	0	11	2¾	56·188	12
13	9·39357	8·76966	·62391	0	12	5¾	62·391	13
14	9·89864	9·21435	·68429	0	13	8¼	68·429	14
15	10·37966	9·63701	·74265	0	14	10	74·265	15
16	10·83777	10·03907	·79870	0	15	11½	79·870	16
17	11·27407	10·42182	·85225	0	17	0½	85·225	17
18	11·68959	10·78647	·90312	0	18	0¾	90·312	18
19	12·08532	11·13414	·95118	0	19	0¼	95·118	19
20	12·46221	11·46582	·99639	0	19	11¼	99·639	**20**
21	12·82115	11·78248	1·03867	1	0	9¼	103·867	21
22	13·16300	12·08497	1·07803	1	1	6¾	107·803	22
23	13·48857	12·37411	1·11446	1	2	3½	111·446	23
24	13·79864	12·65063	1·14801	1	2	11½	114·801	24
25	14·09395	12·91525	1·17870	1	3	6¾	117·870	25
26	14·37519	13·16859	1·20660	1	4	1½	120·660	26
27	14·64303	13·41126	1·23177	1	4	7½	123·177	27
28	14·89813	13·64382	1·25431	1	5	1	125·431	28
29	15·14107	13·86680	1·27427	1	5	5¾	127·427	29
30	15·37245	14·08069	1·29176	1	5	10	129·176	**30**
31	15·59281	14·28593	1·30688	1	6	1½	130·688	31
32	15·80268	14·48297	1·31971	1	6	4¾	131·971	32
33	16·00255	14·67220	1·33035	1	6	7¼	133·035	33
34	16·19290	14·85399	1·33891	1	6	9¼	133·891	34
35	16·37419	15·02871	1·34548	1	6	11	134·548	35
36	16·54685	15·19669	1·35016	1	7	0	135·016	36
37	16·71129	15·35824	1·35305	1	7	0¾	135·305	37
38	16·86789	15·51366	1·35423	1	7	1	135·423	38
39	17·01704	15·66322	1·35382	1	7	0¾	135·382	39
40	17·15909	15·80718	1·35191	1	7	0½	135·191	**40**
41	17·29437	15·94581	1·34856	1	6	11½	134·856	41
42	17·42321	16·07932	1·34389	1	6	10½	134·389	42
43	17·54591	16·20795	1·33796	1	6	9	133·796	43
44	17·66277	16·33190	1·33087	1	6	7¼	133·087	44
45	17·77407	16·45138	1·32269	1	6	5½	132·269	45
46	17·88007	16·56657	1·31350	1	6	3¼	131·350	46
47	17·98101	16·67764	1·30337	1	6	0¾	130·337	47
48	18·07716	16·78478	1·29238	1	5	10¼	129·238	48
49	18·16872	16·88814	1·28058	1	5	6¼	128·058	49
50	18·25593	16·98788	1·26805	1	5	4½	126·805	**50**

Comparison of the Difference in Value between the old or ordinary Table of Present Values, and those calculated at one rate of interest on Capital, and at another rate for its Redemption, at the following rates.

Years	The ordinary or old Table of Present Values, Interest on Capital and for Redemption being 8 per cent.	The new Table of Present Values, Interest for Redemption being 3 per cent. and on Capital 8 per cent.	Difference in Excess of True Value on every £1 Annuity purchased by old Table, in Decimals of a Pound	Difference in Excess of True Value on every £1 Annuity purchased by old Table, in Pounds, Shillings, and Pence £	s.	d.	Rate per Cent. lost on the Purchase of every £1 Annuity by the old Table	Years
1	·92593	·92593	·00000	0	0	0	·000	1
2	1·78327	1·74639	·03688	0	0	9	3·688	2
3	2·57710	2·47813	·09897	0	1	11¾	9·897	3
4	3·31213	3·13453	·17760	0	3	6¾	17·760	4
5	3·99271	3·72641	·26630	0	5	4	26·630	5
6	4·62288	4·26262	·36026	0	7	2½	36·026	6
7	5·20637	4·75045	·45592	0	9	1½	45·592	7
8	5·74664	5·19598	·55066	0	11	0¼	55·066	8
9	6·24689	5·60432	·64257	0	12	10¼	64·257	9
10	6·71008	5·97977	·73031	0	14	7¼	73·031	**10**
11	7·13896	6·32601	·81295	0	16	3	81·295	11
12	7·53608	6·64619	·88989	0	17	9½	88·989	12
13	7·90378	6·94302	·96076	0	19	2½	96·076	13
14	8·24424	7·21884	1·02540	1	0	6	102·540	14
15	8·55948	7·47571	1·08377	1	1	8	108·377	15
16	8·85137	7·71540	1·13597	1	2	8½	113·597	16
17	9·12164	7·93950	1·18214	1	3	7¾	118·214	17
18	9·37189	8·14938	1·22251	1	4	5½	122·251	18
19	9·60360	8·34628	1·25732	1	5	1¾	125·732	19
20	9·81815	8·53128	1·28687	1	5	8¾	128·687	**20**
21	10·01680	8·70536	1·31144	1	6	2¾	131·144	21
22	10·20074	8·86938	1·33136	1	6	7½	133·136	22
23	10·37106	9·02414	1·34692	1	6	11¼	134·692	23
24	10·52876	9·17032	1·35844	1	7	2	135·844	24
25	10·67478	9·30857	1·36621	1	7	4	136·621	25
26	10·80998	9·43946	1·37052	1	7	5	137·052	26
27	10·93517	9·56350	1·37167	1	7	5¼	137·167	27
28	11·05108	9·68118	1·36990	1	7	4¾	136·990	28
29	11·15841	9·79291	1·36550	1	7	3¾	136·550	29
30	11·25778	9·89910	1·35868	1	7	2	135·868	**30**
31	11·34980	10·00011	1·34969	1	6	11¾	134·969	31
32	11·43500	10·09626	1·33874	1	6	9¼	133·874	32
33	11·51389	10·18785	1·32604	1	6	6¼	132·604	33
34	11·58693	10·27517	1·31176	1	6	2¾	131·176	34
35	11·65457	10·35848	1·29609	1	5	11	129·609	35
36	11·71719	10·43800	1·27919	1	5	7	127·919	36
37	11·77518	10·51396	1·26122	1	5	2¾	126·122	37
38	11·82887	10·58657	1·24230	1	4	10¼	124·230	38
39	11·87858	10·65600	1·22258	1	4	5½	122·258	39
40	11·92461	10·72244	1·20217	1	4	0½	120·217	**40**
41	11·96724	10·78604	1·18120	1	3	7½	118·120	41
42	12·00670	10·84697	1·15973	1	3	2½	115·973	42
43	12·04324	10·90535	1·13789	1	2	9	113·789	43
44	12·07707	10·96132	1·11575	1	2	3¾	111·575	44
45	12·10840	11·01501	1·09339	1	1	10½	109·339	45
46	12·13741	11·06653	1·07088	1	1	5	107·088	46
47	12·16427	11·11599	1·04828	1	0	11½	104·828	47
48	12·18914	11·16348	1·02566	1	0	6	102·566	48
49	12·21216	11·20911	1·00305	1	0	0½	100·305	49
50	12·23349	11·25296	·98053	0	19	7½	98·053	**50**

Comparison of the Difference in Value between the old or ordinary Table of Present Values, and those calculated at one rate of interest on Capital, and at another rate for its Redemption, at the following rates.

Years	The ordinary or old Table of Present Values, Interest on Capital and for Redemption being 10 per cent.	The new Table of Present Values, Interest for Redemption being 3 per cent. and on Capital 10 per cent.	Difference in Excess of True Value on every £1 Annuity purchased by old Table, in Decimals of a Pound	Difference in Excess of True Value on every £1 Annuity purchased by old Table, in Pounds, Shillings, and Pence			Rate per Cent. lost on the Purchase of every £1 Annuity by the old Table	Years
				£	s.	d.		
1	·90909	·90909	·00000	0	0	0	·000	1
2	1·73554	1·68745	·04809	0	0	11½	4·809	2
3	2·48685	2·36111	·12574	0	2	6¼	12·574	3
4	3·16987	2·94962	·22025	0	4	4¾	22·025	4
5	3·79079	3·46795	·32284	0	6	5½	32·284	5
6	4·35526	3·92777	·42749	0	8	6½	42·749	6
7	4·86842	4·33828	·53014	0	10	7¼	53·014	7
8	5·33493	4·70685	·62808	0	12	6¾	62·808	8
9	5·75902	5·03946	·71956	0	14	4¾	71·956	9
10	6·14457	5·34101	·80356	0	16	1	80·356	**10**
11	6·49506	5·61553	·87953	0	17	7¼	87·953	11
12	6·81369	5·86641	·94728	0	18	11¼	94·728	12
13	7·10336	6·09646	1·00690	1	0	1½	100·690	13
14	7·36669	6·30810	1·05859	1	1	2	105·859	14
15	7·60608	6·50336	1·10272	1	2	0¾	110·272	15
16	7·82371	6·68401	1·13970	1	2	9½	113·970	16
17	8·02155	6·85154	1·17001	1	3	4¾	117·001	17
18	8·20141	7·00728	1·19413	1	3	10½	119·413	18
19	8·36492	7·15237	1·21255	1	4	3	121·255	19
20	8·51356	7·28780	1·22576	1	4	6	122·576	**20**
21	8·64869	7·41445	1·23424	1	4	8¼	123·424	21
22	8·77154	7·53310	1·23844	1	4	9¼	123·844	22
23	8·88322	7·64445	1·23877	1	4	9½	123·877	23
24	8·98474	7·74909	1·23565	1	4	8½	123·565	24
25	9·07704	7·84758	1·22946	1	4	7¼	122·946	25
26	9·16095	7·94040	1·22055	1	4	5	122·055	26
27	9·23722	8·02799	1·20923	1	4	2¼	120·923	27
28	9·30657	8·11075	1·19582	1	3	11	119·582	28
29	9·36961	8·18902	1·18059	1	3	7¼	118·059	29
30	9·42691	8·26315	1·16376	1	3	3¼	116·376	**30**
31	9·47901	8·33341	1·14560	1	2	11	114·560	31
32	9·52638	8·40007	1·12631	1	2	6¼	112·631	32
33	9·56943	8·46338	1·10605	1	2	1½	110·605	33
34	9·60858	8·52355	1·08503	1	1	8½	108·503	34
35	9·64416	8·58080	1·06336	1	1	3¼	106·336	35
36	9·67651	8·63530	1·04121	1	0	10	104·121	36
37	9·70592	8·68722	1·01870	1	0	4½	101·870	37
38	9·73265	8·73673	·99592	0	19	11	99·592	38
39	9·75696	8·78396	·97300	0	19	5½	97·300	39
40	9·77905	8·82906	·94999	0	19	0	94·999	**40**
41	9·79914	8·87214	·92710	0	18	6½	92·710	41
42	9·81740	8·91332	·90408	0	18	1	90·408	42
43	9·83400	8·95270	·88130	0	17	7½	88·130	43
44	9·84909	8·99039	·85870	0	17	2¼	85·870	44
45	9·86281	9·02648	·83633	0	16	8¾	83·633	45
46	9·87528	9·06105	·81423	0	16	3¼	81·423	46
47	9·88662	9·09417	·79245	0	15	10¼	79·245	47
48	9·89693	9·12594	·77099	0	15	5	77·099	48
49	9·90630	9·15641	·74989	0	14	11¾	74·989	49
50	9·91481	9·18565	·72916	0	14	7	72·916	**50**

Comparison of the Difference in Value between the old or ordinary Table of Present Values, and those calculated at one rate of interest on Capital, and at another rate for its Redemption, at the following rates.

Years	The ordinary or old Table of Present Values, Interest on Capital and for Redemption being 12 per cent.	The new Table of Present Values, Interest for Redemption being 3 per cent. and on Capital 12 per cent.	Difference in Excess of True Value on every £1 Annuity purchased by old Table, in Decimals of a Pound	Difference in Excesses of True Value on every £1 Annuity purchased by old Table, in Pounds, Shillings, and Pence £	s.	d.	Rate per Cent. lost on the Purchase of every £1 Annuity by the old Table	Years
1	·89286	·89286	·00000	0	0	0	·000	1
2	1·69005	1·63236	·05769	0	1	1¾	5·769	2
3	2·40183	2·25464	·14719	0	2	11¼	14·719	3
4	3·03735	2·78531	·25204	0	5	0½	25·204	4
5	3·60478	3·24302	·36176	0	7	2¾	36·176	5
6	4·06131	3·64169	·41962	0	8	4¾	41·962	6
7	4·56376	3·99191	·57185	0	11	5¼	57·185	7
8	4·96764	4·30188	·66576	0	13	3¾	66·576	8
9	5·32825	4·57804	·75021	0	15	0	75·021	9
10	5·65022	4·82554	·82468	0	16	6	82·468	**10**
11	5·93770	5·04853	·88917	0	17	9¼	88·917	11
12	6·19437	5·25039	·94398	0	18	10½	94·398	12
13	6·42355	5·43391	·98964	0	19	9½	98·964	13
14	6·62817	5·60141	1·02676	1	0	6½	102·676	14
15	6·81086	5·75485	1·05601	1	1	1½	105·601	15
16	6·97399	5·89585	1·07814	1	1	6¾	107·814	16
17	7·11963	6·02582	1·09381	1	1	10½	109·381	17
18	7·24967	6·14595	1·10372	1	2	0¾	110·372	18
19	7·36578	6·25728	1·10850	1	2	2	110·850	19
20	7·46944	6·36069	1·10875	1	2	2¼	110·875	**20**
21	7·56200	6·45695	1·10505	1	2	1¼	110·505	21
22	7·64465	6·54676	1·09789	1	1	11½	109·789	22
23	7·71843	6·63469	1·08374	1	1	8	108·374	23
24	7·78432	6·70927	1·07505	1	1	6	107·505	24
25	7·84314	6·78298	1·06016	1	1	2¼	106·016	25
26	7·89566	6·85221	1·04345	1	0	10½	104·345	26
27	7·94255	6·91734	1·02521	1	0	6	102·521	27
28	7·98442	6·97870	1·00572	1	0	1¼	100·572	28
29	8·02181	7·03657	·98524	0	19	8½	98·524	29
30	8·05518	7·09123	·96395	0	19	3¼	96·395	**30**
31	8·08499	7·14291	·94208	0	18	10	94·208	31
32	8·11159	7·19183	·91976	0	18	4¾	91·976	32
33	8·13535	7·23819	·89716	0	17	11¼	89·716	33
34	8·15656	7·28216	·87440	0	17	5¾	87·440	34
35	8·17550	7·32390	·85160	0	17	0½	85·160	35
36	8·19241	7·36356	·82885	0	16	6¾	82·885	36
37	8·20751	7·40129	·80622	0	16	1½	80·622	37
38	8·22099	7·43719	·78380	0	15	8	78·380	38
39	8·23303	7·47139	·76164	0	15	2¾	76·164	39
40	8·24378	7·50399	·73979	0	14	9½	73·979	**40**
41	8·25337	7·53509	·71828	0	14	4¼	71·828	41
42	8·26194	7·56477	·69717	0	13	11¼	69·717	42
43	8·26959	7·59312	·67647	0	13	6¼	67·647	43
44	8·27642	7·62018	·65624	0	13	1½	65·624	44
45	8·28252	7·64613	·63639	0	12	8½	63·639	45
46	8·28796	7·67092	·61704	0	12	4	61·704	46
47	8·29282	7·69465	·59817	0	11	11½	59·817	47
48	8·29716	7·71737	·57979	0	11	7	57·979	48
49	8·30104	7·73915	·56189	0	11	2¾	56·189	49
50	8·30450	7·76003	·54447	0	10	10½	54·447	**50**

Comparison of the Difference in Value between the old or ordinary Table of Present Values, and those calculated at one rate of interest on Capital, and at another rate for its Redemption, at the following rates.

Years	The ordinary or old Table of Present Values, Interest on Capital and for Redemption being 15 per cent.	The new Table of Present Values, Interest for Redemption being 3 per cent. and on Capital 15 per cent.	Difference in Excess of True Value on every £1 Annuity purchased by old Table, in Decimals of a Pound	Difference in Excess of True Value on every £1 Annuity purchased by old Table, in Pounds, Shillings, and Pence £ s. d.	Rate per Cent. lost on the Purchase of every £1 Annuity by the old Table	Years
1	·86957	·86957	·00000	0 0 0	·000	1
2	1·62571	1·55615	·06956	0 1 4½	6·956	2
3	2·28323	2·11180	·17143	0 3 5	17·143	3
4	2·85498	2·57052	·28446	0 5 8¼	28·446	4
5	3·35216	2·95548	·39668	0 7 11	39·668	5
6	3·78448	3·28302	·50146	0 10 0¼	50·146	6
7	4·16042	3·56498	·59544	0 11 10¾	59·544	7
8	4·48732	3·81016	·67716	0 13 6½	67·716	8
9	4·76287	4·02522	·73765	0 14 9	73·765	9
10	5·01877	4·21531	·80346	0 16 9¾	80·346	**10**
11	5·23371	4·38448	·84923	0 16 11¾	84·923	11
12	5·42062	4·53593	·88469	0 17 8¼	88·469	12
13	5·58315	4·67225	·91090	0 18 2½	91·090	13
14	5·72448	4·79556	·92892	0 18 6¾	92·892	14
15	5·84737	4·90758	·93979	0 18 9½	93·979	15
16	5·95423	5·00975	·94448	0 18 10¾	94·448	16
17	6·04716	5·10328	·94388	0 18 10½	94·388	17
18	6·12797	5·18918	·93879	0 18 9¼	93·879	18
19	6·19631	5·26832	·92799	0 18 6½	92·799	19
20	6·25933	5·34143	·91790	0 18 4¼	91·790	**20**
21	6·31246	5·40915	·90331	0 18 0¾	90·331	21
22	6·35866	5·47203	·88663	0 17 8¾	88·663	22
23	6·39884	5·53055	·86829	0 17 4½	86·829	23
24	6·43377	5·58511	·84866	0 16 11¾	84·866	24
25	6·46415	5·63609	·82806	0 16 6¾	82·806	25
26	6·49056	5·68381	·80675	0 16 1½	80·675	26
27	6·51353	5·72855	·78498	0 15 8½	78·498	27
28	6·53351	5·77057	·76294	0 15 3	76·294	28
29	6·55088	5·81008	·74080	0 14 9¾	74·080	29
30	6·56684	5·84729	·71955	0 14 4¾	71·955	**30**
31	6·57911	5·88239	·69672	0 13 11	69·672	31
32	6·59053	5·91528	·67525	0 13 6	67·525	32
33	6·60046	5·94685	·65761	0 13 1¾	65·761	33
34	6·60910	5·97650	·63260	0 12 7¾	63·260	34
35	6·61661	6·00459	·61202	0 12 2¾	61·202	35
36	6·62314	6·03123	·59191	0 11 10	59·191	36
37	6·62881	6·05651	·57230	0 11 5¼	57·230	37
38	6·63375	6·08053	·55322	0 11 0¾	55·322	38
39	6·63805	6·10337	·53468	0 10 8¼	53·468	39
40	6·64178	6·12511	·51667	0 10 4	51·667	**40**
41	6·64502	6·14581	·49921	0 9 11¾	49·921	41
42	6·64785	6·16554	·48231	0 9 7¾	48·231	42
43	6·65030	6·18436	·46544	0 9 3¾	46·594	43
44	6·65244	6·20233	·45011	0 9 0	45·011	44
45	6·65429	6·21948	·43481	0 8 8¼	43·481	45
46	6·65591	6·23587	·42004	0 8 4¾	42·004	46
47	6·65731	6·25154	·40577	0 8 1¼	40·577	47
48	6·65853	6·26654	·39199	0 7 10	39·199	48
49	6·65959	6·28089	·37870	0 7 6¾	37·870	49
50	6·66051	6·29463	·36588	0 7 3¾	36·588	**50**

Comparison of the Difference in Value between the old or ordinary Table of Present Values, and those calculated at one rate of interest on Capital, and at another rate for its Redemption, at the following rates.

Years	The ordinary or old Table of Present Values, Interest on Capital and for Redemption being 18 per cent.	The new Table of Present Values, Interest for Redemption being 3 per cent. and on Capital 18 per cent.	The Difference in Excess of True Value on every £1 Annuity purchased by old Table, in Decimals of a Pound	The Difference in Excess of True Value on every £1 Annuity purchased by old Table, in Pounds, Shillings, and Pence			Rate per Cent. lost on the Purchase of every £1 Annuity by the old Table	Years
				£	s.	d.		
1	·84746	·84746	·00000	0	0	0	·000	1
2	1·56564	1·48674	·07890	0	1	6¾	7·890	2
3	2·17427	1·98598	·18829	0	3	9	18·829	3
4	2·65006	2·38648	·26358	0	5	3¼	26·358	4
5	3·12717	2·71478	·41239	0	8	2¾	41·239	5
6	3·49760	2·98867	·50893	0	10	2	50·893	6
7	3·81153	3·22055	·59098	0	11	9¾	59·098	7
8	4·07757	3·41931	·65826	0	13	1¾	65·826	8
9	4·30302	3·59152	·71150	0	14	2¾	71·150	9
10	4·49409	3·74209	·75200	0	15	0¼	75·200	10
11	4·65601	3·87481	·78120	0	15	7¼	78·120	11
12	4·79322	3·99262	·80060	0	16	0	80·060	12
13	4·90951	4·09786	·81165	0	16	2¾	81·165	13
14	5·00806	4·19241	·81565	0	16	3¾	81·565	14
15	5·09158	4·27777	·81381	0	16	3¼	81·381	15
16	5·16235	4·35519	·80716	0	16	1½	80·716	16
17	5·22233	4·42571	·79662	0	15	11	79·662	17
18	5·27177	4·49017	·78160	0	15	7½	78·160	18
19	5·31624	4·54930	·76694	0	15	4	76·694	19
20	5·35275	4·60372	·74903	0	14	11¾	74·903	20
21	5·38368	4·65394	·72974	0	14	7	72·974	21
22	5·40990	4·70041	·70949	0	14	2¼	70·949	22
23	5·43212	4·74352	·68860	0	13	9¼	68·860	23
24	5·45095	4·78360	·66735	0	13	4	66·735	24
25	5·46691	4·82095	·64596	0	12	11	64·596	25
26	5·48043	4·85582	·62461	0	12	5¾	62·461	26
27	5·49189	4·88844	·60345	0	12	0¾	60·345	27
28	5·50160	4·91900	·58260	0	11	7¾	58·260	28
29	5·50983	4·94769	·56214	0	11	2¼	56·214	29
30	5·51681	4·97465	·54216	0	10	10	54·216	30
31	5·52272	5·00003	·52269	0	10	5¼	52·269	31
32	5·52773	5·02395	·50378	0	10	0½	50·378	32
33	5·53197	5·04653	·48544	0	9	8½	48·544	33
34	5·53557	5·06786	·46771	0	9	4¼	46·771	34
35	5·53862	5·08804	·45058	0	9	0	45·058	35
36	5·54089	5·10715	·43374	0	8	8	43·374	36
37	5·54327	5·12527	·41800	0	8	4¼	41·800	37
38	5·54525	5·14246	·40279	0	8	0½	40·279	38
39	5·54682	5·15879	·38803	0	7	9	38·803	39
40	5·54815	5·17431	·37384	0	7	5½	37·384	40
41	5·54928	5·18908	·36020	0	7	2¼	36·020	41
42	5·55024	5·20314	·34710	0	6	11¼	34·710	42
43	5·55105	5·21654	·33451	0	6	8¼	33·451	43
44	5·55174	5·22931	·32243	0	6	5¼	32·243	44
45	5·55232	5·24150	·31082	0	6	2½	31·082	45
46	5·55281	5·25313	·29978	0	5	11¾	29·978	46
47	5·55322	5·26425	·28897	0	5	9¼	28·897	47
48	5·55359	5·27488	·27871	0	5	6¾	27·871	48
49	5·55389	5·28505	·26884	0	5	4½	26·884	49
50	5·55414	5·29477	·25937	0	5	2¼	25·937	50

Comparison of the Difference in Value between the old or ordinary Table of Present Values, and those calculated at one rate of interest on Capital, and at another rate for its Redemption, at the following rates.

Years	The ordinary or old Table of Present Values, Interest on Capital and for Redemption being 20 per cent.	The new Table of Present Values, Interest for Redemption being 3 per cent. and on Capital 20 per cent.	The Difference in Excess of True Value on every £1 Annuity purchased by old Table, in Decimals of a Pound	The Difference in Excess of True Value on every £1 Annuity purchased by old Table, in Pounds, Shillings, and Pence	Rate per Cent. lost on the Purchase of every £1 Annuity by the old Table	Years
				£ s. d.		
1	·83333	·83333	·00000	0 0 0	·000	1
2	1·52778	1·44381	·08397	0 1 8	8·397	2
3	2·10648	1·91011	·19637	0 3 11	19·637	3
4	2·58873	2·27776	·31097	0 6 2½	31·097	4
5	2·99061	2·57497	·41564	0 8 3¾	41·564	5
6	3·32551	2·82010	·50541	0 10 1¼	50·541	6
7	3·60459	3·02566	·57893	0 11 6¾	57·893	7
8	3·83716	3·20045	·63671	0 12 8¾	63·671	8
9	4·03097	3·35083	·68014	0 13 7	68·014	9
10	4·19247	3·48152	·71095	0 14 2½	71·095	**10**
11	4·32706	3·59612	·73094	0 14 7¼	73·094	11
12	4·43922	3·69738	·74184	0 14 10	74·184	12
13	4·53268	3·78745	·74523	0 14 10¾	74·523	13
14	4·61057	3·86808	·74249	0 14 10	74·249	14
15	4·67547	3·94063	·73484	0 14 8¼	73·484	15
16	4·72956	4·00624	·72332	c 14 5½	72·332	16
17	4·77463	4·06583	·70880	0 14 2	70·880	17
18	4·81219	4·12017	·69202	0 13 10	69·202	18
19	4·84584	4·16990	·67594	0 13 6	67·594	19
20	4·86887	4·21557	·65330	0 13 0¾	65·330	**20**
21	4·89132	4·25764	·63368	0 12 8	63·368	21
22	4·90943	4·29650	·61293	0 12 3	61·293	22
23	4·92453	4·33249	·59204	0 11 10	59·204	23
24	4·93710	4·36591	·57119	0 11 5	57·119	24
25	4·94759	4·39700	·55059	0 11 0	55·059	25
26	4·95632	4·42599	·53033	0 10 7¼	53·033	26
27	4·96360	4·45307	·51053	0 10 2½	51·053	27
28	4·96967	4·47842	·49125	0 9 9¾	49·125	28
29	4·97472	4·50218	·47254	0 9 5¼	47·254	29
30	4·97894	4·52449	·45445	0 9 1	45·445	**30**
31	4·98245	4·54548	·43697	0 8 8¾	43·697	31
32	4·98537	4·56524	·42013	0 8 4¾	42·013	32
33	4·98784	4·58387	·40397	0 8 0¾	40·397	33
34	4·98984	4·60147	·38837	0 7 9	38·837	34
35	4·99154	4·61810	·37344	0 7 5½	37·344	35
36	4·99295	4·63384	·35911	0 7 2	35·911	36
37	4·99412	4·64875	·34537	0 6 10¾	34·537	37
38	4·99510	4·66289	·33221	0 6 7½	33·221	38
39	4·99592	4·67631	·31961	0 6 4½	31·961	39
40	4·99660	4·68906	·30754	0 6 1¾	30·754	**40**
41	4·99717	4·70118	·29597	0 5 11	29·597	41
42	4·99764	4·71272	·28492	0 5 8¼	28·492	42
43	4·99803	4·72371	·27432	0 5 5¾	27·432	43
44	4·99836	4·73418	·26418	0 5 3¼	26·418	44
45	4·99863	4·74417	·25446	0 5 1	25·446	45
46	4·99886	4·75370	·24516	0 4 10¾	24·516	46
47	4·99905	4·76280	·23625	0 4 8½	23·625	47
48	4·99921	4·77150	·22771	0 4 6½	22·771	48
49	4·99934	4·77981	·21953	0 4 4½	21·953	49
50	4·99945	4·78777	·21168	0 4 2¾	21·618	**50**

TABLE XIII.

The Present Value (or Years' Purchase) of £1 per annum in n years; Redemption of Capital being at 2, 2½, 3, and 3½ per cent., with interest allowed to a present purchaser upon his purchase money, or capital invested, at the same rates per cent.

Calculated to 5 places of decimals, and to 100 years for each percentage.

TABLE XIII.   clxxxv

Present Value of £1 per Annum in n years, Redemption of Capital being at 2 and 2½ per cent., with Interest allowed to a Purchaser at the same rates per cent.

n Years	2 per cent.	n Years	2 per cent.	n Years	2½ per cent.	n Years	2½ per cent.
1	·98039	51	31·78785	1	·97561	51	28·64616
2	1·94156	52	32·14495	2	1·92742	52	28·92308
3	2·88388	53	32·49505	3	2·85602	53	29·19325
4	3·80773	54	32·83828	4	3·76197	54	29·45683
5	4·71346	55	33·17479	5	4·64583	55	29·71398
6	5·60143	56	33·50469	6	5·50813	56	29·96456
7	6·47199	57	33·82813	7	6·34939	57	30·20962
8	7·32548	58	34·15523	8	7·17014	58	30·44841
9	8·16224	59	34·45610	9	7·97087	59	30·68137
10	8·98259	60	34·76089	10	8·75206	60	30·90866
11	9·78685	61	35·05969	11	9·51421	61	31·13040
12	10·57534	62	35·35264	12	10·25777	62	31·34673
13	11·34837	63	35·63984	13	10·98319	63	31·55778
14	12·10625	64	35·92142	14	11·69091	64	31·76369
15	12·84926	65	36·19747	15	12·38138	65	31·96458
16	13·57771	66	36·46810	16	13·05500	66	32·16056
17	14·29187	67	36·73344	17	13·71220	67	32·35177
18	14·99203	68	36·99356	18	14·35336	68	32·53831
19	15·67846	69	37·24859	19	14·97889	69	32·72031
20	16·35143	70	37·49862	20	15·58916	70	32·89786
21	17·01121	71	37·74374	21	16·18455	71	33·07108
22	17·65805	72	37·98406	22	16·76541	72	33·24008
23	18·29220	73	38·21967	23	17·32211	73	33·40495
24	18·91393	74	38·45066	24	17·88499	74	33·56581
25	19·52346	75	38·67711	25	18·42438	75	33·72274
26	20·12104	76	38·89913	26	18·95061	76	33·87584
27	20·70690	77	39·11680	27	19·46401	77	34·02521
28	21·28127	78	39·33019	28	19·96489	78	34·17094
29	21·84439	79	39·53940	29	20·45355	79	34·31311
30	22·39646	80	39·74451	30	20·93029	80	34·45182
31	22·93770	81	39·94560	31	21·39541	81	34·58714
32	23·46834	82	40·14275	32	21·84918	82	34·71916
33	23·98856	83	40·33603	33	22·29188	83	34·84796
34	24·49859	84	40·52552	34	22·72379	84	34·97362
35	24·99862	85	40·71129	35	23·14516	85	35·09622
36	25·48884	86	40·89342	36	23·55625	86	35·21582
37	25·96945	87	41·07198	37	23·95732	87	35·33251
38	26·44064	88	41·24704	38	24·34860	88	35·44635
39	26·90259	89	41·41867	39	24·73034	89	35·55741
40	27·35548	90	41·58693	40	25·10278	90	35·66577
41	27·79949	91	41·75189	41	25·46612	91	35·77148
42	28·23479	92	41·91362	42	25·82061	92	35·87462
43	28·66156	93	42·07218	43	26·16645	93	35·97524
44	29·07996	94	42·22762	44	26·50385	94	36·07340
45	29·49016	95	42·38002	45	26·83302	95	36·16917
46	29·89231	96	42·52943	46	27·15417	96	36·26261
47	30·28658	97	42·67592	47	27·46748	97	36·35376
48	30·67312	98	42·81953	48	27·77315	98	36·44269
49	31·05208	99	42·96032	49	28·07137	99	36·52946
50	31·42361	100	43·09835	50	28·36231	100	36·61411

Present Value of £1 per Annum in n years, Redemption of Capital being at 3 and 3½ per cent., with Interest allowed to a Purchaser at the same rates per cent.

n Years	3 per cent.	n Years	3 per cent.	n Years	3½ per cent.	n Years	3½ per cent.
1	·97087	51	25·95123	1	·96618	51	23·62862
2	1·91347	52	26·16624	2	1·89969	52	23·79577
3	2·82861	53	26·37499	3	2·80164	53	23·95726
4	3·71710	54	26·57766	4	3·67308	54	24·11330
5	4·57971	55	26·77443	5	4·51505	55	24·26405
6	5·41719	56	26·96546	6	5·32855	56	24·40971
7	6·23028	57	27·15094	7	6·11454	57	24·55045
8	7·01969	58	27·33101	8	6·87396	58	24·68642
9	7·78611	59	27·50583	9	7·60769	59	24·81780
10	8·53020	60	27·67556	10	8·31661	60	24·94473
11	9·25262	61	27·84035	11	9·00155	61	25·06738
12	9·95400	62	28·00034	12	9·66333	62	25·18587
13	10·63496	63	28·15567	13	10·30274	63	25·30036
14	11·29607	64	28·30648	14.	10·92052	64	25·41097
15	11·93794	65	28·45289	15	11·51741	65	25·51785
16	12·56100	66	28·59504	16	12·09412	66	25·62111
17	13·16612	67	28·73305	17	12·65132	67	25·72088
18	13·75351	68	28·86704	18	13·18968	68	25·81726
19	14·32380	69	28·99712	19	13·70984	69	25·91041
20	14·87748	70	29·12342	20	14·21240	70	26·00040
21	15·41502	71	29·24604	21	14·69797	71	26·08734
22	15·93692	72	29·36509	22	15·16713	72	26·17134
23	16·44361	73	29·48067	23	15·62041	73	26·25251
24	16·93554	74	29·59288	24	16·05837	74	26·33092
25	17·41315	75	29·70183	25	16·48152	75	26·40669
26	17·87684	76	29·80760	26	16·89035	76	26·47989
27	18·32703	77	29·91029	27	17·28537	77	26·55062
28	18·76412	78	30·01000	28	17·66702	78	26·61896
29	19·18856	79	30·10679	29	18·03577	79	26·68498
30	19·60044	80	30·20076	30	18·39205	80	26·74878
31	20·00043	81	30·29200	31	18·73628	81	26·81041
32	20·38877	82	30·38059	32	19·06887	82	26·86996
33	20·76579	83	30·46659	33	19·39021	83	26·92750
34	21·13184	84	30·55009	34	19·70068	84	26·98309
35	21·48722	85	30·63115	35	20·00066	85	27·03680
36	21·83225	86	30·70986	36	20·29049	86	27·08870
37	22·16724	87	30·78627	37	20·57053	87	27·13884
38	22·49246	88	30·86045	38	20·84109	88	27·18729
39	22·80822	89	30·93248	39	21·10258	89	27·23409
40	23·11477	90	31·00241	40	21·35507	90	27·27932
41	23·41240	91	31·07030	41	21·59910	91	27·32301
42	23·70136	92	31·13621	42	21·83488	92	27·36523
43	23·98199	93	31·20021	43	22·06269	93	27·40602
44	24·25427	94	31·26234	44	22·28279	94	27·45543
45	24·51871	95	31·32266	45	22·49545	95	27·48350
46	24·77545	96	31·38122	46	22·70092	96	27·52029
47	25·02471	97	31·43808	47	22·89944	97	27·55584
48	25·26671	98	31·49328	48	23·09124	98	27·59018
49	25·50166	99	31·54687	49	23·27656	99	27·62337
50	25·72976	100	31·59891	50	23·45562	100	27·65543

TABLE XIV.

Multiples of the Present Value of £1 per annum in n years. Available Interest on Capital, 2½ per cent., Redemption being at the rate of 3 per cent.

Calculated to 9 places of decimals.

TABLE XIV. clxxxix

Multiples of the Present Value of £1 per Annum in n years. Interest on Capital 21 per cent., Redemption 3 per cent.

Years	Annuity £1, £10, £100, £1000 or £100,000,000	Years	Annuity £1, £10, £100, £1000 or £100,000,000	Years	Annuity £2, £20, £200, £2000 or £200,000,000	Years	Annuity £2, £20, £200, £2000 or £200,000,000
1	·826446280	51	4·575950694	1	1·652892560	51	9·151901388
2	1·423262988	52	4·582590532	2	2·846525976	52	9·165181064
3	1·874304059	53	4·588951424	3	3·748608118	53	9·177902848
4	2·227037348	54	4·595047977	4	4·454074696	54	9·190095954
5	2·510326407	55	4·600893831	5	5·020652814	55	9·201787662
6	2·742750563	56	4·606501736	6	5·485501126	56	9·213003472
7	2·936802761	57	4·611883630	7	5·873605522	57	9·223767260
8	3·101194563	58	4·617050690	8	6·202389126	58	9·234101380
9	3·242186216	59	4·622013405	9	6·484372432	59	9·244026810
10	3·364392206	60	4·626781616	10	6·728784412	60	9·253563232
11	3·471288737	61	4·631364575	11	6·942577474	61	9·262729150
12	3·565544335	62	4·635770791	12	7·131088670	62	9·271541582
13	3·649241557	63	4·640009037	13	7·298483114	63	9·280018074
14	3·724029470	64	4·644086453	14	7·448058940	64	9·288172906
15	3·791230860	65	4·648010514	15	7·582461720	65	9·296021028
16	3·851919142	66	4·651788088	16	7·703838284	66	9·303576176
17	3·906974478	67	4·655425670	17	7·813948956	67	9·310851340
18	3·957125402	68	4·658929397	18	7·914250804	68	9·317858794
19	4·002980129	69	4·662305077	19	8·005960258	69	9·324610154
20	4·045050412	70	4·665558205	20	8·090100824	70	9·331116410
21	4·083769939	71	4·668693992	21	8·167539879	71	9·337387984
22	4·119508680	72	4·671717378	22	8·239017360	72	9·343434756
23	4·152584168	73	4·674633046	23	8·305168336	73	9·349266092
24	4·183270487	74	4·677445445	24	8·366540974	74	9·354890890
25	4·211805445	75	4·680158806	25	8·423610890	75	9·360317612
26	4·238396399	76	4·682777136	26	8·476792798	76	9·365554272
27	4·263224976	77	4·685304265	27	8·526449952	77	9·370608530
28	4·286450941	78	4·687743826	28	8·572901882	78	9·375487652
29	4·308215397	79	4·690099281	29	8·616430794	79	9·380198562
30	4·328643434	80	4·692373931	30	8·657286868	80	9·384747862
31	4·347846336	81	4·694570917	31	8·695692672	81	9·389141834
32	4·365923441	82	4·696693240	32	8·731846882	82	9·393386480
33	4·382963699	83	4·698743762	33	8·765927398	83	9·397487524
34	4·399046995	84	4·700725212	34	8·798093990	84	9·401450424
35	4·414245285	85	4·702640204	35	8·828490570	85	9·405280408
36	4·428623547	86	4·704491229	36	8·857247094	86	9·408982458
37	4·442240611	87	4·706280668	37	8·884481222	87	9·412561336
38	4·455149870	88	4·708010803	38	8·910299740	88	9·416021606
39	4·467399898	89	4·709683816	39	8·934799796	89	9·419367632
40	4·479034978	90	4·711301793	40	8·958069956	90	9·422603586
41	4·490095567	91	4·712866734	41	8·980191134	91	9·425733468
42	4·500618704	92	4·714380553	42	9·001237408	92	9·428761106
43	4·510638357	93	4·715845084	43	9·021276714	93	9·431693168
44	4·520185743	94	4·717262091	44	9·040371486	94	9·434524182
45	4·529289599	95	4·718633256	45	9·058579198	95	9·437266512
46	4·537976418	96	4·719960199	46	9·075952836	96	9·439920398
47	4·546270673	97	4·721244475	47	9·092541346	97	9·442488950
48	4·554195002	98	4·722487574	48	9·108390004	98	9·444975148
49	4·561770374	99	4·723690927	49	9·123540748	99	9·447381854
50	4·569016247	100	4·724855909	50	9·138032494	100	9·449711818

Multiples of the Present Value of £1 per Annum in n years. Interest on Capital 2½ per cent., Redemption 3 per cent.

Years	Annuity £3, £30, £300, £3000 or £300,000,000	Years	Annuity £3, £30, £300, £3000 or £300,000,000	Years	Annuity £4, £40, £400, £4000 or £400,000,000	Years	Annuity £4, £40, £400, £4000 or £400,000,000
1	2·479338840	51	13·727852082	1	3·305785120	51	18·303802776
2	4·269788964	52	13·747771596	2	5·693051952	52	18·330362128
3	5·622912177	53	13·766854272	3	7·497216236	53	18·355805696
4	6·681112044	54	13·785143931	4	8·908149392	54	18·380191908
5	7·530979221	55	13·802681493	5	10·041305628	55	18·403575324
6	8·228251689	56	13·819505208	6	10·971002252	56	18·426006944
7	8·810408283	57	13·835650890	7	11·747211044	57	18·447534520
8	9·303583689	58	13·851152070	8	12·404778252	58	18·468202760
9	9·726558648	59	13·866040215	9	12·968744864	59	18·488053620
10	10·093176618	60	13·880344848	10	13·457568824	60	18·507126464
11	10·413866211	61	13·894093725	11	13·885154948	61	18·525458300
12	10·696633005	62	13·907312373	12	14·262177340	62	18·543083164
13	10·947724671	63	13·920027111	13	14·596966228	63	18·560036148
14	11·172088410	64	13·932259359	14	14·896117880	64	18·576345812
15	11·373692580	65	13·944031542	15	15·164923440	65	18·592042056
16	11·555757426	66	13·955364264	16	15·407676568	66	18·607152352
17	11·720923434	67	13·966277010	17	15·627897912	67	18·621702680
18	11·871376206	68	13·976788191	18	15·828501608	68	18·635717588
19	12·008940387	69	13·986915231	19	16·011920516	69	18·649220308
20	12·135151236	70	13·996674615	20	16·180201648	70	18·662232820
21	12·251309817	71	14·006081976	21	16·335079756	71	18·674775968
22	12·358526040	72	14·015152134	22	16·478034720	72	18·686869512
23	12·457752504	73	14·023899138	23	16·610336672	73	18·698532184
24	12·549811461	74	14·032336335	24	16·733081948	74	18·709781780
25	12·635416335	75	14·040476418	25	16·847221780	75	18·720635224
26	12·715189197	76	14·048331408	26	16·953585596	76	18·731108544
27	12·789674928	77	14·055912795	27	17·052899904	77	18·741217060
28	12·859352823	78	14·063231478	28	17·145803764	78	18·750975304
29	12·924646191	79	14·070297843	29	17·232861588	79	18·760397124
30	12·985930302	80	14·077121793	30	17·314573736	80	18·769495724
31	13·043539008	81	14·083712751	31	17·391385344	81	18·778288668
32	13·097770323	82	14·090079720	32	17·463693764	82	18·786772960
33	13·148891097	83	14·096231286	33	17·531854796	83	18·794975048
34	13·197140985	84	14·102175636	34	17·596187980	84	18·802900848
35	13·242735855	85	14·107920612	35	17·656981140	85	18·810560816
36	13·285870641	86	14·113473687	36	17·714494188	86	18·817964916
37	13·326721833	87	14·118842004	37	17·768962444	87	18·825122672
38	13·365449610	88	14·124032409	38	17·820599480	88	18·832043212
39	13·402199694	89	14·129051448	39	17·869599592	89	18·838735264
40	13·437104934	90	14·133905379	40	17·916139912	90	18·845207172
41	13·470286701	91	14·138600202	41	17·960382268	91	18·851466936
42	13·501856112	92	14·143141659	42	18·002474816	92	18·857522212
43	13·531915071	93	14·147535252	43	18·042553428	93	18·863380336
44	13·560557229	94	14·151786273	44	18·080742972	94	18·869048364
45	13·587868797	95	14·155899768	45	18·117158396	95	18·874533024
46	13·613929254	96	14·159880597	46	18·151905672	96	18·879840796
47	13·638812019	97	14·163733425	47	18·185082662	97	18·884977900
48	13·662585006	98	14·167462722	48	18·216780008	98	18·889950296
49	13·685311122	99	14·171072781	49	18·247081496	99	18·894763708
50	13·707048741	100	14·174567727	50	18·276064988	100	18·899423636

TABLE XIV.   cxci

Multiples of the Present Value of £1 per Annum in n years. Interest on Capital 2½ per cent., Redemption 3 per cent.

Years	Annuity £5, £50, £500, £5000 or £500,000,000	Years	Annuity £5, £50, £500, £5000 or £500,000,000	Years	Annuity £6, £60, £600, £6000 or £600,000,000	Years	Annuity £6, £60, £600, £6000 or £600,000,000
1	4·132231400	51	22·879753470	1	4·958677680	51	27·455704164
2	7·116314940	52	22·912952660	2	8·539577928	52	27·495543192
3	9·371520295	53	22·944757120	3	11·245824354	53	27·533708544
4	11·135186740	54	22·975239885	4	13·362224088	54	27·570287862
5	12·551632035	55	23·004469155	5	15·061958442	55	27·605362986
6	13·713752815	56	23·032508680	6	16·456503378	56	27·639010416
7	14·684013805	57	23·059418150	7	17·620816566	57	27·671301780
8	15·505972815	58	23·085253450	8	18·607167378	58	27·702304140
9	16·210931080	59	23·110067025	9	19·453117296	59	27·732080430
10	16·821961030	60	23·133908080	10	20·186353236	60	27·760689696
11	17·356443685	61	23·156822875	11	20·827732422	61	27·788187450
12	17·827721675	62	23·178853955	12	21·393266010	62	27·814624746
13	18·246207785	63	23·200045185	13	21·895449342	63	27·840054222
14	18·620147350	64	23·220432265	14	22·344176820	64	27·864518718
15	18·956154300	65	23·240052570	15	22·747385160	65	27·888063084
16	19·259595710	66	23·258940440	16	23·111514852	66	27·910728528
17	19·534872390	67	23·277128350	17	23·441846868	67	27·932554020
18	19·785627010	68	23·294646985	18	23·742752412	68	27·953576382
19	20·014900645	69	23·311525385	19	24·017880774	69	27·973830462
20	20·225252060	70	23·327791025	20	24·270302472	70	27·993349230
21	20·418849695	71	23·343469960	21	24·502619634	71	28·012163952
22	20·597543400	72	23·358586890	22	24·717052080	72	28·030304268
23	20·762920840	73	23·373165230	23	24·915505008	73	28·047798276
24	20·916352435	74	23·387227225	24	25·099622922	74	28·064672670
25	21·059027225	75	23·400794030	25	25·270832670	75	28·080952836
26	21·191981995	76	23·413885680	26	25·430378394	76	28·096662816
27	21·316124880	77	23·426521325	27	25·579349856	77	28·111825590
28	21·432254705	78	23·438719130	28	25·718705645	78	28·126462956
29	21·541076985	79	23·450496405	29	25·849292382	79	28·140595686
30	21·643217170	80	23·461869655	30	25·971860604	80	28·154243586
31	21·739231680	81	23·472854585	31	26·087078016	81	28·167425502
32	21·829617205	82	23·483466200	32	26·195540646	82	28·180159440
33	21·914818495	83	23·493718810	33	26·297782194	83	28·192462572
34	21·995234975	84	23·503626060	34	26·394281970	84	28·204351272
35	22·071226425	85	23·513201020	35	26·485471710	85	28·215841224
36	22·143117735	86	23·522456146	36	26·571741282	86	28·226947374
37	22·211203055	87	23·531403340	37	26·653443666	87	28·237684008
38	22·275749350	88	23·540054015	38	26·730899220	88	28·248064818
39	22·336999490	89	23·548419080	39	26·804399388	89	28.258102896
40	22·395174890	90	23·556508965	40	26·874209868	90	28·267810758
41	22·450477835	91	23·564333670	41	26·940573402	91	28·277200404
42	22·503093520	92	23·571902765	42	27·003712224	92	28·286283318
43	22·553191785	93	23·579225420	43	27·063830142	93	28·295070504
44	22·600928715	94	23·586310455	44	27·121114458	94	28·303572546
45	22·646447995	95	23·593166280	45	27·175737594	95	28·311799536
46	22·689882090	96	23·599800995	46	27·227858508	96	28·319761194
47	22·731353365	97	23·606222375	47	27·277624038	97	28·327466850
48	22·770975010	98	23·612437870	48	27·325170012	98	28·334925444
49	22·808851870	99	23·618454635	49	27·370622244	99	28·342145562
50	22·845081235	100	23·624279545	50	27·414097482	100	28·349135454

Multiples of the Present Value of £1 per Annum in n years. Interest on Capital 2½ per cent., Redemption 3 per cent.

Years	Annuity £7, £70, £700, £7000 or £700,000,000	Years	Annuity £7, £70, £700, £7000 or £700,000,000	Years	Annuity £8, £80, £800, £8000 or £800,000,000	Years	Annuity £8, £80, £800, £8000 or £800,000,000
1	5·785123960	51	32·031654858	1	6·611570240	51	36·607605552
2	9·962840916	52	32·078133724	2	11·386103904	52	36·660724256
3	13·120128413	53	32·122659968	3	14·994432472	53	36·711611392
4	15·589261436	54	32·165335839	4	17·816298784	54	36·760383816
5	17·572284849	55	32·206256817	5	20·082611256	55	36·807150648
6	19·199253941	56	32·245512152	6	21·942004504	56	36·852013888
7	20·557619327	57	32·283185410	7	23·494422088	57	36·895069040
8	21·708361941	58	32·319354830	8	24·809556504	58	36·936405520
9	22·695303512	59	32·354093835	9	25·937489728	59	36·976107240
10	23·550745442	60	32·387471312	10	26·915137648	60	37·014252928
11	24·299021159	61	32·419552025	11	27·770309896	61	37·050916600
12	24·958810345	62	32·450395537	12	28·524354680	62	37·086166328
13	25·544690899	63	32·480063259	13	29·193932456	63	37·120072296
14	26·068206290	64	32·508605171	14	29·792235760	64	37·152691624
15	26·538616020	65	32·536073598	15	30·329846880	65	37·184084112
16	26·963433994	66	32·562516616	16	30·815353136	66	37·214304704
17	27·348821346	67	32·587979690	17	31·255795824	67	37·243405360
18	27·699877814	68	32·612505779	18	31·657003216	68	37·271435176
19	28·020860903	69	32·636135539	19	32·023841032	69	37·298440616
20	28·315352884	70	32·658907435	20	32·360403296	70	37·324465640
21	28·586389573	71	32·680857944	21	32·670159512	71	37·349551936
22	28·836560760	72	32·702021646	22	32·956069440	72	37·373739024
23	29·068089176	73	32·722431322	23	33·220673343	73	37·397064368
24	29·282893409	74	32·742118115	24	33·466163896	74	37·419563560
25	29·482638115	75	32·761111642	25	33·694443560	75	37·441270448
26	29·668774793	76	32·779439952	26	33·907171192	76	37·462217088
27	29·842574832	77	32·797129855	27	34·105799808	77	37·482434120
28	30·005156587	78	32·814206782	28	34·291607528	78	37·501970608
29	30·157507779	79	32·830694967	29	34·465723176	79	37·520794248
30	30·300504038	80	32·846617517	30	34·629147472	80	37·538991448
31	30·434924352	81	32·861996419	31	34·782770688	81	37·556567336
32	30·561464087	82	32·876852680	32	34·927387528	82	37·573545920
33	30·680745893	83	32·891206334	33	35·063709592	83	37·589950096
34	30·793328965	84	32·905076484	34	35·192375960	84	37·605801696
35	30·899716995	85	32·918481428	35	35·313962280	85	37·621121632
36	31·000364829	86	32·931438603	36	35·428988376	86	37·635929832
37	31·095684277	87	32·943964676	37	35·537924888	87	37·650245344
38	31·186049090	88	32·956075621	38	35·641198960	88	37·664086424
39	31·271799286	89	32·967786712	39	35·739199184	89	37·677470528
40	31·353244846	90	32·979112551	40	35·832279824	90	37·690414344
41	31·430668969	91	32·990067138	41	35·920764536	91	37·702933872
42	31·504330928	92	33·000663871	42	36·004949632	92	37·715044424
43	31·574468499	93	33·010915588	43	36·085106856	93	37·726760672
44	31·641300201	94	33·020834637	44	36·161485944	94	37·738096728
45	31·705027193	95	33·030432792	45	36·234316792	95	37·749066048
46	31·765834926	96	33·039721393	46	36·303811344	96	37·759681592
47	31·823894711	97	33·048711325	47	36·370165384	97	37·769955800
48	31·879635014	98	33·057413018	48	36·433560616	98	37·779900592
49	31·932392618	99	33·065836489	49	36·494162992	99	37·789527416
50	31·983113729	100	33·073991363	50	36·552129976	100	37·798847272

TABLE XIV.

Multiples of the Present Value of £1 per Annum in n years.   Interest on Capital
21 per cent., Redemption 3 per cent.

Years	Annuity £9, £90, £900, £9000 or £900,000.000	Years	Annuity £9, £90, £900, £9000 or £900,000,000	Years	Annuity £10, £100, £1000, £10,000 or £1000,000,000	Years	Annuity £10, £100, £1000, £10,000 or £1000,000,000
1	7·438016520	51	41·183556246	1	8·264462800	51	45·759506940
2	12·809366892	52	41·243314788	2	14·232629880	52	45·825905320
3	16·868736531	53	41·300562816	3	18·743040590	53	45·889514240
4	20·043336132	54	41·355431793	4	22·270373480	54	45·950479770
5	22·592937663	55	41·408044479	5	25·103264070	55	46·008938310
6	24·684755067	56	41·458515624	6	27·427505630	56	46·065017360
7	26·431224849	57	41·506952670	7	29·368027610	57	46·118836300
8	27·910751067	58	41·553456210	8	31·011945630	58	46·170506900
9	29·179675944	59	41·598120645	9	32·421862160	59	46·220134050
10	30·279529854	60	41·641034544	10	33·643922060	60	46·267816160
11	31·241598633	61	41·682281175	11	34·712887370	61	46·313645750
12	32·089899015	62	41·721937119	12	35·655443350	62	46·357707910
13	32·843174013	63	41·760081333	13	36·492415570	63	46·400090370
14	33·516265230	64	41·796778077	14	37·240294700	64	46·440864530
15	34·121077740	65	41·832094626	15	37·912308600	65	46·480105140
16	34·667272278	66	41·866092792	16	38·519191420	66	46·517880880
17	35·162770302	67	41·898831030	17	39·069744780	67	46·554265700
18	35·614128618	68	41·930964573	18	39·571254020	68	46·589293970
19	36·026821161	69	41·960745693	19	40·029801290	69	46·623050770
20	36·405453708	70	41·990023845	20	40·450504120	70	46·655582050
21	36·753929451	71	42·018245928	21	40·837699390	71	46·686939920
22	37·075578120	72	42·045456402	22	41·195086800	72	46·717173780
23	37·373257512	73	42·071697414	23	41·525841680	73	46·746330460
24	37·649434383	74	42·097009005	24	41·832704870	74	46·774454450
25	37·906249005	75	42·121429254	25	42·118054450	75	46·801588060
26	38·145567591	76	42·144994224	26	42·383963990	76	46·827771360
27	38·369024784	77	42·167738385	27	42·632249760	77	46·853042650
28	38·578058469	78	42·189694434	28	42·864509410	78	46·877438260
29	38·773938573	79	42·210893529	29	43·082153970	79	46·900992810
30	38·957790906	80	42·231365379	30	43·286434340	80	46·923739310
31	39·130617024	81	42·251138253	31	43·478463360	81	46·945709170
32	39·293310969	82	42·270239160	32	43·659234410	82	46·966932400
33	39·446673291	83	42·288693858	33	43·829636990	83	46·987437620
34	39·591422955	84	42·306526908	34	43·990469950	84	47·007252120
35	39·728207565	85	42·323761836	35	44·143452850	85	47·026420240
36	39·857611923	86	42·340421061	36	44·286235470	86	47·044912290
37	39·980165499	87	42·356526012	37	44·422406110	87	47·062806680
38	40·096348830	88	42·372097227	38	44·551498700	88	47·080108030
39	40·206599082	89	42·387154344	39	44·673998980	89	47·096838160
40	40·311314802	90	42·401716137	40	44·790349780	90	47·113017930
41	40·410860103	91	42·415800606	41	44·900955670	91	47·128667340
42	40·505568336	92	42·429424977	42	45·006187040	92	47·143805530
43	40·595745213	93	42·442605756	43	45·106383570	93	47·158450840
44	40·681671687	94	42·455358819	44	45·201857430	94	47·172620910
45	40·763606391	95	42·467699304	45	45·292895990	95	47·186332560
46	40·841787762	96	42·479641791	46	45·379764180	96	47·199601990
47	40·916436057	97	42·491200275	47	45·462706730	97	47·212444750
48	40·987755018	98	42·502388166	48	45·541950020	98	47·224875740
49	41·055933366	99	42·513218343	49	45·617703740	99	47·236909270
50	41·121146223	100	42·523703181	50	45·690162470	100	47·248559090

TABLE XV.

MULTIPLES OF REDEMPTION FUNDS,

*At the rate of **3** per cent. per annum, necessary to produce £1, £2, £3, £4, £5, £6, £7, £8, £9, and £10; or from £1 to £100,000,000, £2 to £200,000,000, £3 to £300,000,000, etc., up to £10 or £1,000,000,000; and by employing the decimal system of notation for any intermediate sum.*

*Calculated to **10** places of decimals.*

Multiples of Redemption Funds, at 3 per cent. per Annum, necessary to produce the following sums in n years.

Years	Sum to produce £1, £10, £100, £1000, or £100,000,000	Years	Sum to produce £1, £10, £100, £1000, or £100,000,000	Years	Sum to produce £2, £20, £200, £2000, or £200,000,000	Years	Sum to produce £2, £20, £200, £2000, or £200,000,000
1	1·0000000000	51	·0085338232	1	2·0000000000	51	·0170676464
2	·4926108374	52	·0082171837	2	·9852216748	52	·0164343674
3	·3235303633	53	·0079147059	3	·6470607266	53	·0158294118
4	·2390270452	54	·0076255841	4	·4780540904	54	·0152511682
5	·1883545714	55	·0073490710	5	·3767091428	55	·0146981420
6	·1545975005	56	·0070844726	6	·3091950010	56	·0141689452
7	·1305063538	57	·0068311432	7	·2610127076	57	·0136622864
8	·1124563888	58	·0065884819	8	·2249127776	58	·0131769638
9	·0984338570	59	·0063559281	9	·1968677140	59	·0127118562
10	·0872305066	60	·0061329587	10	·1744610132	60	·0122659174
11	·0780774478	61	·0059190847	11	·1561548956	61	·0118381694
12	·0704620855	62	·0057138575	12	·1409241710	62	·0114277150
13	·0640295440	63	·0055168216	13	·1280590880	63	·0110336432
14	·0585263390	64	·0053276021	14	·1170526780	64	·0106552042
15	·0537665805	65	·0051458128	15	·1075331610	65	·0102916256
16	·0496108493	66	·0049710095	16	·0992216986	66	·0099421990
17	·0459525294	67	·0048031288	17	·0919050588	67	·0096062576
18	·0427086959	68	·0046415871	18	·0854173918	68	·0092831742
19	·0398138806	69	·0044861787	19	·0796277612	69	·0089723574
20	·0372157076	70	·0043366251	20	·0744314152	70	·0086732502
21	·0348717765	71	·0041926632	21	·0697435530	71	·0083853264
22	·0327473948	72	·0040540446	22	·0654947896	72	·0081080892
23	·0308139027	73	·0039205345	23	·0616278054	73	·0078410690
24	·0290474159	74	·0037919109	24	·0580948318	74	·0075838218
25	·0274278710	75	·0036679633	25	·0548557420	75	·0073359266
26	·0259382903	76	·0035484929	26	·0518765806	76	·0070969858
27	·0245642103	77	·0034333105	27	·0491284206	77	·0068666210
28	·0232932334	78	·0033222371	28	·0465864668	78	·0066444742
29	·0221146711	79	·0032151027	29	·0442293422	79	·0064302054
30	·0210192593	80	·0031117457	30	·0420385186	80	·0062234914
31	·0199989288	81	·0030120127	31	·0399978576	81	·0060240254
32	·0190466183	82	·0029157577	32	·0380932366	82	·0058315154
33	·0181561219	83	·0028228417	33	·0363122438	83	·0056456834
34	·0173219634	84	·0027331326	34	·0346439268	84	·0054662652
35	·0165392916	85	·0026465042	35	·0330785832	85	·0052930084
36	·0158037942	86	·0025628365	36	·0316075884	86	·0051256730
37	·0151116244	87	·0024820151	37	·0302232488	87	·0049640302
38	·0144593401	88	·0024039306	38	·0289186802	88	·0048078612
39	·0138438516	89	·0023284787	39	·0276877032	89	·0046569574
40	·0132623779	90	·0022555599	40	·0265247558	90	·0045111198
41	·0127124089	91	·0021850789	41	·0254248178	91	·0043701578
42	·0121916731	92	·0021169449	42	·0243833462	92	·0042338898
43	·0116981103	93	·0020510708	43	·0233962206	93	·0041021416
44	·0112298469	94	·0019873733	44	·0224596938	94	·0039747464
45	·0107851757	95	·0019257729	45	·0215703514	95	·0038515458
46	·0103625378	96	·0018661933	46	·0207250756	96	·0037323866
47	·0099605065	97	·0018085613	47	·0199210130	97	·0036171226
48	·0095777738	98	·0017528070	48	·0191555476	98	·0035056140
49	·0092131383	99	·0016988633	49	·0184262766	99	·0033977266
50	·0088654944	100	·0016466659	50	·0177309888	100	·0032933318

Multiples of Redemption Funds, at **3** per cent. per Annum, necessary to produce the following sums in **n** years.

Years	Sum to produce £3, £30, £300, £3000, or £300.000.000	Years	Sum to produce £3, £30, £300, £3000, or £300,000,000	Years	Sum to produce £4, £40, £400, £4000, or £400,000,000	Years	Sum to produce £4, £40, £400, £4000, or £400,000,000
1	3·0000000000	51	·0256014696	1	4·0000000000	51	·0341352928
2	1·4778325122	52	·0246515511	2	1·9704433496	52	·0328687348
3	·9705910899	53	·0237441177	3	1·2941214532	53	·0316588236
4	·7170811356	54	·0228767523	4	·9561081808	54	·0305023364
5	·5650637142	55	·0220472130	5	·7534182856	55	·0293962840
6	·4637925015	56	·0212534178	6	·6183900020	56	·0283378904
7	·3915190614	57	·0204934296	7	·5220254152	57	·0273245728
8	·3373691664	58	·0197654457	8	·4498255552	58	·0263539276
9	·2953015710	59	·0190677843	9	·3937354280	59	·0254237124
10	·2616915198	60	·0183988761	10	·3489220264	60	·0245318348
11	·2342323434	61	·0177572541	11	·3123097912	61	·0236763388
12	·2113862565	62	·0171415725	12	·2818483420	62	·0228554300
13	·1920886320	63	·0165504648	13	·2561181760	63	·0220672864
14	·1755790170	64	·0159828063	14	·2341053560	64	·0213104084
15	·1612997415	65	·0154374384	15	·2150663220	65	·0205832512
16	·1488325479	66	·0149132985	16	·1984433972	66	·0198843980
17	·1375575882	67	·0144093864	17	·1838101176	67	·0192125152
18	·1281260877	68	·0139247613	18	·1708347836	68	·0185663484
19	·1194416418	69	·0134585361	19	·1592555224	69	·0179447148
20	·1116471228	70	·0130098753	20	·1488628304	70	·0173465004
21	·1046153295	71	·0125779896	21	·1394871060	71	·0167706528
22	·0982421844	72	·0121621338	22	·1309895792	72	·0162161784
23	·0924417081	73	·0117616035	23	·1232556108	73	·0156821380
24	·0871422477	74	·0113757327	24	·1161896636	74	·0151676436
25	·0822836130	75	·0110038899	25	·1097114840	75	·0146718532
26	·0778148709	76	·0106454787	26	·1037531612	76	·0141939716
27	·0736926309	77	·0102999315	27	·0982568412	77	·0137332420
28	·0698797002	78	·0099667113	28	·0931729336	78	·0132889484
29	·0663440133	79	·0096453081	29	·0884586844	79	·0128604108
30	·0630577779	80	·0093352371	30	·0840770372	80	·0124469828
31	·0599967864	81	·0090360381	31	·0799957152	81	·0120480508
32	·0571398549	82	·0087472731	32	·0761864732	82	·0116630308
33	·0544683657	83	·0084685251	33	·0726244876	83	·0112913668
34	·0519658902	84	·0081993978	34	·0692878536	84	·0109325304
35	·0496178748	85	·0079395126	35	·0661571664	85	·0105860168
36	·0474113826	86	·0076885095	36	·0632151768	86	·0102513460
37	·0453348732	87	·0074460453	37	·0604464976	87	·0099280604
38	·0433780203	88	·0072117918	38	·0578373064	88	·0096157224
39	·0415315548	89	·0069854361	39	·0553754064	89	·0095139148
40	·0397871337	90	·0067666797	40	·0530495116	90	·0090222396
41	·0381372267	91	·0065552367	41	·0508496356	91	·0087403156
42	·0365750193	92	·0063508347	42	·0487666924	92	·0084677796
43	·0350943309	93	·0061532124	43	·0467924412	93	·0082042832
44	·0336895407	94	·0059621199	44	·0449193876	94	·0079494933
45	·0323555271	95	·0057773187	45	·0431407028	95	·0077030916
46	·0310876134	96	·0055985799	46	·0414501512	96	·0074667732
47	·0298815195	97	·0054256839	47	·0398420260	97	·0072342452
48	·0287333214	98	·0052584210	48	·0383110952	98	·0070112280
49	·0276394149	99	·0050965899	49	·0368525532	99	·0067954532
50	·0265964832	100	·0049399777	50	·0354619776	100	·0065866636

Multiples of Redemption Funds, at **3** per cent. per Annum, necessary to produce the following sums in **n** years.

Years	Sum to produce £5, £50, £500, £5000, or £500,000,000	Years	Sum to produce £5, £50, £500, £5000, or £500,000,000	Years	Sum to produce £6, £60, £600, £6000, or £600,000,000	Years	Sum to produce £6, £60, £600, £6000, or £600,000,000
1	5·0000000000	51	·0426691160	1	6·0000000000	51	·0512029392
2	2·4630541870	52	·0410859185	2	2·9556550244	52	·0493031022
3	1·6176518165	53	·0395735295	3	1·9411821798	53	·0474882354
4	1·1951352260	54	·0381279205	4	1·4341622712	54	·0457535046
5	·9417728570	55	·0367453550	5	1·1301274284	55	·0440944260
6	·7729875025	56	·0354223630	6	·9275850030	56	·0425068356
7	·6525317690	57	·0341557160	7	·7830381228	57	·0409868592
8	·5622819440	58	·0329424095	8	·6747383328	58	·0395308914
9	·4921692850	59	·0317796405	9	·5906031420	59	·0381355686
10	·4361525330	60	·0306647935	10	·5233830396	60	·0367977522
11	·3903872390	61	·0295954235	11	·4684646868	61	·0355145082
12	·3523104275	62	·0285692875	12	·4227725130	62	·0342831450
13	·3201477200	63	·0275841080	13	·3841772640	63	·0331009296
14	·2926316950	64	·0266380105	14	·3511580340	64	·0319656126
15	·2688329025	65	·0257290640	15	·3225994830	65	·0308748768
16	·2480542465	66	·0248554975	16	·2976650958	66	·0298265970
17	·2297626470	67	·0240156440	17	·2751151764	67	·0288187728
18	·2135434795	68	·0232079355	18	·2562521754	68	·0278495226
19	·1990694030	69	·0224308935	19	·2388832836	69	·0269170722
20	·1860785380	70	·0216831255	20	·2232942446	70	·0260197506
21	·1743588825	71	·0209633160	21	·2092306590	71	·0251559792
22	·1637369740	72	·0202702230	22	·1964843688	72	·0243242676
23	·1540695135	73	·0196026725	23	·1848834162	73	·0235232070
24	·1452370795	74	·0189595545	24	·1742844954	74	·0227514654
25	·1371393550	75	·0183398165	25	·1645672260	75	·0220077798
26	·1296914515	76	·0177424645	26	·1556297418	76	·0212909574
27	·1228210515	77	·0171665525	27	·1473852618	77	·0205998630
28	·1164661670	78	·0166111855	28	·1397594004	78	·0199334226
29	·1105733555	79	·0160755135	29	·1226880266	79	·0192906162
30	·1050962965	80	·0155587285	30	·1261155558	80	·0186704742
31	·0999946440	81	·0150600635	31	·1199935728	81	·0180720762
32	·0952330915	82	·0145787885	32	·1142797098	82	·0174945462
33	·0907806095	83	·0141142085	33	·1089367314	83	·0169370502
34	·0866098170	84	·0136656630	34	·1039317804	84	·0163987956
35	·0826964580	85	·0132325210	35	·0992357496	85	·0158790252
36	·0790189710	86	·0128141825	36	·0948227652	86	·0153770190
37	·0755581220	87	·0124100755	37	·0906697464	87	·0148920906
38	·0722967005	88	·0120196530	38	·0867560406	88	·0144235836
39	·0692192580	89	·0116423935	39	·0830631096	89	·0139708722
40	·0663118895	90	·0112777995	40	·0795742674	90	·0135333594
41	·0635620445	91	·0109253945	41	·0762744534	91	·0131104734
42	·0609583655	92	·0105847245	42	·0731500386	92	·0127016694
43	·0584905515	93	·0102553540	43	·0701886618	93	·0123064248
44	·0561492345	94	·0099368665	44	·0673790814	94	·0119242398
45	·0539258785	95	·0096288645	45	·0647110542	95	·0115546374
46	·0518126890	96	·0093309665	46	·0621752268	96	·0111971598
47	·0498025325	97	·0090428065	47	·0597630390	97	·0108513678
48	·0478888690	98	·0087640350	48	·0574666428	98	·0105168420
49	·0460656915	99	·0084943165	49	·0552788298	99	·0101931798
50	·0443274320	100	·0082333295	50	·0531929664	100	·0098799954

Multiples of Redemption Funds, at **3** per cent. per Annum, necessary to produce the following sums in n years.

Years	Sum to produce £7, £70, £700, £7000, or £·00,000,000	Years	Sum to produce £7, £70, £700, £7000, or £700 000,000	Years	Sum to produce £8, £80, £800, £8000, or £800,000,000	Years	Sum to produce £8, £80, £800, £8000, or £800,000,000
1	7·0000000000	51	·0597367624	1	8·0000000000	51	·0682705856
2	3·4482758618	52	·0575202859	2	3·9408866992	52	·0657374696
3	2·2647125431	53	·0554029413	3	2·5882429064	53	·0633176472
4	1·6731893164	54	·0533790887	4	1·9122163616	54	·0610046728
5	1·3184819998	55	·0514434970	5	1·5068365712	55	·0587925680
6	1·0821825035	56	·0495913082	6	1·2367800040	56	·0566757808
7	·9135444766	57	·0478180024	7	1·0440508304	57	·0546491456
8	·7871947216	58	·0461183733	8	·8996511104	58	·0527078552
9	·6890369990	59	·0444914967	9	·7874708560	59	·0508474248
10	·6106135462	60	·0429307109	10	·6978440528	60	·0490636696
11	·5465421346	61	·0414335929	11	·6246195824	61	·0473526776
12	·4932345985	62	·0399970025	12	·5636966840	62	·0457108600
13	·4482068080	63	·0386177512	13	·5122363520	63	·0441345728
14	·4096843730	64	·0372932147	14	·4682107120	64	·0426208168
15	·3763660635	65	·0360206896	15	·4301326440	65	·0411665024
16	·3472759451	66	·0347976965	16	·3968867944	66	·0397687960
17	·3216677058	67	·0336219016	17	·3676202352	67	·0384250304
18	·2989608713	68	·0324911097	18	·3416695672	68	·0371326968
19	·2786971642	69	·0314032509	19	·3185110448	69	·0358894296
20	·2605099532	70	·0303563757	20	·2977256608	70	·0346930008
21	·2441024355	71	·0293486424	21	·2789742120	71	·0335413056
22	·2292317636	72	·0283783122	22	·2619791584	72	·0324323568
23	·2156973189	73	·0274437415	23	·2465112216	73	·0313542760
24	·2033319113	74	·0265433763	24	·2323793272	74	·0303352872
25	·1919950970	75	·0256757431	25	·2194229680	75	·0293437064
26	·1815680321	76	·0248394503	26	·2075063224	76	·0283879432
27	·1719494721	77	·0240431735	27	·1965136824	77	·0274664840
28	·1630526338	78	·0232556597	28	·1863458672	78	·0265778968
29	·1548026977	79	·0225057189	29	·1769173688	79	·0257208216
30	·1471348151	80	·0217822199	30	·1681540744	80	·0248939656
31	·1399925016	81	·0210840889	31	·1599914304	81	·0240961016
32	·1333263281	82	·0204103039	32	·1523729464	82	·0233260616
33	·1270928533	83	·0197598919	33	·1452489752	83	·0225827336
34	·1212537438	84	·0191319282	34	·1385757072	84	·0218650608
35	·1157750412	85	·0185255294	35	·1323143328	85	·0211720336
36	·1106265594	86	·0179398555	36	·1264303536	86	·0205026920
37	·1057813708	87	·0173741059	37	·1208929952	87	·0198561208
38	·1012153807	88	·0168275142	38	·1156742208	88	·0192314448
39	·0969069612	89	·0162993509	39	·1107508128	89	·0186278296
40	·0928366453	90	·0157889193	40	·1060990232	90	·0180444792
41	·0889868623	91	·0152955523	41	·1016992712	91	·0174806312
42	·0853417117	92	·0148186143	42	·0975333848	92	·0169355592
43	·0818867721	93	·0143574956	43	·0935848824	93	·0164085664
44	·0786089283	94	·0139116131	44	·0898387752	94	·0158989864
45	·0754962299	95	·0134804103	45	·0862814056	95	·0154061832
46	·0725377646	96	·0130633531	46	·0829003024	96	·0149295464
47	·0697235455	97	·0126599291	47	·0796840520	97	·0144684904
48	·0670444166	98	·0122696490	48	·0766221904	98	·0140224560
49	·0644919681	99	·0118920431	49	·0737051064	99	·0135909064
50	·0620584608	100	·0115266613	50	·0709239552	100	·0131733272

Multiples of Redemption Funds, at 3 per cent. per Annum, necessary to produce the following sums in n years.

Years	Sum to produce £9, £90, £900, £9000, or £900,000,000	Years	Sum to produce £9, £90, £900, £9000, or £900,000,000	Years	Sum to produce £10, £100, £1000, £10,000, or £1000,000,000	Years	Sum to produce £10, £100, £1000, £10,000, or £1000,000,000
1	8·0000000000	51	·0768044088	1	10·0000000000	51	·0853382320
2	4·4334975366	52	·0739546533	2	4·9261083740	52	·0821718370
3	2·9117732697	53	·0712323531	3	3·2353036330	53	·0791470590
4	2·1512434068	54	·0686302569	4	2·3902704520	54	·0762558410
5	1·6951911426	55	·0661416390	5	1·8835547140	55	·0734907100
6	1·3913775045	56	·0637602534	6	1·5459750050	56	·0708447260
7	1·1745571842	57	·0614802888	7	1·3050635380	57	·0683114320
8	1·0121074992	58	·0592963371	8	1·1245638880	58	·0658848190
9	·8859047130	59	·0572033529	9	·9843385700	59	·0635592810
10	·7850745594	60	·0551966283	10	·8723050660	60	·0613295870
11	·7026970302	61	·0532717623	11	·7807744780	61	·0591908470
12	·6341587695	62	·0514247175	12	·7046208550	62	·0571385750
13	·5762658960	63	·0496513944	13	·6402954400	63	·0551682160
14	·5267370510	64	·0479484189	14	·5852633900	64	·0532760210
15	·4838992245	65	·0463123152	15	·5376658050	65	·0514581280
16	·4464976437	66	·0447398955	16	·4961084930	66	·0497109950
17	·4135727646	67	·0432281592	17	·4595252940	67	·0480312880
18	·3843782631	68	·0417742839	18	·4270869590	68	·0464158710
19	·3583249254	69	·0403756083	19	·3981388060	69	·0448617870
20	·3349413684	70	·0390296259	20	·3721570760	70	·0433662510
21	·3138459885	71	·0377339688	21	·3487177650	71	·0419266320
22	·2947265532	72	·0364864014	22	·3274739480	72	·0405404460
23	·2773251243	73	·0352848105	23	·3081390270	73	·0392053450
24	·2614267431	74	·0341271981	24	·2904741590	74	·0379191020
25	·2468508390	75	·0330116697	25	·2742787100	75	·0366796330
26	·2334446127	76	·0319364361	26	·2593829030	76	·0354849290
27	·2210778927	77	·0308997945	27	·2456421030	77	·0343331050
28	·2096391006	78	·0299001339	28	·2329323340	78	·0332222710
29	·1990320399	79	·0289359243	29	·2211467110	79	·0321510270
30	·1891733337	80	·0280057113	30	·2101925930	80	·0311174570
31	·1799903592	81	·0271081143	31	·1998928880	81	·0301201270
32	·1714195647	82	·0262418193	32	·1904661830	82	·0291575770
33	·1634050971	83	·0254055755	33	·1815612190	83	·0282284170
34	·1558976706	84	·0245981934	34	·1732196340	84	·0273313260
35	·1488536244	85	·0238185378	35	·1653929160	85	·0264650420
36	·1422341478	86	·0230655285	36	·1580379420	86	·0256283650
37	·1360046196	87	·0223381359	37	·1511162440	87	·0248201510
38	·1301340609	88	·0216353754	38	·1445934010	88	·0240393060
39	·1245946644	89	·0209563083	39	·1384385160	89	·0232847870
40	·1193614011	90	·0203000391	40	·1326237790	90	·0225555990
41	·1144116801	91	·0196657101	41	·1271240890	91	·0218507890
42	·1097250579	92	·0190525041	42	·1219167310	92	·0211694490
43	·1052829927	93	·0184596372	43	·1169811030	93	·0205107080
44	·1010686221	94	·0178863597	44	·1122984690	94	·0198737330
45	·0970665813	95	·0173319561	45	·1078517570	95	·0192577290
46	·0932628402	96	·0167957397	46	·1036253980	96	·0186619330
47	·0896445585	97	·0162770517	47	·0996050650	97	·0180856130
48	·0861999642	98	·0157752630	48	·0957777380	98	·0175280700
49	·0829182447	99	·0152897697	49	·0921313830	99	·0169886330
	·0797894496	100	·0148199931	50	·0886549440	100	·0164666590

c c

TABLE XVI.

Present Value of a Perpetuity of £1, receivable once in every nth year, the first payment due n years hence; also of a Perpetuity of £1 deferred n years, at the rates of 3, 3½, 4, 4½, 5, 6, 7, and 8 per cent.

Calculated to 4 places of decimals.

TABLE XVI. ccv

Present Value of a Perpetuity of £1, receivable once in every nth Year, the first payment due n years hence; also Present Value of a Perpetuity of £1 Deferred n Years.

3 Per Cent.

Years	Present Values of a Perpetuity of £1 Deferred n Years	Years	Present Values of a Perpetuity of £1 Deferred n Years	Years	Present Value of £1 receivable once in every nth Year, the first due n Years hence	Years	Present Value of £1 receivable once in every nth Year, the first due n Years hence
1	32·3625	51	7·3821	1	33·3333	51	·2845
2	31·4199	52	7·1671	2	16·4204	52	·2739
3	30·5047	53	6·9583	3	10·7843	53	·2638
4	29·6162	54	6·7557	4	7·9676	54	·2542
5	28·7536	55	6·5589	5	6·2785	55	·2450
6	27·9161	56	6·3679	6	5·1533	56	·2361
7	27·1031	57	6·1824	7	4·3502	57	·2277
8	26·3136	58	6·0023	8	3·7485	58	·2196
9	25·5472	59	5·8275	9	3·2811	59	·2119
10	24·8031	60	5·6578	10	2·9077	60	·2044
11	24·0807	61	5·4930	11	2·6026	61	·1973
12	23·3793	62	5·3330	12	2·3487	62	·1905
13	22·6984	63	5·1777	13	2·1343	63	·1839
14	22·0373	64	5·0269	14	1·9509	64	·1776
15	21·3954	65	4·8804	15	1·7922	65	·1715
16	20·7722	66	4·7383	16	1·6537	66	·1657
17	20·1672	67	4·6003	17	1·5318	67	·1601
18	19·5798	68	4·4663	18	1·4236	68	·1547
19	19·0095	69	4·3362	19	1·3271	69	·1495
20	18·4559	70	4·2099	20	1·2405	70	·1446
21	17·9183	71	4·0873	21	1·1624	71	·1398
22	17·3964	72	3·9682	22	1·0916	72	·1351
23	16·8897	73	3·8527	23	1·0271	73	·1307
24	16·3978	74	3·7405	24	·9682	74	·1264
25	15·9202	75	3·6315	25	·9143	75	·1223
26	15·4565	76	3·5257	26	·8646	76	·1183
27	15·0063	77	3·4230	27	·8188	77	·1144
28	14·5692	78	3·3233	28	·7764	78	·1107
29	14·1449	79	3·2265	29	·7372	79	·1072
30	13·7329	80	3·1326	30	·7006	80	·1037
31	13·3329	81	3·0413	31	·6666	81	·1004
32	12·9446	82	2·9527	32	·6349	82	·0972
33	12·5675	83	2·8667	33	·6052	83	·0941
34	12·2015	84	2·7832	34	·5774	84	·0911
35	11·8461	85	2·7022	35	·5513	85	·0882
36	11·5011	86	2·6235	36	·5268	86	·0854
37	11·1661	87	2·5471	37	·5037	87	·0827
38	10·8409	88	2·4729	38	·4820	88	·0801
39	10·5251	89	2·4009	39	·4615	89	·0776
40	10·2186	90	2·3309	40	·4421	90	·0752
41	9·9209	91	2·2630	41	·4237	91	·0728
42	9·6320	92	2·1971	42	·4064	92	·0706
43	9·3514	93	2·1331	43	·3899	93	·0684
44	9·0791	94	2·0710	44	·3743	94	·0662
45	8·8146	95	2·0107	45	·3595	95	·0642
46	8·5579	96	1·9521	46	·3454	96	·0622
47	8·3086	97	1·8953	47	·3320	97	·0603
48	8·0666	98	1·8401	48	·3193	98	·0584
49	7·8317	99	1·7865	49	·3071	99	·0566
50	7·6036	100	1·7344	50	·2955	100	·0549

Present Value of a Perpetuity of £1, receivable once in every nth Year,
the first payment due n years hence; also Present Value of a Perpetuity of £1
Deferred n Years.

3½ PER CENT.

Years	Present Values of a Perpetuity of £1 Deferred n Years	Years	Present Values of a Perpetuity of £1 Deferred n Years	Years	Present Value of £1 receivable once in every nth Year, the first due n Years hence	Years	Present Value of £1 receivable once in every nth Year, the first due n Years hence
1	27·6052	51	4·9428	1	28·5714	51	·2092
2	26·6717	52	4·7757	2	14·0400	52	·2007
3	25·7698	53	4·6142	3	9·1981	53	·1926
4	24·8983	54	4·4581	4	6·7786	54	·1849
5	24·0564	55	4·3074	5	5·3280	55	·1775
6	23·2429	56	4·1617	6	4·3619	56	·1705
7	22·4569	57	4·0210	7	3·6727	57	·1638
8	21·6975	58	3·8850	8	3·1565	58	·1574
9	20·9637	59	3·7536	9	2·7556	59	·1512
10	20·2548	60	3·6267	10	2·4355	60	·1454
11	19·5699	61	3·5041	11	2·1741	61	·1398
12	18·9081	62	3·3856	12	1·9567	62	·1344
13	18·2687	63	3·2711	13	1·7732	63	·1293
14	17·6509	64	3·1605	14	1·6163	64	·1244
15	17·0540	65	3·0536	15	1·4807	65	·1197
16	16·4773	66	2·9503	16	1·3624	66	·1152
17	15·9201	67	2·8505	17	1·2584	67	·1108
18	15·3817	68	2·7542	18	1·1662	68	·1067
19	14·8616	69	2·6610	19	1·0840	69	·1027
20	14·3590	70	2·5710	20	1·0103	70	·0989
21	13·8735	71	2·4841	21	·9439	71	·0952
22	13·4043	72	2·4001	22	·8838	72	·0917
23	12·9510	73	2·3189	23	·8291	73	·0883
24	12·5131	74	2·2405	24	·7792	74	·0851
25	12·0899	75	2·1647	25	·7335	75	·0820
26	11·6811	76	2·0915	26	·6916	76	·0790
27	11·2861	77	2·0208	27	·6529	77	·0761
28	10·9044	78	1·9525	28	·6172	78	·0733
29	10·5357	79	1·8864	29	·5842	79	·0707
30	10·1794	80	1·8227	30	·5535	80	·0681
31	9·8352	81	1·7610	31	·5249	81	·0657
32	9·5026	82	1·7015	32	·4983	82	·0633
33	9·1812	83	1·6439	33	·4735	83	·0611
34	8·8707	84	1·5883	34	·4503	84	·0589
35	8·5708	85	1·5346	35	·4285	85	·0568
36	8·2809	86	1·4827	36	·4081	86	·0547
37	8·0009	87	1·4326	37	·3890	87	·0528
38	7·7303	88	1·3841	38	·3709	88	·0509
39	7·4689	89	1·3373	39	·3539	89	·0491
40	7·2164	90	1·2921	40	·3379	90	·0474
41	6·9723	91	1·2484	41	·3228	91	·0457
42	6·7365	92	1·2062	42	·3085	92	·0441
43	6·5087	93	1·1654	43	·2950	93	·0425
44	6·2886	94	1·1260	44	·2822	94	·0410
45	6·0760	95	1·0879	45	·2701	95	·0396
46	5·8705	96	1·0511	46	·2586	96	·0382
47	5·6720	97	1·0156	47	·2477	97	·0369
48	5·4802	98	·9812	48	·2373	98	·0356
49	5·2949	99	·9481	49	·2275	99	·0343
50	5·1158	100	·9160	50	·2181	100	·0331

Present Value of a Perpetuity of £1, receivable once in every nth Year, the first payment due n years hence; also Present Value of a Perpetuity of £1 Deferred n Years.

4 Per Cent.

Years	Present Values of a Perpetuity of £1 Deferred n Years	Years	Present Values of a Perpetuity of £1 Deferred n Years	Years	Present Value of £1 receivable once in every nth Year, the first due n Years hence	Years	Present Value of £1 receivable once in every nth Year, the first due n Years hence
1	24·0385	51	3·3825	1	25·0000	51	·1565
2	23·1139	52	3·2524	2	12·2549	52	·1496
3	22·2249	53	3·1273	3	8·0087	53	·1430
4	21·3701	54	3·0070	4	5·8873	54	·1367
5	20·5482	55	2·8914	5	4·6157	55	·1308
6	19·7579	56	2·7802	6	3·7690	56	·1251
7	18·9979	57	2·6733	7	3·1652	57	·1197
8	18·2673	58	2·5704	8	2·7132	58	·1146
9	17·5647	59	2·4716	9	2·3623	59	·1097
10	16·8891	60	2·3765	10	2·0823	60	·1050
11	16·2395	61	2·2851	11	1·8537	61	·1006
12	15·6149	62	2·1972	12	1·6638	62	·0964
13	15·0144	63	2·1127	13	1·5036	63	·0923
14	14·4369	64	2·0315	14	1·3667	64	·0884
15	13·8816	65	1·9533	15	1·2485	65	·0848
16	13·3477	66	1·8782	16	1·1455	66	·0812
17	12·8343	67	1·8060	17	1·0550	67	·0779
18	12·3407	68	1·7365	18	·9748	68	·0746
19	11·8661	69	1·6697	19	·9045	69	·0716
20	11·4097	70	1·6055	20	·8395	70	·0686
21	10·9708	71	1·5437	21	·7820	71	·0658
22	10·5489	72	1·4844	22	·7300	72	·0631
23	10·1432	73	1·4273	23	·6827	73	·0605
24	9·7530	74	1·3724	24	·6397	74	·0581
25	9·3779	75	1·3196	25	·6003	75	·0557
26	9·0172	76	1·2688	26	·5642	76	·0535
27	8·6704	77	1·2200	27	·5310	77	·0513
28	8·3369	78	1·1731	28	·5003	78	·0492
29	8·0163	79	1·1280	29	·4720	79	·0473
30	7·7080	80	1·0846	30	·4458	80	·0454
31	7·4115	81	1·0429	31	·4214	81	·0435
32	7·1264	82	1·0028	32	·3987	82	·0418
33	6·8524	83	·9642	33	·3776	83	·0401
34	6·5888	84	·9271	34	·3579	84	·0385
35	6·3354	85	·8915	35	·3394	85	·0370
36	6·0917	86	·8572	36	·3222	86	·0355
37	5·8574	87	·8242	37	·3060	87	·0341
38	5·6321	88	·7925	38	·2908	88	·0327
39	5·4155	89	·7620	39	·2765	89	·0314
40	5·2072	90	·7327	40	·2631	90	·0302
41	5·0069	91	·7045	41	·2504	91	·0290
42	4·8144	92	·6774	42	·2385	92	·0279
43	4·6292	93	·6514	43	·2272	93	·0268
44	4·4512	94	·6263	44	·2166	94	·0257
45	4·2800	95	·6022	45	·2066	95	·0247
46	4·1153	96	·5791	46	·1971	96	·0237
47	3·9571	97	·5568	47	·1880	97	·0228
48	3·8049	98	·5354	48	·1795	98	·0219
49	3·6585	99	·5148	49	·1714	99	·0210
50	3·5178	100	·4950	50	·1638	100	·0202

Present Value of a Perpetuity of £1, receivable once in every nth Year,
the first payment due n years hence; also Present Value of a Perpetuity of £1
Deferred n Years.

2½ Per Cent.

Years	Present Values of a Perpetuity of £1 Deferred n Years	Years	Present Values of a Perpetuity of £1 Deferred n Years	Years	Present Value of £1 receivable once in every nth Year, the first due n Years hence	Years	Present Value of £1 receivable once in every nth Year, the first due n Years hence
1	21·2653	51	2·3543	1	22·2222	51	·1185
2	20·3496	52	2·2529	2	10·8666	52	·1128
3	19·4733	53	2·1559	3	7·0839	53	·1074
4	18·6347	54	2·0630	4	5·1943	54	·1023
5	17·8322	55	1·9742	5	4·0620	55	·0975
6	17·0643	56	1·8892	6	3·3084	56	·0929
7	16·3295	57	1·8078	7	2·7711	57	·0886
8	15·6263	58	1·7300	8	2·3691	58	·0844
9	14·9534	59	1·6555	9	2·0572	59	·0805
10	14·3095	60	1·5842	10	1·8084	60	·0770
11	13·6933	61	1·5160	11	1·6055	61	·0732
12	13·1036	62	1·4507	12	1·4370	62	·0698
13	12·5394	63	1·3882	13	1·2950	63	·0666
14	11·9994	64	1·3284	14	1·1738	64	·0636
15	11·4827	65	1·2712	15	1·0692	65	·0607
16	10·9882	66	1·2165	16	·9781	66	·0579
17	10·5150	67	1·1641	17	·8982	67	·0553
18	10·0622	68	1·1140	18	·8275	68	·0528
19	9·6289	69	1·0660	19	·7646	69	·0504
20	9·2143	70	1·0201	20	·7084	70	·0481
21	8·8175	71	·9762	21	·6578	71	·0459
22	8·4378	72	·9341	22	·6121	72	·0439
23	8·0744	73	·8938	23	·5707	73	·0419
24	7·7267	74	·8554	24	·5330	74	·0400
25	7·3940	75	·8188	25	·4986	75	·0382
26	7·0756	76	·7833	26	·4672	76	·0365
27	6·7709	77	·7496	27	·4382	77	·0349
28	6·4793	78	·7173	28	·4116	78	·0334
29	6·2003	79	·6864	29	·3870	79	·0319
30	5·9333	80	·6569	30	·3643	80	·0305
31	5·6778	81	·6286	31	·3432	81	·0291
32	5·4333	82	·6015	32	·3236	82	·0278
33	5·1994	83	·5756	33	·3054	83	·0266
34	4·9755	84	·5508	34	·2885	84	·0254
35	4·7612	85	·5271	35	·2727	85	·0243
36	4·5562	86	·5044	36	·2579	86	·0232
37	4·3600	87	·4827	37	·2441	87	·0222
38	4·1722	88	·4619	38	·2312	88	·0212
39	3·9926	89	·4420	39	·2190	89	·0203
40	3·8206	90	·4230	40	·2076	90	·0194
41	3·6561	91	·4048	41	·1969	91	·0186
42	3·4987	92	·3873	42	·1869	92	·0177
43	3·3480	93	·3707	43	·1774	93	·0170
44	3·2038	94	·3547	44	·1685	94	·0162
45	3·0659	95	·3394	45	·1600	95	·0155
46	2·9339	96	·3248	46	·1521	96	·0148
47	2·8075	97	·3108	47	·1446	97	·0142
48	2·6866	98	·2974	48	·1375	98	·0136
49	2·5709	99	·2846	49	·1308	99	·0130
50	2·4602	100	·2724	50	·1245	100	·0124

TABLE XVI.   ccix

Present Value of a Perpetuity of £1, receivable once in every nth Year, the first payment due n years hence; also Present Value of a Perpetuity of £1 Deferred n Years.

5 Per Cent.

Years	Present Values of a Perpetuity of £1 Deferred n Years	Years	Present Values of a Perpetuity of £1 Deferred n Years	Years	Present Value of £1 receivable once in every nth Year, the first due n Years hence	Years	Present Value of £1 receivable once in every nth Year, the first due n Years hence
1	19·0476	51	1·6610	1	20·0000	51	·0906
2	18·1406	52	1·5819	2	9·7561	52	·0859
3	17·2768	53	1·5066	3	6·3442	53	·0815
4	16·4540	54	1·4349	4	4·6402	54	·0773
5	15·6705	55	1·3665	5	3·6195	55	·0733
6	14·9243	56	1·3015	6	2·9403	56	·0696
7	14·2136	57	1·2395	7	2·4564	57	·0661
8	13·5368	58	1·1805	8	2·0944	58	·0627
9	12·8922	59	1·1242	9	1·8138	59	·0596
10	12·2783	60	1·0707	10	1·5901	60	·0566
11	11·6936	61	1·0197	11	1·4078	61	·0537
12	11·1367	62	·9712	12	1·2565	62	·0510
13	10·6064	63	·9249	13	1·1291	63	·0485
14	10·1014	64	·8809	14	1·0205	64	·0461
15	9·6203	65	·8389	15	·9268	65	·0438
16	9·1622	66	·7990	16	·8454	66	·0416
17	8·7259	67	·7609	17	·7740	67	·0396
18	8·3104	68	·7247	18	·7109	68	·0376
19	7·9147	69	·6902	19	·6549	69	·0357
20	7·5378	70	·6573	20	·6049	70	·0340
21	7·1788	71	·6260	21	·5599	71	·0323
22	6·8370	72	·5962	22	·5194	72	·0307
23	6·5114	73	·5678	23	·4827	73	·0292
24	6·2014	74	·5408	24	·4494	74	·0278
25	5·9061	75	·5150	25	·4190	75	·0264
26	5·6248	76	·4905	26	·3913	76	·0251
27	5·3570	77	·4671	27	·3658	77	·0238
28	5·1019	78	·4449	28	·3425	78	·0228
29	4·8589	79	·4237	29	·3209	79	·0216
30	4·6275	80	·4035	30	·3010	80	·0206
31	4·4072	81	·3843	31	·2826	81	·0196
32	4·1973	82	·3660	32	·2656	82	·0186
33	3·9975	83	·3486	33	·2498	83	·0177
34	3·8071	84	·3320	34	·2351	84	·0169
35	3·6258	85	·3162	35	·2214	85	·0161
36	3·4531	86	·3011	36	·2087	86	·0153
37	3·2887	87	·2868	37	·1968	87	·0145
38	3·1321	88	·2731	38	·1857	88	·0138
39	2·9830	89	·2601	39	·1753	89	·0132
40	2·8409	90	·2477	40	·1656	90	·0125
41	2·7056	91	·2359	41	·1564	91	·0119
42	2·5768	92	·2247	42	·1479	92	·0114
43	2·4541	93	·2140	43	·1399	93	·0108
44	2·3372	94	·2038	44	·1323	94	·0103
45	2·2259	95	·1941	45	·1252	95	·0098
46	2·1199	96	·1849	46	·1186	96	·0093
47	2·0190	97	·1761	47	·1123	97	·0089
48	1·9228	98	·1677	48	·1064	98	·0085
49	1·8313	99	·1597	49	·1008	99	·0080
50	1·7441	100	·1521	50	·0955	100	·0077

Present Value of a Perpetuity of £1, receivable once in every nth Year, the first payment due n years hence; also Present Value of a Perpetuity of £1 Deferred n Years.

6 PER CENT.

Years	Present Values of a Perpetuity of £1 Deferred n Years	Years	Present Values of a Perpetuity of £1 Deferred n Years	Years	Present Value of £1 receivable once in every nth Year, the first due n Years hence	Years	Present Value of £1 receivable once in every nth Year, the first due n Years hence
1	15·7233	51	·8536	1	16·6667	51	·0540
2	14·8333	52	·8053	2	8·0906	52	·0508
3	13·9937	53	·7597	3	5·2352	53	·0478
4	13·2016	54	·7167	4	3·8099	54	·0449
5	12·5432	55	·6761	5	2·8233	55	·0423
6	11·7493	56	·6379	6	2·3894	56	·0398
7	11·0843	57	·6017	7	1·9856	57	·0375
8	10·4569	58	·5677	8	1·6839	58	·0353
9	9·8650	59	·5356	9	1·4504	59	·0332
10	9·3066	60	·5052	10	1·2645	60	·0313
11	8·7798	61	·4766	11	1·1132	61	·0294
12	8·2828	62	·4497	12	·9880	62	·0277
13	7·8140	63	·4242	13	·8827	63	·0261
14	7·3717	64	·4002	14	·7931	64	·0246
15	6·9544	65	·3775	15	·7160	65	·0232
16	6·5608	66	·3562	16	·6492	66	·0218
17	6·1894	67	·3360	17	·5907	67	·0206
18	5·8391	68	·3170	18	·5393	68	·0194
19	5·5086	69	·2991	19	·4937	69	·0183
20	5·1967	70	·2821	20	·4531	70	·0172
21	4·9026	71	·2662	21	·4167	71	·0162
22	4·6251	72	·2511	22	·3841	72	·0153
23	4·3633	73	·2369	23	·3546	73	·0144
24	4·1163	74	·2235	24	·3280	74	·0136
25	3·8833	75	·2108	25	·3038	75	·0128
26	3·6635	76	·1989	26	·2817	76	·0121
27	3·4561	77	·1876	27	·2616	77	·0114
28	3·2605	78	·1770	28	·2432	78	·0107
29	3·0759	79	·1670	29	·2263	79	·0101
30	2·9018	80	·1575	30	·2108	80	·0095
31	2·7376	81	·1486	31	·1965	81	·0090
32	2·5826	82	·1402	32	·1834	82	·0085
33	2·4364	83	·1323	33	·1712	83	·0080
34	2·2985	84	·1248	34	·1600	84	·0075
35	2·1684	85	·1177	35	·1496	85	·0071
36	2·0457	86	·1111	36	·1399	86	·0067
37	1·9299	87	·1048	37	·1310	87	·0063
38	1·8206	88	·0988	38	·1226	88	·0060
39	1·7176	89	·0932	39	·1149	89	·0056
40	1·6204	90	·0880	40	·1077	90	·0053
41	1·5287	91	·0830	41	·1010	91	·0050
42	1·4421	92	·0783	42	·0947	92	·0047
43	1·3605	93	·0739	43	·0888	93	·0045
44	1·2835	94	·0697	44	·0834	94	·0042
45	1·2108	95	·0657	45	·0783	95	·0040
46	1·1423	96	·0620	46	·0736	96	·0037
47	1·0776	97	·0585	47	·0691	97	·0035
48	1·0166	98	·0552	48	·0650	98	·0033
49	·9591	99	·0521	49	·0611	99	·0031
50	·9048	100	·0491	50	·0574	100	·0030

TABLE XVI. ccxi

Present Value of a Perpetuity of £1, receivable once in every nth Year, the first payment due n years hence; also Present Value of a Perpetuity of £1 Deferred n Years.

7 PER CENT.

Years	Present Values of a Perpetuity of £1 Deferred n Years	Years	Present Values of a Perpetuity of £1 Deferred n Years	Years	Present Value of £1 receivable once in every nth Year, the first due n Years hence	Years	Present Value of £1 receivable once in every nth Year, the first due n Years hence
1	13·3511	51	·4532	1	14·2857	51	·0328
2	12·4777	52	·4236	2	6·9013	52	·0306
3	11·6614	53	·3959	3	4·4436	53	·0285
4	10·8985	54	·3700	4	3·2175	54	·0266
5	10·1855	55	·3458	5	2·4842	55	·0248
6	9·5192	56	·3232	6	1·9971	56	·0231
7	8·8964	57	·3020	7	1·6508	57	·0216
8	8·3144	58	·2823	8	1·3924	58	·0202
9	7·7705	59	·2638	9	1·1927	59	·0188
10	7·2621	60	·2465	10	1·0340	60	·0176
11	6·7870	61	·2304	11	·9051	61	·0164
12	6·3430	62	·2153	12	·7986	62	·0153
13	5·9281	63	·2012	13	·7093	63	·0143
14	5·5402	64	·1881	14	·6335	64	·0133
15	5·1778	65	·1758	15	·5685	65	·0125
16	4·8391	66	·1643	16	·5123	66	·0116
17	4·5225	67	·1535	17	·4632	67	·0109
18	4·2266	68	·1435	18	·4202	68	·0101
19	3·9501	69	·1341	19	·3822	69	·0095
20	3·6917	70	·1253	20	·3485	70	·0089
21	3·4502	71	·1171	21	·3184	71	·0083
22	3·2245	72	·1095	22	·2915	72	·0077
23	3·0135	73	·1023	23	·2673	73	·0072
24	2·8164	74	·0956	24	·2456	74	·0067
25	2·6321	75	·0894	25	·2259	75	·0063
26	2·4599	76	·0835	26	·2080	76	·0059
27	2·2990	77	·0780	27	·1918	77	·0055
28	2·1486	78	·0729	28	·1770	78	·0051
29	2·0080	79	·0682	29	·1636	79	·0048
30	1·8767	80	·0637	30	·1512	80	·0045
31	1·7539	81	·0595	31	·1400	81	·0042
32	1·6392	82	·0556	32	·1296	82	·0039
33	1·5319	83	·0520	33	·1201	83	·0037
34	1·4317	84	·0486	34	·1114	84	·0034
35	1·3380	85	·0454	35	·1033	85	·0032
36	1·2505	86	·0425	36	·0959	86	·0030
37	1·1687	87	·0397	37	·0891	87	·0028
38	1·0922	88	·0371	38	·0828	88	·0026
39	1·0208	89	·0347	39	·0770	89	·0024
40	·9540	90	·0324	40	·0716	90	·0023
41	·8916	91	·0303	41	·0666	91	·0021
42	·8333	92	·0283	42	·0619	92	·0020
43	·7788	93	·0264	43	·0577	93	·0019
44	·7278	94	·0247	44	·0537	94	·0017
45	·6802	95	·0231	45	·0500	95	·0016
46	·6357	96	·0216	46	·0466	96	·0015
47	·5941	97	·0202	47	·0434	97	·0014
48	·5552	98	·0188	48	·0404	98	·0013
49	·5189	99	·0176	49	·0307	99	·0012
50	·4850	100	·0165	50	·0351	100	·0012

Present Value of a Perpetuity of £1, receivable once in every nth Year, the first payment due n years hence; also Present Value of a Perpetuity of £1 Deferred n Years.

8 Per Cent.

Years	Present Values of a Perpetuity of £1 Deferred n Years	Years	Present Values of a Perpetuity of £1 Deferred n Years	Years	Present Value of £1 receivable once in every nth Year, the first due n Years hence	Years	Present Value of £1 receivable once in every nth Year, the first due n Years hence
1	11·5741	51	·2468	1	12·5000	51	·0201
2	10·7167	52	·2285	2	6·0096	52	·0186
3	9·9229	53	·2116	3	3·8504	53	·0172
4	9·1879	54	·1959	4	2·7740	54	·0159
5	8·5073	55	·1814	5	2·1307	55	·0147
6	7·8771	56	·1680	6	1·7039	56	·0136
7	7·2936	57	·1555	7	1·4009	57	·0126
8	6·7534	58	·1440	8	1·1752	58	·0117
9	6·2531	59	·1333	9	1·0010	59	·0108
10	5·7899	60	·1234	10	·8629	60	·0100
11	5·3610	61	·1143	11	·7510	61	·0092
12	4·9639	62	·1058	12	·6587	62	·0085
13	4·5962	63	·0980	13	·5815	63	·0079
14	4·2558	64	·0907	14	·5162	64	·0073
15	3·9405	65	·0840	15	·4604	65	·0068
16	3·6486	66	·0778	16	·4122	66	·0063
17	3·3784	67	·0720	17	·3704	67	·0058
18	3·1281	68	·0667	18	·3338	68	·0054
19	2·8964	69	·0618	19	·3016	69	·0050
20	2·6819	70	·0572	20	·2732	70	·0046
21	2·4832	71	·0529	21	·2479	71	·0043
22	2·2993	72	·0490	22	·2254	72	·0039
23	2·1289	73	·0454	23	·2053	73	·0036
24	1·9712	74	·0420	24	·1872	74	·0034
25	1·8259	75	·0389	25	·1710	75	·0031
26	1·6900	76	·0360	26	·1563	76	·0029
27	1·5643	77	·0334	27	·1431	77	·0027
28	1·4489	78	·0309	28	·1311	78	·0025
29	1·3416	79	·0286	29	·1202	79	·0023
30	1·2422	80	·0265	30	·1103	80	·0021
31	1·1502	81	·0245	31	·1013	81	·0020
32	1·0650	82	·0227	32	·0931	82	·0018
33	·9861	83	·0210	33	·0856	83	·0017
34	·9131	84	·0195	34	·0788	84	·0016
35	·8454	85	·0180	35	·0725	85	·0014
36	·7828	86	·0167	36	·0668	86	·0013
37	·7248	87	·0155	37	·0616	87	·0012
38	·6711	88	·0143	38	·0567	88	·0011
39	·6214	89	·0132	39	·0523	89	·0011
40	·5754	90	·0123	40	·0483	90	·0010
41	·5328	91	·0114	41	·0445	91	·0009
42	·4933	92	·0105	42	·0411	92	·0008
43	·4568	93	·0097	43	·0379	93	·0008
44	·4229	94	·0090	44	·0350	94	·0007
45	·3916	95	·0083	45	·0323	95	·0007
46	·3626	96	·0077	46	·0299	96	·0006
47	·3357	97	·0072	47	·0276	97	·0006
48	·3109	98	·0066	48	·0255	98	·0005
49	·2878	99	·0061	49	·0236	99	·0005
50	·2665	100	·0057	50	·0218	100	·0005

TABLE XVII.

SINGLE LIFE ANNUITIES (CARLISI.E).

Available Interest on Capital being at the rates of 4, 5, 6, 8, *and* 10 *per cent. per annum, Redemption of Capital being at the rate of* **3** *per cent.*

Calculated to **3** *places of decimals.*

TABLE XVII.   CCXV

Single Life Annuities (Carlisle).   Available Interest on Capital at the following
rates, Redemption being at **3** per cent.

Age	4 Per Cent.	Age	4 Per Cent.	Age	5 Per Cent.	Age	5 Per Cent.
0	14·763	52	11·939	0	12·864	52	10·666
1	16·725	53	11·645	1	14·329	53	10·430
2	17·696	54	11·346	2	15·035	54	10·190
3	18·489	55	11·039	3	15·604	55	9·941
4	18·887	56	10·726	4	15·886	56	9·687
5	19·154	57	10·406	5	16·075	57	9·425
6	19·254	58	10·086	6	16·146	58	9·162
7	19·268	59	9·781	7	16·155	59	8·909
8	19·225	60	9·495	8	16·125	60	8·672
9	19·144	61	9·240	9	16·068	61	8·458
10	19·036	**62**	8·988	**10**	15·990	**62**	8·247
11	18·914	63	8·731	11	15·906	63	8·030
12	18·793	64	8·463	12	15·820	64	7·803
13	18·671	65	8·187	13	15·733	65	7·568
14	18·546	66	7·901	14	15·645	66	7·322
15	18·422	67	7·602	15	15·556	67	7·065
16	18·304	68	7·295	16	15·472	68	6·799
17	18·189	69	6·976	17	15·389	69	6·521
18	18·072	70	6·650	18	15·306	70	6·235
19	17·952	71	6·311	19	15·220	71	5·937
20	17·828	**72**	5·991	**20**	15·130	**72**	5·652
21	17·697	73	5·699	21	15·037	73	5·392
22	17·563	74	5·439	22	14·939	74	5·159
23	17·423	75	5·224	23	14·838	75	4·965
24	17·277	76	5·012	24	14·732	76	4·773
25	17·127	77	4·816	25	14·623	77	4·594
26	16·973	78	4·615	26	14·510	78	4·411
27	16·814	79	4·390	27	14·394	79	4·206
28	16·654	80	4·182	28	14·277	80	4·015
29	16·501	81	3·956	29	14·164	81	3·805
30	16·358	**82**	3·751	**30**	14·058	**82**	3·616
31	16·212	83	3·542	31	13·951	83	3·421
32	16·062	84	3·339	32	13·839	84	3·231
33	15·903	85	3·128	33	13·721	85	3·033
34	15·737	86	2·944	34	13·597	86	2·860
35	15·565	87	2·793	35	13·468	87	2·718
36	15·386	88	2·701	36	13·334	88	2·630
37	15·203	89	2·596	37	13·197	89	2·530
38	15·016	90	2·438	38	13·056	90	2·380
39	14·825	91	2·421	39	12·911	91	2·364
40	14·634	**92**	2·512	**40**	12·766	**92**	2·451
41	14·449	93	2·616	41	12·625	93	2·550
42	14·265	94	2·663	42	12·484	94	2·594
43	14·081	95	2·683	43	12·343	95	2·613
44	13·889	96	2·633	44	12·195	96	2·565
45	13·691	97	2·495	45	12·042	97	2·434
46	13·483	98	2·333	46	11·881	98	2·279
47	13·265	99	2·086	47	11·711	99	2·044
48	13·033	100	1·655	48	11·530	100	1·628
49	12·781	101	1·213	49	11·332	101	1·199
50	12·513	102	·765	50	11·121	102	·759
51	12·228	**103**	·323	**51**	10·896	**103**	·322

Single Life Annuities (Carlisle).   Available Interest on Capital at the following rates, Redemption being at **3** per cent.

Age	6 Per Cent.	Age	6 Per Cent.	Age	8 Per Cent.	Age	8 Per Cent.
0	11·398	52	9·638	0	9·282	52	8·080
1	12·533	53	9·445	1	10·021	53	7·945
2	13·070	54	9·248	2	10·362	54	7·804
3	13·498	55	9·043	3	10·629	55	7·658
4	13·709	56	8·831	4	10·759	56	7·506
5	13·849	57	8·613	5	10·845	57	7·348
6	13·901	58	8·393	6	10·877	58	7·187
7	13·908	59	8·181	7	10·881	59	7·030
8	13·886	60	7·980	8	10·868	60	6·882
9	13·844	61	7·799	9	10·842	61	6·746
10	13·787	62	7·618	10	10·807	62	6·611
11	13·723	63	7·433	11	10·768	63	6·471
12	13·659	64	7·238	12	10·728	64	6·323
13	13·594	65	7·035	13	10·688	65	6·167
14	13·528	66	6·823	14	10·647	66	6·003
15	13·462	67	6·599	15	10·606	67	5·830
16	13·399	68	6·366	16	10·567	68	5·647
17	13·337	69	6·122	17	10·529	69	5·454
18	13·274	70	5·869	18	10·490	70	5·252
19	13·209	71	5·604	19	10·449	71	5·039
20	13·142	72	5·350	20	10·407	72	4·833
21	13·072	73	5·116	21	10·363	73	4·641
22	12·998	74	4·906	22	10·316	74	4·467
23	12·921	75	4·730	23	10·267	75	4·321
24	12·840	76	4·556	24	10·217	76	4·175
25	12·757	77	4·392	25	10·164	77	4·038
26	12·672	78	4·225	26	10·110	78	3·896
27	12·583	79	4·036	27	10·053	79	3·734
28	12·493	80	3·859	28	9·995	80	3·583
29	12·407	81	3·666	29	9·940	81	3·415
30	12·325	82	3·490	30	9·888	82	3·262
31	12·243	83	3·308	31	9·835	83	3·102
32	12·157	84	3·130	32	9·779	84	2·945
33	12·066	85	2·944	33	9·720	85	2·780
34	11·970	86	2·780	34	9·658	86	2·634
35	11·870	87	2·645	35	9·593	87	2·512
36	11·765	88	2·563	36	9·524	88	2·438
37	11·658	89	2·468	37	9·454	89	2·352
38	11·548	90	2·324	38	9·381	90	2·221
39	11·434	91	2·309	39	9·306	91	2·207
40	11·321	92	2·392	40	9·231	92	2·283
41	11·210	93	2·486	41	9·157	93	2·368
42	11·099	94	2·528	42	9·083	94	2·407
43	10·987	95	2·546	43	9·007	95	2·423
44	10·870	96	2·501	44	8·929	96	2·382
45	10·748	97	2·376	45	8·846	97	2·268
46	10·619	98	2·229	46	8·759	98	2·134
47	10·484	99	2·003	47	8·666	99	1·926
48	10·338	100	1·602	48	8·567	100	1·552
49	10·179	101	1·185·	49	8·457	101	1·157
50	10·008	102	·754	50	8·339	102	·742
51	9·825	103	·321	51	8·212	103	·318

TABLE XVII. ccxvii

Single Life Annuities (Carlisle).   Available Interest on Capital at the following rates, Redemption being at **3** per cent.

Age	10 per cent.	Age	10 per cent.	Age	10 per cent.	Age	10 per cent.
0	7·829	26	8·409	52	6·956	78	3·614
1	8·348	27	8·370	53	6·855	79	3·475
2	8·583	28	8·330	54	6·751	80	3·343
3	8·765	29	8·292	55	6·641	81	3·197
4	8·854	30	8·255	56	6·526	82	3·062
5	8·912	31	8·218	57	6·406	83	2·921
6	8·933	32	8·179	58	6·284	84	2·782
7	8·937	33	8·138	59	6·164	85	2 634
8	8·927	34	8·094	60	6·049	86	2·502
9	8·910	35	8·048	61	5·944	87	2·392
10	8·886	36	8·000	62	5·839	88	2·324
11	8·860	37	7·951	63	5·730	89	2·246
12	8·833	38	7·899	64	5·613	90	2·127
13	8·806	39	7·846	65	5·490	91	2·114
14	8·778	40	7·792	66	5·360	92	2·183
15	8·750	41	7·739	67	5·221	93	2·261
16	8·723	42	7·686	68	5·074	94	2·296
17	8·697	43	7·632	69	4·918	95	2·311
18	8·671	44	7·576	70	4·753	96	2·274
19	8·643	45	7·517	71	4·578	97	2·170
20	8·614	46	7·453	72	4·407	98	2·046
21	8·584	47	7·386	73	4·247	99	1·854
22	8·552	48	7·314	74	4·101	100	1·505
23	8·518	49	7·234	75	3·977	101	1·131
24	8·483	50	7·147	76	3·853	102	·732
25	8·447	51	7·053	77	3·736	103	·316

TABLE XVIII.

Decimal Equivalents for every Farthing in the Pound. Calculated to 8 places of decimals.

By removing the decimal point two places to the right, the rate per cent. corresponding to the s. d. columns will be obtained :—Thus 10¾d. = 0·4479167, and by observing the rule referred to above, the number will be converted into 04·479167, which is the rate per cent.

Decimal Equivalents for every Farthing in the Pound.

s.	d.	Decimal	s.	d.	Decimal	s.	d.	Decimal	s.	d.	Decimal
	¼	·00104167	1	0¼	·05104167	2	0¼	·10104167	3	0¼	·15104167
	½	·00208333	1	0½	·05208333	2	0½	·10208333	3	0½	·15208333
	¾	·003125	1	0¾	·053125	2	0¾	·103125	3	0¾	·153125
	1	·00416667	1	1	·05416667	2	1	·10416667	3	1	·15416667
	1¼	·00520833	1	1¼	·05520833	2	1¼	·10520833	3	1¼	·15520833
	1½	·00625	1	1½	·05625	2	1½	·10625	3	1½	·15625
	1¾	·00729167	1	1¾	·05729167	2	1¾	·10729167	3	1¾	·15729167
	2	·00833333	1	2	·05833333	2	2	·10833333	3	2	·15833333
	2¼	·009375	1	2¼	·059375	2	2¼	·109375	3	2¼	·159375
	2½	·01041667	1	2½	·06041667	2	2½	·11041667	3	2½	·16041667
	2¾	·01145883	1	2¾	·06145883	2	2¾	·11145883	3	2¾	·16145883
	3	·0125	1	3	·0625	2	3	·1125	3	3	·1625
	3¼	·01354167	1	3¼	·06354167	2	3¼	·11354167	3	3¼	·16354167
	3½	·01458333	1	3½	·06458333	2	3½	·11458333	3	3½	·16458333
	3¾	·015625	1	3¾	·065625	2	3¾	·115625	3	3¾	·165625
	4	·01666667	1	4	·06666667	2	4	·11666667	3	4	·16666667
	4¼	·01770833	1	4¼	·06770833	2	4¼	·11770833	3	4¼	·16770833
	4½	·01875	1	4½	·06875	2	4½	·11875	3	4½	·16875
	4¾	·01979167	1	4¾	·06979167	2	4¾	·11979167	3	4¾	·16979167
	5	·02083333	1	5	·07083333	2	5	·12083333	3	5	·17083333
	5¼	·021875	1	5¼	·071875	2	5¼	·121875	3	5¼	·171875
	5½	·02291667	1	5½	·07291667	2	5½	·12291667	3	5½	·17291667
	5¾	·02395833	1	5¾	·07395833	2	5¾	·12395833	3	5¾	·17395833
	6	·025	1	6	·075	2	6	·125	3	6	·175
	6¼	·02604167	1	6¼	·07604167	2	6¼	·12604167	3	6¼	·17604167
	6½	·02708333	1	6½	·07708333	2	6½	·12708333	2	6½	·17708333
	6¾	·028125	1	6¾	·078125	2	6¾	·128125	3	6¾	·178125
	7	·02916667	1	7	·07916667	2	7	·12916667	3	7	·17916667
	7¼	·03020833	1	7¼	·08020833	2	7¼	·13020833	3	7¼	·18020833
	7½	·03125	1	7½	·08125	2	7½	·13125	3	7½	·18125
	7¾	·03229167	1	7¾	·08229167	2	7¾	·13229167	3	7¾	·18229167
	8	·03333333	1	8	·08333333	2	8	·13333333	3	8	·18333333
	8¼	·034375	1	8¼	·084375	2	8¼	·134375	3	8¼	·184375
	8½	·03541667	1	8½	·08541667	2	8½	·13541667	3	8½	·18541667
	8¾	·03645833	1	8¾	·08645833	2	8¾	·13645833	3	8¾	·18645833
	9	·0375	1	9	·0875	2	9	·1375	3	9	·1875
	9¼	·03854167	1	9¼	·08854167	2	9¼	·13854167	3	9¼	·18854167
	9½	·03958333	1	9½	·08958333	2	9½	·13958333	3	9½	·18958333
	9¾	·040625	1	9¾	·090625	2	9¾	·140625	3	9¾	·190625
	10	·04166667	1	10	·09166667	2	10	·14166667	3	10	·19166667
	10¼	·04270833	1	10¼	·09270833	2	10¼	·14270833	3	10¼	·19270833
	10½	·04375	1	10½	·09375	2	10½	·14375	3	10½	·19375
	10¾	·04479167	1	10¾	·09479167	2	10¾	·14479167	3	10¾	·19479167
	11	·04583333	1	11	·09583333	2	11	·14583333	3	11	·19583333
	11¼	·046875	1	11¼	·096875	2	11¼	·146875	3	11¼	·196875
	11½	·04791667	1	11½	·09791667	2	11½	·14791667	3	11½	·19791667
	11¾	·04895833	1	11¾	·09895833	2	11¾	·14895833	3	11¾	·19895833
1	0	·05	2	0	·1	3	0	·15	4	0	·20

Decimal Equivalents for every Farthing in the Pound.

s.	d.	Decimal	s.	d.	Decimal	s.	d.	Decimal	s.	d.	Decimal
4	0¼	·20104167	5	0¼	·25104167	6	0¼	·30104167	7	0¼	·35104167
4	0½	·20208333	5	0½	·25208333	6	0½	·30208333	7	0½	·35208333
4	0¾	·203125	5	0¾	·253125	6	0¾	·303125	7	0¾	·353125
4	1	·20416667	5	1	·25416667	6	1	·30416667	7	1	·35416667
4	1¼	·20520833	5	1¼	·25520833	6	1¼	·30520833	7	1¼	·35520833
4	1½	·20625	5	1½	·25625	6	1½	·30625	7	1½	·35625
4	1¾	·20729167	5	1¾	·25729167	6	1¾	·30729167	7	1¾	·35729167
4	2	·20833333	5	2	·25833333	6	2	·30833333	7	2	·35833333
4	2¼	·209375	5	2¼	·259375	6	2¼	·309375	7	2¼	·359375
4	2½	·21041667	5	2½	·26041667	6	2½	·31045967	7	2½	·36041667
4	2¾	·21145883	5	2¾	·26145883	6	2¾	·31145883	7	2¾	·36145883
4	3	·2125	5	3	·2625	6	3	·3125	7	3	·3625
4	3¼	·21354167	5	3¼	·26354167	6	3¼	·3135416	7	3¼	·36354167
4	3½	·21458333	5	3½	·26458333	6	3½	·31458333	7	3½	·36458333
4	3¾	·215625	5	3¾	·265625	6	3¾	·315625	7	3¾	·365625
4	4	·21666667	5	4	·26666667	6	4	·31666667	7	4	·36666667
4	4¼	·21770833	5	4¼	·26770833	6	4¼	·31770833	7	4¼	·36770833
4	4½	·21875	5	4½	·26875	6	4½	·31875	7	4½	·36875
4	4¾	·21979167	5	4¾	·26979167	6	4¾	·31979167	7	4¾	·36979167
4	5	·22083333	5	5	·27083333	6	5	·32083333	7	5	·37083333
4	5¼	·221875	5	5¼	·271875	6	5¼	·321875	7	5¼	·371875
4	5½	·22291667	5	5½	·27291667	6	5½	·32291667	7	5½	·37291667
4	5¾	·22395833	5	5¾	·27395833	6	5¾	·32395833	7	5¾	·37395833
4	6	·225	5	6	·275	6	6	·325	7	6	·375
4	6¼	·22604167	5	6¼	·27604167	6	6¼	·32604167	7	6¼	·37604167
4	6½	·22708333	5	6½	·27708333	6	6½	·37708333	7	6½	·37708333
4	6¾	·228125	5	6¾	·278125	6	6¾	·328125	7	6¾	·378125
4	7	·22916667	5	7	·27916667	6	7	·32916667	7	7	·37916667
4	7¼	·23020833	5	7¼	·28020833	6	7¼	·33020833	7	7¼	·38020833
4	7½	·23125	5	7½	·28125	6	7½	·33125	7	7½	·38125
4	7¾	·23229167	5	7¾	·28229167	6	7¾	·33229167	7	7¾	·38229167
4	8	·23333333	5	8	·28333333	6	8	·33333333	7	8	·38333333
4	8¼	·234375	5	8¼	·284375	6	8¼	·334375	7	8¼	·384375
4	8½	·23541667	5	8½	·28541667	6	8½	·33541667	7	8½	·38541667
4	8¾	·23645833	5	8¾	·28645833	6	8¾	·33645833	7	8¾	·38645833
4	9	·2375	5	9	·3875	6	9	·3375	7	·9	·3875
4	9¼	·23854167	5	9¼	·28854167	6	9¼	·33854167	7	9¼	·38854167
4	9½	·23958333	5	9½	·28958333	6	9½	·33958333	7	9½	·38958333
-4	9¾	·240625	5	9¾	·290625	6	9¾	·340625	7	9¾	·390625
4	10	·24166667	5	10	·29166667	6	10	·34166667	7	10	·39166667
4	10¼	·24270833	5	10¼	·29270833	6	10¼	·34270833	7	10¼	·39270833
4	10½	·24375	5	10½	·29375	6	10½	·34375	7	10½	·39375
4	10¾	·24479167	5	10¾	·29479167	6	10¾	·34479167	7	10¾	·39479167
4	11	·24583333	5	11	·29583333	6	11	·34583333	7	11	·39583333
4	11¼	·246875	5	11¼	·296875	6	11¼	·346875	7	11¼	·396875
4	11½	·24791667	5	11½	·29791667	6	11½	·34791667	7	11½	·39791667
4	11¾	·24895833	5	11¾	·29895833	6	11¾	·34895833	7	11¾	·39895833
5	0	·25	6	0	·30	7	0	·35	8	0	·40

Decimal Equivalents for every Farthing in the Pound.

s.	d.	Decimal	s.	d.	Decimal	s.	d.	Decimal	s.	d.	Decimal
8	0¼	·40104167	9	0¼	·45104167	10	0¼	·50104167	11	0¼	·55104167
8	0½	·40208333	9	0½	·45208333	10	0½	·50208333	11	0½	·55208333
8	0¾	·403125	9	0¾	·453125	10	0¾	·503125	11	0¾	·553125
8	1	·40416667	9	1	·45416667	10	1	·50416667	11	1	·55416667
8	1¼	·40520833	9	1¼	·45520833	10	1¼	·50520833	11	1¼	·55520833
8	1½	·40625	9	1½	·45625	10	1½	·50625	11	1½	·55625
8	1¾	·40729167	9	1¾	·45729167	10	1¾	·50729167	11	1¾	·55729167
8	2	·40833333	9	2	·45833333	10	2	·50833333	11	2	·55833333
8	2¼	·409375	9	2¼	·459375	10	2¼	·509375	11	2¼	·559375
8	2½	·41041667	9	2½	·46041667	10	2½	·51041667	11	2½	·56041667
8	2¾	·41145883	9	2¾	·46145883	10	2¾	·51145883	11	2¾	·56145883
8	3	·4125	9	3	·4625	10	3	·5125	11	3	·5625
8	3¼	·41354167	9	3¼	·46354167	10	3¼	·51354167	11	3¼	·56354167
8	3½	·41458333	9	3½	·46458333	10	3½	·51458333	11	3½	·56458333
8	3¾	·415625	9	3¾	·465625	10	3¾	·515625	11	3¾	·565625
8	4	·41666667	9	4	·46666667	10	4	·51666667	11	4	·56666667
8	4¼	·41770833	9	4¼	·46770833	10	4¼	·51770833	11	4¼	·56770833
8	4½	·41875	9	4½	·46875	10	4½	·51875	11	4½	·56875
8	4¾	·41979167	9	4¾	·46979167	10	4¾	·51979167	11	4¾	·56979167
8	5	·42083333	9	5	·47083333	10	5	·52083333	11	5	·57083333
8	5¼	·421875	9	5¼	·471875	10	5¼	·521875	11	5¼	·571875
8	5½	·42291667	9	5½	·47291667	10	5½	·52291667	11	5½	·57291667
8	5¾	·42395833	9	5¾	·47395833	10	5¾	·52395833	11	5¾	·57395833
8	6	·425	9	6	·475	10	6	·525	11	6	·575
8	6¼	·42604167	9	6¼	·47604167	10	6¼	·52604167	11	6¼	·57604167
8	6½	·42708333	9	6½	·47708333	10	6½	·52708333	11	6½	·57708333
8	6¾	·428125	9	6¾	·478125	10	6¾	·528125	11	6¾	·578125
8	7	·42916667	9	7	·47916667	10	7	·52916667	11	7	·57916667
8	7¼	·43020833	9	7¼	·48020833	10	7¼	·53020833	11	7¼	·58020833
8	7½	·43125	9	7½	·48125	10	7½	·53125	11	7½	·58125
8	7¾	·43229167	9	7¾	·48229167	10	7¾	·53229167	11	7¾	·58229167
8	8	·43333333	9	8	·48333333	10	8	·53333333	11	8	·58333333
8	8¼	·434375	9	8¼	·484375	10	8¼	·534375	11	8¼	·584375
8	8½	·43541667	9	8½	·48541667	10	8½	·53541667	11	8½	·58541667
8	8¾	·43645833	9	8¾	·48645833	10	8¾	·53645833	11	8¾	·58645833
8	9	·4375	9	9	·4875	10	9	·5375	11	9	·5875
8	9¼	·43854167	9	9¼	·48854167	10	9¼	·53854167	11	9¼	·58854167
8	9½	·43958333	9	9½	·48958333	10	9½	·53958333	11	9½	·58958333
8	9¾	·440625	9	9¾	·490625	10	9¾	·540625	11	9¾	·590625
8	10	·44166667	9	10	·49166667	10	10	·54166667	11	10	·59166667
8	10¼	·44270833	9	10¼	·49270833	10	10¼	·54270833	11	10¼	·59270833
8	10½	·44375	9	10½	·49375	10	10½	·54375	11	10½	·59375
8	10¾	·44479167	9	10¾	·49479167	10	10¾	·54479167	11	10¾	·59479167
8	11	·44583333	9	11	·49583333	10	11	·54583333	11	11	·59583333
8	11¼	·446875	9	11¼	·496875	10	11¼	·546875	11	11¼	·596875
8	11½	·44791667	9	11½	·49791667	10	11½	·54791667	11	11½	·59791667
8	11¾	·44895833	9	11¾	·49895833	10	11¾	·54895833	11	11¾	·59895833
9	0	·45	10	0	·50	11	0	·55	12	0	·60

THE ENGINEER'S VALUING ASSISTANT.

Decimal Equivalents for every Farthing in the Pound.

s.	d.	Decimal	s.	d.	Decimal	s.	d.	Decimal	s.	d.	Decimal
12	0¼	·60104167	13	0¼	·65104167	14	0¼	·70104167	15	0¼	·75104167
12	0½	·60208333	13	0½	·65208333	14	0½	·70208333	15	0½	·75208333
12	0¾	·603125	13	0¾	·653125	14	0¾	·703125	15	0¾	·753125
12	1	·60416667	13	1	·65416667	14	1	·70416667	15	1	·75416667
12	1¼	·60520833	13	1¼	·65520833	14	1¼	·70520833	15	1¼	·75520833
12	1½	·60625	13	1½	·65625	14	1½	·70625	15	1½	·75625
12	1¾	·60729167	13	1¾	·65729167	14	1¾	·70729167	15	1¾	·75729167
12	2	·60833333	13	2	·65833333	14	2	·70833333	15	2	·75833333
12	2¼	·60559375	13	2¼	·65559375	14	2¼	·70559375	15	2¼	·75559375
12	2½	·61041667	13	2½	·66041667	14	2½	·71041667	15	2½	·76041667
12	2¾	·61145883	13	2¾	·66145883	14	2¾	·71145883	15	2¾	·76145883
12	3	·6125	13	3	·6625	14	3	·7125	15	3	·7625
12	3¼	·61354167	13	3¼	·66354167	14	3¼	·71354167	15	3¼	·76354167
12	3½	·61458333	13	3½	·66458333	14	3½	·71458333	15	3½	·76458333
12	3¾	·615625	13	3¾	·665625	14	3¾	·715625	15	3¾	·765625
12	4	·61666667	13	4	·66666667	14	4	·71666667	15	4	·76666667
12	4¼	·61770833	13	4¼	·66770833	14	4¼	·71770833	15	4¼	·76770833
12	4½	·61875	13	4½	·66875	14	4½	·71875	15	4½	·76875
12	4¾	·61979167	13	4¾	·66979167	14	4¾	·71979167	15	4¾	·76979167
12	5	·62083333	13	5	·67083333	14	5	·72083333	15	5	·77083333
12	5¼	·621875	13	5¼	·671875	14	5¼	·721875	15	5¼	·771875
12	5½	·62291667	13	5½	·67291667	14	5½	·72291667	15	5½	·77291667
12	5¾	·62395833	13	5¾	·67395833	14	5¾	·72395833	15	5¾	·77395833
12	6	·625	13	6	·675	14	6	·725	15	6	·775
12	6¼	·62604167	13	6¼	·67604167	14	6¼	·72604167	15	6¼	·77604167
12	6½	·62708333	13	6½	·67708333	14	6½	·72708333	15	6½	·77708333
12	6¾	·628125	13	6¾	·678125	14	6¾	·728125	15	6¾	·778125
12	7	·62916667	13	7	·67916667	14	7	·72916667	15	7	·77916667
12	7¼	·63020833	13	7¼	·68020833	14	7¼	·73020833	15	7¼	·78020833
12	7½	·63125	13	7½	·68125	14	7½	·73125	15	7½	·78125
12	7¾	·63229167	13	7¾	·68229167	14	7¾	·73229167	15	7¾	·78229167
12	8	·63333333	13	8	·68333333	14	8	·73333333	15	8	·78333333
12	8¼	·634375	13	8¼	·684375	14	8¼	·734375	15	8¼	·784375
12	8½	·63541667	13	8½	·68541667	14	8½	·73541667	15	8½	·78541667
12	8¾	·63645833	13	8¾	·68645833	14	8¾	·73645833	15	8¾	·78645833
12	9	·6375	13	9	·6875	14	9	·7375	15	9	·7875
12	9¼	·63854167	13	9¼	·68854167	14	9¼	·73854167	15	9¼	·78854167
12	9½	·63958333	13	9½	·68958333	14	9½	·73958333	15	9½	·78958333
12	9¾	·640625	13	9¾	·690625	14	9¾	·740625	15	9¾	·790625
12	10	·64166667	13	10	·69166667	14	10	·74166667	15	10	·79166667
12	10¼	·64270833	13	10¼	·69270833	14	10¼	·74270833	15	10¼	·79270833
12	10½	·64375	13	10½	·69375	14	10½	·74375	15	10½	·79375
12	10¾	·64478167	13	10¾	·69479167	14	10¾	·74479167	15	10¾	·79479167
12	11	·64583333	13	11	·69583333	14	11	·74583333	15	11	·79583333
12	11¼	·646875	13	11¼	·696875	14	11¼	·746875	15	11¼	·796875
12	11½	·64791667	13	11½	·69791667	14	11½	·74791667	15	11½	·79791667
12	11¾	·64895833	13	11¾	·69895833	14	11¾	·74895833	15	11¾	·79895833
13	0	·65	14	0	·70	15	0	·75	16	0	·80

TABLE XVIII.   CCXXV

Decimal Equivalents for every Farthing in the Pound.

s.	d.	Decimal	s.	d.	Decimal	s.	d.	Decimal	s.	d.	Decimal
16	0¼	·80104167	17	0¼	·85104167	18	0¼	·90104167	19	0¼	·95104167
16	0½	·80208333	17	0½	·85208333	18	0½	·90208333	19	0½	·95208333
16	0¾	·803125	17	0¾	·853125	18	0¾	·903125	19	0¾	·953125
16	1	·80416667	17	1	·85416667	18	1	·90416667	19	1	·95416667
16	1¼	·80520833	17	1¼	·85520833	18	1¼	·90520833	19	1¼	·95520833
16	1½	·80625	17	1½	·85625	18	1½	·90625	19	1½	·95625
16	1¾	·80729167	17	1¾	·85729167	18	1¾	·90729167	19	1¾	·95729167
16	2	·80833333	17	2	·85833333	18	2	·90833333	19	2	·95833333
16	2¼	·80559375	17	2¼	·85559375	18	2¼	·90559375	19	2¼	·95559375
16	2½	·81041667	17	2½	·86041667	18	2½	·91041667	19	2½	·96041667
16	2¾	·81145883	17	2¾	·86145883	18	2¾	·91145883	19	2¾	·96145883
16	3	·8125	17	3	·8625	18	3	·9125	19	3	·9625
16	3¼	·81354167	17	3¼	·86354167	18	3¼	·91354167	19	3¼	·96354167
16	3½	·81458333	17	3½	·86458333	18	3½	·91458333	19	3½	·96458333
16	3¾	·815625	17	3¾	·865625	18	3¾	·915625	19	3¾	·965625
16	4	·81666667	17	4	·86666667	18	4	·91666667	19	4	·96666667
16	4¼	·81770833	17	4¼	·86770833	18	4¼	·91770833	19	4¼	·96770833
16	4½	·81875	17	4½	·86875	18	4½	·91875	19	4½	·96875
16	4¾	·81979167	17	4¾	·86979167	18	4¾	·91979167	19	4¾	·96979167
16	5	·82083333	17	5	·87083333	18	5	·92083333	19	5	·97083333
16	5¼	·821875	17	5¼	·871875	18	5¼	·921875	19	5¼	·971875
16	5½	·82291667	17	5½	·87291667	18	5½	·92291667	19	5½	·97291667
16	5¾	·82395833	17	5¾	·87395833	18	5¾	·92395833	19	5¾	·97395833
16	6	·825	17	6	·875	18	6	·925	19	6	·975
16	6¼	·82604167	17	6¼	·87604167	18	6¼	·92604167	19	6¼	·97604167
16	6½	·82708333	17	6½	·87708333	18	6½	·92708333	19	6½	·97708333
16	6¾	·828125	17	6¾	·878125	18	6¾	·928125	19	6¾	·978125
16	7	·82916667	17	7	·87916667	18	7	·92916667	19	7	·97916667
16	7¼	·83020833	17	7¼	·88020833	18	7¼	·93020833	19	7¼	·98020833
16	7½	·83125	17	7½	·88125	18	7½	·93125	19	7½	·98125
16	7¾	·83229167	17	7¾	·88229167	18	7¾	·93229167	19	7¾	·98229167
16	8	·83333333	17	8	·88333333	18	8	·93333333	19	8	·98333333
16	8¼	·834375	17	8¼	·884375	18	8¼	·934375	19	8¼	·984375
16	8½	·83541667	17	8½	·88541667	18	8½	·93541667	19	8½	·98541667
16	8¾	·83645833	17	8¾	·88645833	18	8¾	·93645833	19	8¾	·98645833
16	9	·8375	17	9	·8875	18	9	·9375	19	9	·9875
16	9¼	·83854167	17	9¼	·88854167	18	9¼	·93854167	19	9¼	·98854167
16	9½	·83958333	17	9½	·88958333	18	9½	·93958333	19	9½	·98958333
16	9¾	·840625	17	9¾	·890625	18	9¾	·940625	19	9¾	·990625
16	10	·84166667	17	10	·89166667	18	10	·94166667	19	10	·99166667
16	10¼	·84270833	17	10¼	·89270833	18	10¼	·94270833	19	10¼	·99270833
16	10½	·84375	17	10½	·89375	18	10½	·94375	19	10½	·99375
16	10¾	·84479167	17	10¾	·89479167	18	10¾	·94479167	19	10¾	·99479167
16	11	·84583333	17	11	·89583333	18	11	·04583333	19	11	·99583333
16	11¼	·846875	17	11¼	·896875	18	11¼	·946875	19	11¼	·996875
16	11½	·84791667	17	11½	·89791667	18	11½	·94791667	19	11½	·99791667
16	11¾	·84895833	17	11¾	·89895833	18	11¾	·94895833	19	11¾	·99895833
17	0	·85	18	0	·90	19	0	·95	20	0	1·

INDEX.

———◆———

f f 2

PAGE

LONDON : PRINTED BY
SPOTTISWOODE AND CO., NEW-STREET SQUARE
AND PARLIAMENT STREET